Novel Specialty Optical Fibers and Applications

Novel Specialty Optical Fibers and Applications

Editors

Dora Juanjuan Hu
Georges Humbert

MDPI • Basel • Beijing • Wuhan • Barcelona • Belgrade • Manchester • Tokyo • Cluj • Tianjin

Editors

Dora Juanjuan Hu
Sustainable Built
Environment Division
Institute for Infocomm
Research
Singapore
Singapore

Georges Humbert
Xlim Research Institute
UMR 7252 CNRS/University
of Limoges
Limoges
France

Editorial Office
MDPI
St. Alban-Anlage 66
4052 Basel, Switzerland

This is a reprint of articles from the Special Issue published online in the open access journal *Photonics* (ISSN 2304-6732) (available at: www.mdpi.com/journal/photonics/special_issues/novel_specialty_optical_fibers).

For citation purposes, cite each article independently as indicated on the article page online and as indicated below:

LastName, A.A.; LastName, B.B.; LastName, C.C. Article Title. *Journal Name* **Year**, *Volume Number*, Page Range.

ISBN 978-3-0365-5126-5 (Hbk)
ISBN 978-3-0365-5125-8 (PDF)

Contents

About the Editors

Dora Juanjuan Hu

Dora Juan Juan HU received her B.Eng. (Hons.) and Ph.D. degrees from Nanyang Technological University, Singapore, in 2004 and 2010, respectively. From 2009 to 2012, she worked with Institute for Infocomm Research (I2R) as a Scientist and a Principle Investigator to develop novel fiber devices for sensor applications. In 2012, she received an A*STAR Postdoctoral Fellowship (2012–2014) under which she joined the Wellman Center for Photomedicine, Massachusetts General Hospital, Boston, MA, USA, as a Research Fellow and was involved in the study of novel photonic technologies for endoscopic diagnosis. She joined the Femtosecond Optics Group, Imperial College London, in 2013, where she worked on visible and mid-infrared fiber lasers. She is currently heading the Sustainable Built Environment Division at I2R, A*STAR, Singapore, and driving the innovations in technologies and solutions to improve sustainability, livability and productivity in built environments.

Georges Humbert

Georges Humbert received his PhD degree from the University of Rouen (France) in 2002. During his thesis, he designed and fabricated optical fiber components. Then, he spent two years as a research officer at the University of Bath, where he designed, fabricated and characterized photonic crystal fibers. In 2005, he joined XLIM research institute from the CNRS and the University of Limoges, as a CNRS research officer, and since 2016 he has been a CNRS researcher. His research activities are focused on the development of specialty optical fibers and their applications for sensing (liquid, gas, biochemicals), new light sources and lasers, generating nonlinear effects, and guiding and controlling terahertz waves. His works on the realization of specialty optical fibers include the conception, fabrication, and characterization of original solid-core and hollow-core photonic crystal fibers, multimaterial fibers, glass–ceramic fibers and also the development of novel fabrication processes of fibers or fiber components.

photonics

Editorial

Special Issue "Novel Specialty Optical Fibers and Applications": An Overview

Dora Juan Juan Hu [1,*] and Georges Humbert [2]

1 Institute for Infocomm Research, Agency for Science, Technology and Research, Singapore 138632, Singapore
2 Xlim Research Institute, UMR 7252 CNRS/University of Limoges, 87060 Limoges, France; georges.humbert@xlim.fr
* Correspondence: jjhu@i2r.a-star.edu.sg

Citation: Hu, D.J.J.; Humbert, G. Special Issue "Novel Specialty Optical Fibers and Applications": An Overview. *Photonics* **2022**, *9*, 497. https://doi.org/10.3390/photonics9070497

Received: 2 June 2022
Accepted: 14 July 2022
Published: 15 July 2022

Publisher's Note: MDPI stays neutral with regard to jurisdictional claims in published maps and institutional affiliations.

Novel specialty optical fibers refer to optical fibers that have been engineered in terms of design, material and structure, and have been post-processed for novel functionalities and applications. The optical properties in novel specialty optical fibers can be manipulated to achieve optimum performance, resulting in numerous important applications. In this Special Issue, the recent advancements in specialty fiber technology are highlighted and reviewed, with a focus on design, development, and application, as well as future challenges and emerging opportunities.

Eight review papers and seven original research papers are published in this Special Issue, "Novel Specialty Optical Fibers and Applications". A brief overview of the published work is presented in this section.

Specialty fibers composed of special materials introduce new functionalities and open up new spectral windows for applications due to the presence of special materials. Chalcogenide glass (ChG) microfibers (MFs) have introduced wide opportunities in the mid-infrared (mid-IR) spectral range due to their broadband transparency, high optical nonlinearity, and hospitality to rare-earth dopants. The emerging and fast-growing field of ChG MFs in mid-IR optics and applications is reviewed in [1]. Bismuth-doped fiber power amplifier is developed to cover the spectral range of 1650–1700 nm, which is unavailable with rare-earth-doped fibers, enabling applications in optical telecommunication and optical-based methane sensing [2].

The original properties offered by light propagation within hollow-core fibers leads to novel opportunities in light generation, fiber sensing, communication, and light manipulation. W. Pei et al. report here the development of an all-fiber laser operating at the 1.6–1.7 μm band based on deuterium-filled hollow-core photonic crystal fibers [3]. This all-fiber structure enhances the Raman scattering of deuterium gas, offering tunable laser operating from 1643 nm to 1656 nm. An all-fiber gas cavity based on anti-resonant hollow-core fibers is fabricated and experimentally demonstrated with the stable and efficient pump coupling of 78.1% at a maximum injecting laser power of 260 W [4]. This work demonstrates a feasible and simple method for an all-fiber gas cavity for the development of high-power fiber gas lasers. The opportunities offered by hollow-core fibers for sensing are reviewed in these two contributions. Recent developments related to all-fiber online Raman sensors with hollow-core fibers are reviewed in [5], and their applicability in relation to reusability, easiness to clean, rapid detection for safety inspections, food detection, water quality detection, etc., is emphasized. The development of anti-resonant hollow-core fiber (AR-HCF)-based sensing applications is reviewed in [6]. This novel class of hollow-core fiber offers a unique sensing platform to achieve highly accurate and ultra-compact fiber optic sensors with large measurement ranges.

Different strategies exploiting the functionalization of specialty optical fibers with liquids or metals and/or specific fiber designs, such D-shape multicore fibers, have been investigated for developing more sensitive fiber sensors. Thereby, the refractive index sensing performance of functionalized fiber can be significantly enhanced with gold film (and a thin layer of titanium dioxide) deposited on a D-shaped honeycomb-microstructure optical fiber. This novel fiber design, proposed in [7], is expected to achieve an ultra-high wavelength sensitivity of 20,100 nm/RIU and a minimum resolution as low as 4.98×10^{-7} RIU (from numerical studies). The functionalization of fiber tips with an array of metallic nanodots is experimentally demonstrated to achieve substantially improved large-angle light-collection performance at multiple wavelengths [8]. This concept paves the way for high-performance fiber-based optical devices and is particularly relevant within the context of endoscopic-type applications in life science and light collection within quantum technology.

In addition to metal functionalization, structural engineering in specialty fibers has resulted in the greater flexibility and controllability of optical properties and thus led to versatile applications for physical and biomedical sensing and communications. Multicore fibers that were initially introduced as a promising platform to realize spatial division multiplexed transmission have been attracting growing interest in sensing applications, as reviewed in [9]. Finally, the specific guiding properties at the critical wavelength (CWL) in Mach–Zehnder-based interferometers (MZIs) have been exploited through different designs of few-mode special optical fibers and microfibers for various sensing applications. The recent development of such a configuration, as reviewed in [10], offers desirable sensing advantages, such as large measurement range, high sensitivity, and multi-parameter sensing.

Specialty fibers for distributed fiber sensor technologies are reported in two review papers in the Special Issue. The recent development of specialty fiber-based Brillouin optical time domain analysis (BOTDA) is reviewed in [11]. These specialty fibers present many opportunities for BOTDA systems, including breakthroughs in Brillouin gains, the discriminative sensing of temperature and strains, multiplexed sensing systems, and the distributed sensing of parameters such as curvature, pressure and salinity. In addition, scattering-enhanced specialty-fiber-based distributed acoustic sensing (DAS) systems are reviewed in [12]. These specialty scattering-enhanced optical fibers enable practical applications for DAS systems in many fields, including geological and resource exploration, structural health monitoring, and hydroacoustic exploration.

In addition to spectroscopy and sensing applications, one review paper and two regular papers report specialty fibers for communication applications. Stable orbital angular momentum (OAM) mode generation and transmission in novel fibers are reviewed in [13]. It is worth noting that improvements in fabrication techniques, such as ensuring the uniformity of the longitudinal hollow structures of the novel fibers, remains a critical challenge for OAM mode stability. Through design optimization and simulation investigation, a gold-coated photonic crystal fiber design is proposed to achieve a high loss ratio for polarization filtering at 1550 and 1310 nm [14]. In addition, a dual-core photonic crystal fiber with a liquid–crystal infiltration-based polarization beam splitter is proposed to achieve a splitting bandwidth of 349 nm, covering the entire E + S + C + L + U communication bands [15].

Funding: This research received no external funding.

Acknowledgments: The guest editors would like to thank all authors for their contributions to this Special Issue. We would like to thank all reviewers for providing constructive review comments and recommendations. In particular, we would like to thank editors of *Photonics* journal for the excellent support to this Special Issue.

Conflicts of Interest: The author declares no conflict of interest.

References

1. Cai, D.; Xie, Y.; Guo, X.; Wang, P.; Tong, L. Chalcogenide Glass Microfibers for Mid-Infrared Optics. *Photonics* **2021**, *8*, 497. [CrossRef]
2. Gomółka, G.; Krajewska, M.; Kaleta, M.; Khegai, A.M.; Alyshev, S.V.; Lobanov, A.S.; Firstov, S.V.; Nikodem, M. Operation of a Single-Frequency Bismuth-Doped Fiber Power Amplifier near 1.65 μm. *Photonics* **2020**, *7*, 128. [CrossRef]
3. Pei, W.; Li, H.; Huang, W.; Wang, M.; Wang, Z. All-Fiber Gas Raman Laser by D2-Filled Hollow-Core Photonic Crystal Fibers. *Photonics* **2021**, *8*, 382. [CrossRef]

4. Shi, J.; Ye, X.; Cui, Y.; Huang, W.; Li, H.; Zhou, Z.; Wang, M.; Chen, Z.; Wang, Z. All-Fiber Gas Cavity Based on Anti-Resonant Hollow-Core Fibers Fabricated by Splicing with End Caps. *Photonics* **2021**, *8*, 371. [CrossRef]
5. Ding, H.; Hu, D.J.J.; Yu, X.; Liu, X.; Zhu, Y.; Wang, G. Review on All-Fiber Online Raman Sensor with Hollow Core Microstructured Optical Fiber. *Photonics* **2022**, *9*, 134. [CrossRef]
6. Ni, W.; Yang, C.; Luo, Y.; Xia, R.; Lu, P.; Hu, D.; Danto, S.; Shum, P.; Wei, L. Recent Advancement of Anti-Resonant Hollow-Core Fibers for Sensing Applications. *Photonics* **2021**, *8*, 128. [CrossRef]
7. Bu, W.; Wu, Z.; Shum, P.P.; Shao, X.; Pu, J. Sensitivity Enhanced Refractive Index Fiber Sensor Based on Long-Range Surface Plasmon Resonance in SiO_2-Au-TiO_2 Heterostructure. *Photonics* **2021**, *8*, 379. [CrossRef]
8. Wang, N.; Zeisberger, M.; Hübner, U.; Schmidt, M. Nanograting-Enhanced Optical Fibers for Visible and Infrared Light Collection at Large Input Angles. *Photonics* **2021**, *8*, 295. [CrossRef]
9. Zhao, Z.; Dang, Y.; Tang, M. Advances in Multicore Fiber Grating Sensors. *Photonics* **2022**, *9*, 381. [CrossRef]
10. Lu, C.; Dong, X.; Wu, C. Characteristics of Critical-Wavelength-Existed Fiber-Optic Mach–Zehnder Interferometers and Their Sensing Applications. *Photonics* **2022**, *9*, 378. [CrossRef]
11. Hu, D.J.J.; Humbert, G.; Dong, H.; Zhang, H.; Hao, J.; Sun, Q. Review of Specialty Fiber Based Brillouin Optical Time Domain Analysis Technology. *Photonics* **2021**, *8*, 421. [CrossRef]
12. Sun, Y.; Li, H.; Fan, C.; Yan, B.; Chen, J.; Yan, Z.; Sun, Q. Review of a Specialty Fiber for Distributed Acoustic Sensing Technology. *Photonics* **2022**, *9*, 277. [CrossRef]
13. Wang, Z.; Tu, J.; Gao, S.; Li, Z.; Yu, C.; Lu, C. Transmission and Generation of Orbital ANGULAR Momentum Modes in Optical Fibers. *Photonics* **2021**, *8*, 246. [CrossRef]
14. Wang, C.; Zhang, G.; Wu, Z.; Zhang, Y.; Zhang, Y.; Jiang, L.; Bi, W. A Broadband Gold-Coated Photonic Crystal Fiber Polarization Filter with a High Loss Ratio of Both Polarizations at 1550 and 1310 nm. *Photonics* **2021**, *8*, 488. [CrossRef]
15. Qu, Y.; Han, Y.; Yuan, J.; Zhou, X.; Yan, B.; Wang, K.; Sang, X.; Yu, C. A Novel Liquid Crystal-Filled, Dual-Core Photonic Crystal Fiber Polarization Beam Splitter Covering the E + S + C + L + U Communication Band. *Photonics* **2021**, *8*, 461. [CrossRef]

photonics

MDPI

Review

Chalcogenide Glass Microfibers for Mid-Infrared Optics

Dawei Cai [1,†], Yu Xie [1,†], Xin Guo [1,2], Pan Wang [1,2] and Limin Tong [1,3,*]

1 State Key Laboratory of Modern Optical Instrumentation, College of Optical Science and Engineering, Zhejiang University, Hangzhou 310027, China; caidawei@zju.edu.cn (D.C.); 11730034@zju.edu.cn (Y.X.); guoxin@zju.edu.cn (X.G.); nanopan@zju.edu.cn (P.W.)
2 Intelligent Optics & Photonics Research Center, Jiaxing Institute of Zhejiang University, Jiaxing 314031, China
3 Collaborative Innovation Center of Extreme Optics, Shanxi University, Taiyuan 030006, China
* Correspondence: phytong@zju.edu.cn
† These authors contributed equally to this paper.

Abstract: With diameters close to the wavelength of the guided light, optical microfibers (MFs) can guide light with tight optical confinement, strong evanescent fields and manageable waveguide dispersion and have been widely investigated in the past decades for a variety of applications. Compared to silica MFs, which are ideal for working in visible and near-infrared regions, chalcogenide glass (ChG) MFs are promising for mid-infrared (mid-IR) optics, owing to their easy fabrication, broad-band transparency and high nonlinearity, and have been attracting increasing attention in applications ranging from near-field coupling and molecular sensing to nonlinear optics. Here, we review this emerging field, mainly based on its progress in the last decade. Starting from the high-temperature taper drawing technique for MF fabrication, we introduce basic mid-IR waveguiding properties of typical ChG MFs made of As_2S_3 and As_2Se_3. Then, we focus on ChG-MF-based passive optical devices, including optical couplers, resonators and gratings and active and nonlinear applications of ChG MFs for mid-IR Raman lasers, frequency combs and supercontinuum (SC) generation. MF-based spectroscopy and chemical/biological sensors are also introduced. Finally, we conclude the review with a brief summary and an outlook on future challenges and opportunities of ChG MFs.

Keywords: mid-infrared (mid-IR); chalcogenide glasses (ChGs); optical microfibers (MFs); supercontinuum (SC); molecular sensing

Citation: Cai, D.; Xie, Y.; Guo, X.; Wang, P.; Tong, L. Chalcogenide Glass Microfibers for Mid-Infrared Optics. *Photonics* **2021**, *8*, 497. https://doi.org/10.3390/photonics8110497

Received: 5 October 2021
Accepted: 3 November 2021
Published: 5 November 2021

1. Introduction

In the past decades, optical microfibers (MFs) have brought wide opportunities in renewing and expanding fiber optics and technology at a wavelength scale [1–5]. Fabricated by a high-temperature taper-drawing technique [6], MFs exhibit excellent surface smoothness and diameter uniformity, which bestow them favorable features of low waveguiding loss [1,7], tight optical confinement, high fractional evanescent fields and large manageable waveguide dispersion [8], making them a versatile platform for both scientific research (e.g., optical nonlinearity [9–11] and atom optics [12–14]) and technological applications (e.g., optical sensors [15–17] and microlasers [18–20]). As one of the most widely studied optical materials, MFs based on silica have been employed for a variety of applications [3,4]; however, their uses in the mid-infrared (mid-IR, 2.5–20 μm) spectral range are restricted by strong absorption of silica glass [21]. In recent years, along with the rapid progress in mid-IR photonics, MFs for mid-IR photonics have attracted increasing interest [22,23].

Typically, mid-IR MFs are fabricated from a category of mid-IR-transparent materials, including oxide glasses (e.g., germanates [24], fluorotellurites [25] and tellurites [26]), fluoride glasses (ZBLAN) [27,28], chalcogenide glasses (ChGs, glasses containing one or more chalcogens: sulfur (S), selenium (Se) and tellurium (Te)) [29–31], as well as semiconductors (e.g., cadmium telluride (CdTe) and [32] silicon (Si) [33]). Among these materials, ChGs are mostly investigated for their special merits, including broadband intrinsic transparency

(0.5–25 μm), high optical nonlinearity (about 100–1000 times larger than that of silica glass) and hospitality to rare-earth dopants [29,34,35]. The typical optical loss of ChG fibers in the mid-IR is around 0.1–10 dB/m [35], while the lowest loss, down to 0.012 dB/m at 3 μm, has been reported in multimode As_2S_3 fibers [36]. Additionally, photonic devices based on ChG MFs, including resonators [37,38], gratings [39–41] and sensors [42,43], have been investigated in the visible (VIS) or near-infrared (near-IR) region and have recently extended to the mid-IR region [44–46]. In this review, combined with the latest progress in ChG MFs, we first introduce the fabrication and optical waveguiding properties of ChG MFs. Secondly, we introduce ChG MF-based passive optical devices (e.g., couplers, resonators and gratings), as well as active and nonlinear applications of ChG MFs in the mid-IR region including Raman lasers, frequency combs and supercontinuum (SC) generation, followed by MF-based spectroscopy and chemical/biological sensors. Finally, we summarize this review and present a brief outlook into future challenges and opportunities in this field.

2. Fabrication

Excellent surface smoothness and geometric uniformity are regarded as essential guarantees for achieving low-loss optical waveguiding in MFs. Typically, for MFs made of glass, the high-temperature taper-drawing process is the best technique for fabricating high-quality MFs with extremely low surface roughness (e.g., <0.5 nm [47]), excellent diameter uniformity and circular cross-section [48]. Tapered fibers based on mid-IR materials such as tellurite glass and fluorotellurite glass have been successfully fabricated [25,26]. Besides, owing to the relatively low softening temperature (T_s, e.g., 180 °C for As_2S_3 glass) of ChGs, ChG MFs are typically fabricated from ChG fibers heated by an electric heater (several millimeters in heating zone length) at a temperature slightly higher than the T_s [31,49], as schematically illustrated in Figure 1a. When stretching the fiber bi-directionally at both sides, the fiber is elongated and forms a waist (i.e., the MF) with two transition tapers connected to the untapered fibers. Close-up scanning electron microscope (SEM) images of an as-fabricated ChG (As_2S_3 glass) MF are shown in the inset of Figure 1a, showing the excellent surface smoothness and diameter uniformity.

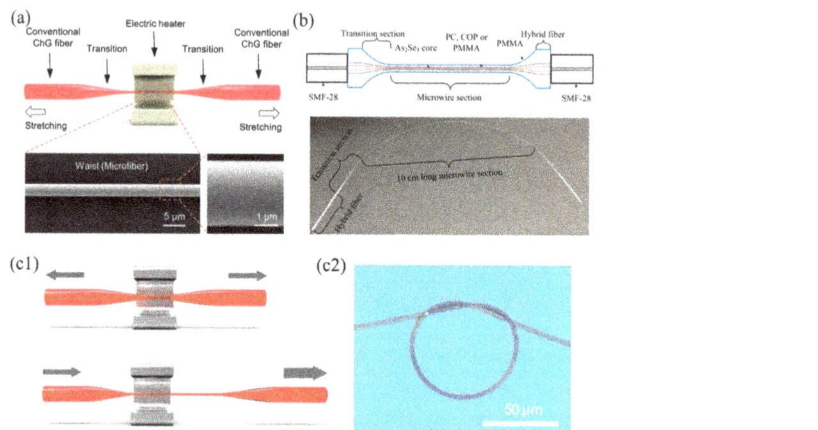

Figure 1. Fabrication of ChG MFs. (**a**) Schematic diagram of taper drawing a ChG MF from a conventional ChG fiber. Insets: SEM images of an as-fabricated As_2S_3 MF. (**b**) Schematic diagram of taper drawing a hybrid ChG-PMMA MF. Inset: optical image of as-fabricated hybrid ChG biconical fiber taper. Reprinted with the permission from [49] © The Optical Society. (**c1**) Schematic diagram of taper drawing a long-waist ChG MF by a two-step taper-drawing process. (**c2**) 62-micrometer diameter As_2S_3 MF-based knot assembled from a 3.5-micrometer diameter MF fabricated by the method in (**c1**). Reprinted with the permission from [45] © CLP Publishing.

Usually, made from soft glass, as-fabricated ChG MF is relatively fragile and easily damaged by mechanical disturbance. To enhance the mechanical strength of ChG MF and reduce its sensitivity to the surrounding environment, in 2010, Baker et al. reported a kind of hybrid ChG-PMMA fiber taper consisting of an As_2Se_3 core and a PMMA cladding [31]. The PMMA cladding possesses a T_s compatible with As_2Se_3 glass and enables taper-drawing fabrication of an MF from the two materials at the same time. A 9.7-cm-length hybrid biconical As_2Se_3 fiber taper with a core diameter down to 0.8 μm and a PMMA cladding diameter of 2.4 μm was fabricated successfully. In 2011, they demonstrated a generalized heat-brush tapering approach, which allowed the ratio of the feed and drawing velocities changes within each tapering sweep [50]. The waist diameter of the as-fabricated As_2Se_3 fiber taper decreased linearly from 15 to 10 μm over a 2.0-centimeter length. In 2016, Li et al. fabricated polymer-cladded (PC, COP or PMMA) As_2Se_3 fiber tapers with a waist diameter down to 1.5 μm and a waist length of 10 cm [49], as illustrated in Figure 1b. Both ends of the ChG fiber tapers were polished and adhered to single-mode silica fibers using UV-cured epoxy for efficient light input and output. Owing to the large waveguide nonlinear parameter and engineerable chromatic dispersion of the MFs, nonlinear optical effects including self-phase modulation, four-wave mixing and Raman scattering with wavelength up to 2.2 μm could be achieved.

To ease the tapering system and obtain long MFs with uniform diameters with a large reduction ratio of fiber diameter between the preform fiber and the final MF, in 2020, Xie et al. demonstrated a two-step taper-drawing process [45]. As schematically shown in Figure 1c1, commercial As_2S_3 fiber was first heated and stretched bi-directionally with a low speed (0.4 $mm \cdot s^{-1}$). Secondly, the pre-drawn MF with a diameter in several micrometers was stretched unidirectionally at a fast rate (6 $mm \cdot s^{-1}$) at one end and a slow rate (0.4 $mm \cdot s^{-1}$) at the other end to draw a long-waist MF with a uniform diameter. Biconical fiber tapers with a waist length longer than 5 cm and diameters of around 3 μm were easily fabricated for assembling mid-IR photonic devices. In addition, as-drawn As_2S_3 MF (3.5-μm diameter) could be assembled into a 62-μm diameter knot structure with an elastic strain of up to 5.6% (Figure 1c2), indicating the excellent structural uniformity and surface-defect-free condition of the MF.

3. Optical Waveguiding Properties

MFs offer a number of attractive optical properties including tight optical confinement, strong evanescent field and tailorable waveguide dispersion [4,8]. The special merits of ChGs bestow the ChG MFs more possibilities. In this section, we discuss the optical properties of As_2S_3 (transmission range: 0.7–6 μm [35]) and As_2Se_3 (transmission range: 1–10 μm [35]) MFs contributed to applications such as optical waveguiding and nonlinearity. We assumed that the MF has a circular cross-section and an infinite air cladding with a step-index profile, and only the fundamental mode (HE_{11}) was considered. The refractive indices of the As_2S_3 (As_2Se_3) glass around 1.55-μm and 4-μm wavelengths are 2.44 (2.73) and 2.41 (2.68), respectively [51,52]. The refractive index of As_2Se_3 glass around 8-μm wavelength is 2.67.

The effective refractive index ($n_{eff} = \beta/k_0$, β is the propagation constant of the guided mode, $k_0 = 2\pi/\lambda$ is the wavenumber of the transmission wavelength λ), effective mode area (A_{eff}) and fractional evanescent fields (η) are essential waveguiding parameters for designing fiber-optic devices, including near-field couplers, gratings, resonators and sensors [31,53]. Figure 2a–c shows the calculated As_2S_3 and As_2Se_3 MF diameter-dependent n_{eff}, A_{eff}, and η in the mid-IR region, respectively. The results in the 1.55-μm wavelength (dash lines) are also shown for comparison. ChG MFs with diameters larger than the wavelengths of the guided light exhibit n_{eff} of HE_{11} mode close to the refractive indices of the bulk materials (Figure 2a), which is independent on the guided wavelength. When the diameter reduces close to or smaller than the wavelength, the n_{eff} decreases dramatically, indicating more fractional evanescent fields exist around the MF (Figure 2b), beneficial for enhancing interactions between the mid-IR light and samples surrounding the MF in applications such as optical trapping, mid-IR spectroscopy and chemical/biological sensing related to molecular fingerprints [3,17,54]. Meanwhile, owing to the high nonlinear refractive index n_2 of As_2S_3 and As_2Se_3 materials and tight optical confinement (i.e., small A_{eff}) of the guided modes in MFs, these MFs usually have large effective nonlinearity γ ($\gamma = k_0\, n_2/A_{eff}$) and relatively low-threshold nonlinear optical effects when their diameters are close to the wavelength of the guided light in the material (i.e., λ/n, λ and n are the wavelength in air and refractive index of material, respectively. See Figure 2c) [31,55].

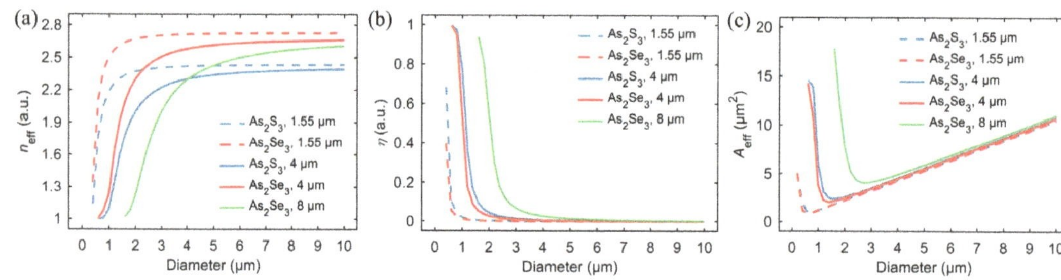

Figure 2. Numerical solutions of diameter-dependent (**a**) effective refractive index (n_{eff}), (**b**) fractional evanescent fields (η), and (**c**) effective mode area (A_{eff}) of As_2S_3 and As_2Se_3 MFs at wavelengths of 1.55 μm, 4 μm, and 8 μm, respectively.

4. Photonic Applications

4.1. Near-Field Optical Couplers

MF near-field couplers are simple fiber-optic components that can be used for efficient in/out-coupling with other optical structures such as micro/nano waveguides [48,56–58], whispering gallery mode (WGM) microcavities [59,60] and photonic crystal structures [61–63]. By precisely adjusting the coupling length or gap distance between the MF and the photonic structure, the phase-matching condition (and thus the coupling efficiency) can be precisely adjusted. MF near-field couplers possess the advantages of easy operation and small footprints.

With increasing demands of in/out coupler at mid-IR range in recent years, mid-IR MFs have been applied in the near-field optical coupling of a variety of photonic structures, including mid-IR waveguides and microcavities. For example, in 2018, to test the waveguiding properties of CdTe microwires (a quasi-one-dimensional single-crystal semiconductor) in the mid-IR region, Xin et al. used a 1.5-μm-diameter As_2S_3 MF as input coupler and a 5-μm-diameter As_2S_3 MF as output coupler for a 1.5-μm-diameter CdTe microwire (Figure 3a) [32]. The input MF diameter was carefully chosen to ensure single-mode operation within the 4.45–4.7 μm range. The relatively large effective refractive index contrast between the As_2S_3 MF and CdTe microwire can decrease the end-face reflection and achieve flat broadband coupling over the 4.45–4.7 μm and 8.36–8.6 μm range.

High-quality factor (Q, defined as the ratio of the resonant wavelength to the full width at half-maximum (FWHM)) resonators in the mid-IR are critical to promote the performances of mid-IR laser sources, cavity-based sensors and mid-IR frequency comb [66]. However, achieving ultra-high Q cavities and measuring their quality factors in the mid-IR remain challenging. In 2016, Lecaplain et al. developed an efficient coupling technique for WGM-based fluoride crystalline microresonators assisted by tapered ChG fibers (Figure 3b1) [64]. The ChG tapered fibers were pulled down to the 1-μm diameter (i.e., a ChG MF) to achieve phase matching with the fundamental WGMs of the resonator. By translating the microresonator along the taper relative to the waist, the phase-matching conditions were tuned precisely. Figure 3b2 shows the normalized transmission as a function of the coupling parameter. The data were consistent with the theoretical model ($K = \kappa_{ex}/\kappa_0 = \left(1 \pm \sqrt{T}\right)/\left(1 \mp \sqrt{T}\right)$, in which κ_0 is the intrinsic loss rate, κ_{ex} is the coupling loss rate and T is the transmission). More recently, Xie et al. used As_2S_3 biconical tapered fibers for the first characterization of ChG microsphere resonators in the mid-IR region (Figure 3c) [65]. A 1.6-μm-diameter As_2S_3 fiber taper was used for near-field coupling with ChG microspheres with the coupling loss less than 3 dB over 4.465–4.705 μm.

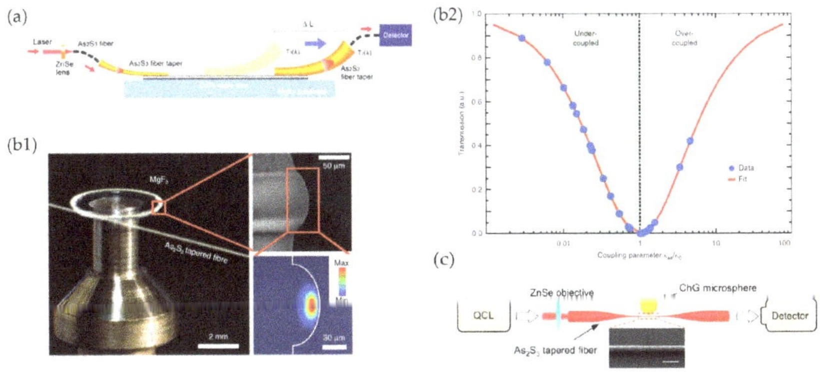

Figure 3. Mid-IR MF near-field couplers. (**a**) Schematic diagram of As_2S_3 tapered fiber coupler used for CdTe microwire transmission spectra measurement. Reprinted with the permission from [32] © The Optical Society. (**b1**) Optical micrograph and SEM image of the MgF_2 microresonator coupled with an As_2S_3 biconical fiber taper. Inset: Finite element model simulation of the optical intensity profile of the fundamental WGM at λ = 4.5 μm. (**b2**) Transmission as a function of the coupling parameter $K = \kappa_{ex}/\kappa_0$ for varying taper waist radius when taper-resonator coupling was achieved. Reprinted with the permission from [64] © Springer Nature. (**c**) Schematic illustration of As_2S_3 biconical fiber taper used for characterization of an As_2S_3 microsphere. Reprinted with the permission from [65] © John Wiley and Sons.

4.2. MF-Based Resonators

The combination of low-loss waveguiding and high-efficiency evanescent coupling between adjacent MFs bestows a kind of competitive high-Q resonators in various structures (e.g., loop [67,68], knot [1,69], ring [70–72] and coil [73,74]), showing the desirable features of easy fabrication, fiber compatibility and resonance tunability. During the past two decades, silica MF-based resonators have attracted wide attention in the VIS and near-IR applications, including optical filters, sensors and lasers [57,74–76]. Additionally, there are several kinds of high-Q (10^4–10^5) resonators based on ChG MFs in the near-IR region, including SU8 polymer-embedded As_2S_3 MF knot resonators (MKRs) (Figure 4a1,a2) [38], self-touching As_2Se_3 MF loop resonators (MLRs) by thermally shaping As_2Se_3 glass into an MF and splicing to cleaved silica fiber tapers (Figure 4b1,b2) [37] and resonators based on photoinduced WGMs localized in the cross-section of the ChG MFs [77]. These studies are beneficial to the further exploitations of ChGs, but they were not extended to the mid-IR region.

In 2020, Xie et al. demonstrated the first ChG MF-based resonator in the mid-IR region [45]. As_2S_3 MFs with uniform diameters and long waists were fabricated via a two-step taper drawing process (Figure 1c1). The MF structure was immersed in a drop of ethyl alcohol to avoid possible surface damage and assembled to a knot with double ends naturally connected to fiber tapers. Figure 4c1 shows an 824-μm diameter MKR assembled from a 3.2-μm-diameter As_2S_3 MF. A Q factor of about 2.84 × 10^4 was obtained at the wavelength of 4469.14 nm (Figure 4c2,c3). The free spectral range (FSR) of the MKR can not only be tuned by tightening the knot structure in liquid (Figure 4c4), but the resonance peaks can also be thermally tuned at a thermal tuning rate of 110 pm·°C^{-1} (Figure 4c5). Furthermore, to increase the long-term stability of the MKR device and isolate it from contaminations, a 551-μm-diameter MKR was embedded in low-index polymer (PMMA) film while a Q factor of 1.1 × 10^4 around 4.5-μm wavelength was retained.

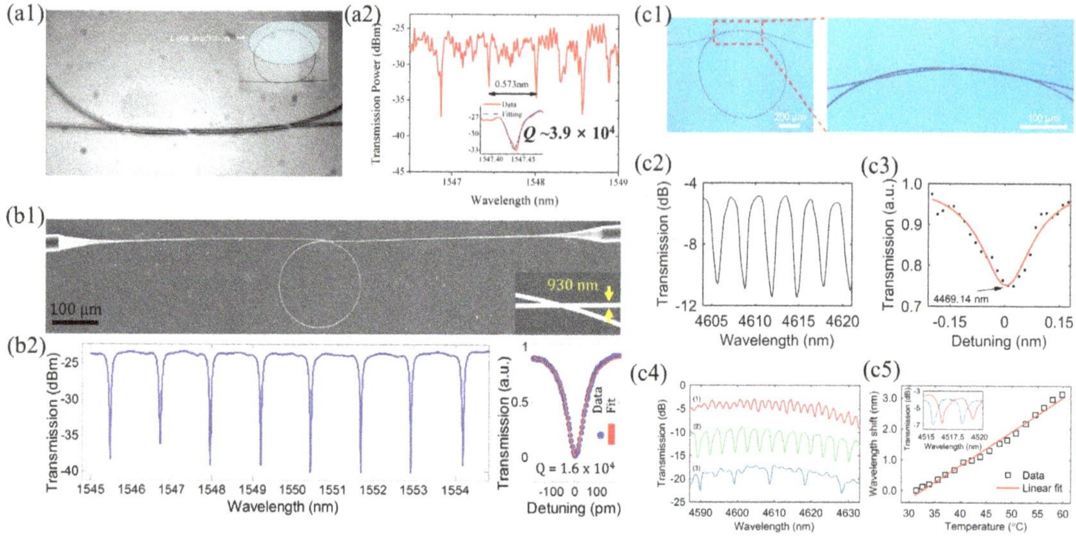

Figure 4. ChG MF-based resonators. (**a1**) Optical micrograph of the intertwisted overlap region of a SU8 polymer-embedded As_2S_3 MKR. (**a2**) Transmission spectrum of the MKR. Reprinted with the permission from [38] © The Optical Society. (**b1**) SEM image of a 178.2-μm diameter As_2Se_3 MLR. The inset shows the self-coupling region. (**b2**) Transmission spectrum of the MLR. Inset: Lorentzian fitting to a resonance mode at 1552.9 nm. Reprinted with the permission from [37] © The Optical Society. (**c1**) Optical micrograph of an 824-μm diameter As_2S_3 MKR. Inset: optical micrograph of the intertwisted overlap area. (**c2**) Transmission spectrum of the MKR. (**c3**) Lorentzian fitting to a resonance mode centered at 4469.14-nm wavelength. (**c4**) Transmission spectra of an As_2S_3 MKR by tightening the knot structure in liquid (from top to bottom). (**c5**) Resonance shift of an MKR with the temperature rising from 31.4 °C to 59.8 °C. Inset: transmission spectra of resonance modes corresponding to 31.4 °C (blue line) and 40.6 °C (red line). Reprinted with the permission from [45] © CLP Publishing.

As MF-based mid-IR resonators are essential for MF-based optical technology, they are expected to attract increasing attention in mid-IR optical applications, including mid-IR microlasers, narrow-wide filters, cavity-enhanced spectroscopy and the label-free detection of molecules [54,78].

4.3. MF Gratings

MF gratings have attracted intense attention owing to their miniaturized footprints, large fraction of evanescent field and high flexibility. Typical grating fabrication techniques include femtosecond (fs) laser inscription [79,80], ultraviolet laser irradiation [81] and focused ion beam milling [82,83]. So far, MF gratings have been employed for photonic sensors for measuring the refractive index, force and temperature, as well as for functionalized devices, including tunable filters and Fabry–Perot cavities [84–87].

Previously, ChG fiber gratings working in the mid-IR region [88] and ChG MF gratings working in the near-IR have been reported [40,89] (Figure 5a), while mid-IR MF gratings have been seldomly studied. In 2020, Cai et al. demonstrated the first mid-IR MF Bragg grating (mFBG) based on an As_2S_3 MF (Figure 5b1) [46]. The transmission spectrum shows a 15 dB depth around 4.5-micrometer wavelength (Figure 5b3), indicating a photo-induced refractive index change of 0.02. The dependence of the grating formation on the accumulated influence of exposure power density and time was also investigated. The interference pattern exposure technique can be applied to other photosensitive ChG MFs.

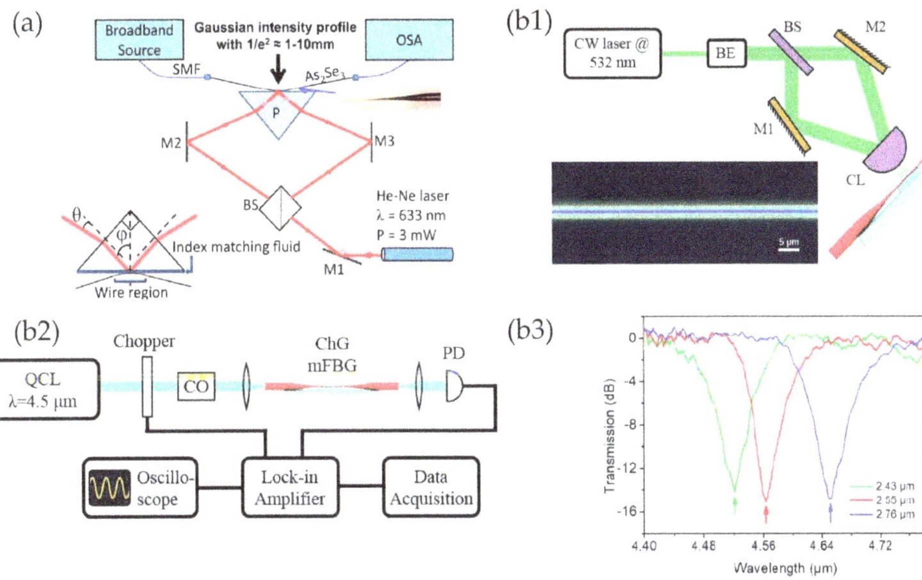

Figure 5. Mid-IR MF gratings. (**a**) Experimental setup for the As_2Se_3 MF Bragg grating photo inscription assisted by prism and in-situ monitoring of the process. Reprinted with the permission from [40] © The Optical Society. (**b1**) Schematic illustration of the As_2S_3 mFBG fabrication setup using interference pattern formed by two 532 nm beams. Inset: optical micrograph of a 2.4-μm diameter ChG mFBG. (**b2**) Schematic illustration of the optical characterization of ChG mid-IR mFBGs. (**b3**) Transmission spectra of mFBGs with different MF diameters. Reprinted with the permission from [46] © The Optical Society.

Compared with conventional long-period grating (LPG), MFs could increase the depth of the evanescent field into the ambient environment, which may be helpful to enhance the performance of MF-based optical sensors. In 2019, Wang et al. reported a temperature sensor based on long-period fiber gratings (LPFGs) inscribed on tapered multimode ChG fibers [90]. Simulation results show that the temperature sensitivity could be increased by reducing the waist diameter of tapered fiber. A resonant wavelength at 3 μm was obtained when the period of LPFGs was 176 μm and the maximum sensitivity of temperature could achieve 12.6 nm/°C at a 3-micrometer wavelength.

4.4. Raman Lasers

The Raman gain provided by silica fibers has led to Raman fiber lasers in the near-IR range for optical imaging and telecommunications-related applications [91]. The mid-IR Raman fiber lasers have been applied in LIDAR, pharmaceutical science and military applications [92,93]. The Raman gain coefficients of typical ChGs (e.g., 4.3~5.7 × 10^{-12} m/W for As_2S_3 and 2~5 × 10^{-11} m/W for As_2Se_3 at 1.5 μm [94]) are more than 900 times larger than that of silica, making them excellent candidates for the mid-IR Raman lasers. In 2017, Abdukerim et al. demonstrated an all-fiber Raman laser based on a PMMA-coated 10-centimeter $As_{38}Se_{62}/As_{38}S_{62}$ MF [95]. The Raman laser emitted at 2.025 μm with pump pulses at 1.938 μm and a threshold pump power of 4.6 W (Figure 6a). To generate a shower of Raman solitons at the same wavelength, in 2021, Guo et al. pumped a 1960-nanometer picosecond laser into a tapered fluorotellurite MF and demonstrated Raman solitons around 3 μm [96]. The group velocity dispersion was controlled by the diameter of the fluorotellurite MF, and all the generated solitons could be put at the same wavelength.

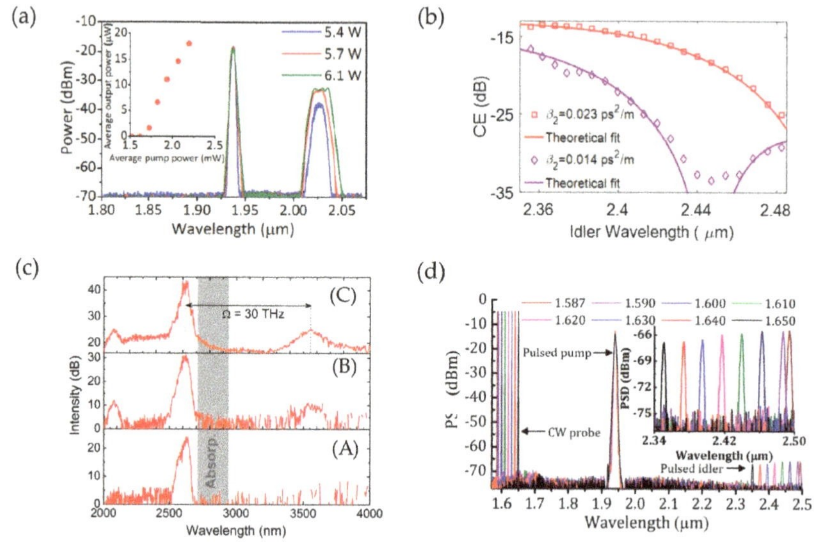

Figure 6. Raman scattering, four-wave mixing and modulation instability in mid-IR MFs. (a) Output spectra of the Raman laser based on a multi material ChG MF with increasing pump power. Inset: Average output power versus average pump power. Reprinted with the permission from [95] © AIP Publishing. (b) Conversion efficiency vs. idler wavelength for different tapering periods of all-fiber wavelength converters based on tapering As_2Se_3 MFs. Reprinted with the permission from [97] © The Optical Society. (c) Experimental observation of MI when pumping an As_2Se_3-polymer MF at λ = 2620 nm in the normal dispersion regime. Input peak powers estimated at (A) 200 mW, (B) 1 W and (C) 10 W. Reprinted with the permission from [98] © The Optical Society. (d) Measured output idler spectra resulting from wavelength tuning of the probe from 1.587 to 1.650 μm with a power of 0.54 mW. Reprinted with the permission from [99] © The Optical Society.

4.5. Frequency Conversion

Besides the large nonlinear optical coefficient and broad mid-IR transparency, ChGs are excellent candidates for mid-IR frequency conversion (e.g., optical parametric oscillation (OPO) and optical parametric amplification (OPA)) for their high laser damage threshold and moderate birefringence [100]. The efficiency of frequency conversion depends on phase matching, nonlinear gain and losses. The tapering of ChG fibers to increase nonlinearity while compensating the normal material dispersion has been demonstrated. In 2016, Abdukerim et al. reported the first fiber optical parametric oscillator (FOPO) using an As_2Se_3-COP MF at 2 μm based on four-wave mixing (FWM) [101]. The MF with a core diameter of 1.47 μm led to a waveguide nonlinearity of 24 $W^{-1}m^{-1}$ and a zero-dispersion wavelength of 1.875 μm. The FOPO had a pump threshold power of 5 W and covered a tunable wavelength range of 290 nm. In the FWM process, where pump and idler are far-detuned, the exact phase-matching wavelength is highly sensitive to chromatic dispersion and hard to control. In 2019, Alamgir et al. demonstrated the far-detuned wavelength converter by in-situ monitoring the output wavelength while tapering down the As_2Se_3 fiber [97]. As the tapering proceeded, the phase-matching condition was satisfied closer to the pump wavelength and the gain spectrum gradually shifted towards shorter wavelengths. The idlers were generated with a spectral range of 2.347–2.481 μm from a pump wavelength of 1.938 μm (Figure 6b).

The modulation instability (MI) is another frequency conversion process where the low-amplitude noise is parametrically amplified on the pump signal together with the growth of symmetric sidebands on both sides of the pump. Although most of the spontaneous MI requires pumping in the anomalous dispersion regime, MI can also be pumped in a normal dispersion regime when the fiber possesses a higher-order group velocity dispersion profile. To prove the broadband frequency conversion due to normal dispersion MI, in 2014, Godin et al. demonstrated a parametric frequency conversion based on MI at 2 and 3.5 μm by pumping an As_2Se_3-polymer MF with an fs OPO at 2.63 μm [98]. The measured frequency shift of 30 THz was the largest reported, using normal dispersion pumped MI in a single-pass configuration (Figure 6c). To prove the potential for improving the technical specifications of mid-IR fiber wavelength converter in terms of tunability and conversion efficiency, in 2017, Li et al. generated tunable parametric sidebands via MI in a 10-centimeter long As_2Se_3-CYTOP tapered MF with a diameter of 1.625 μm (Figure 6d) [99]. The widely spaced Stokes and anti-Stokes bands were generated with a frequency shift as large as 49.3 THz, the largest reported in soft glass materials.

4.6. Supercontinuum Generation

Supercontinuum (SC) generation in optical fibers originates from the interaction between ultrashort laser pulses and fibers with high nonlinearity [107]. ChGs are regarded as good candidates for mid-IR SC generation with a large transparency window and comparatively high optical nonlinearity [22].

The uses of MFs reduce the power consumption and allow a shorter interaction length for SC generation. In 2013, Al-kadry et al. demonstrated SC generation from 1260 to 2200 nm using a 10-centimeter long As_2Se_3 MF [108]. The pump wavelength was self-frequency shifted to 1775 nm in the anomalous dispersion regime of the As_2Se_3 MF to avoid two photons' absorption. In the same year, Rudy et al. reported a generated SC from 1 to 3.7 μm in a 2.1-millimeter length As_2S_3 MF with 300-pJ 2-nanometer ultrafast pump pulses [109]. The spectral bandwidth was limited by the absorption caused by third harmonic generation. In 2014, Al-kadry et al. further used pump pulses of 800 fs at a 1.94-μm wavelength to generate an SC covered bandwidth of two octaves from 1.1 to 4.4 μm [102]. The waveguide nonlinearity of a 1.6-micrometer diameter As_2Se_3 MF was as high as 32.2 $W^{-1}m^{-1}$, resulting in a pumping power of only 500 pJ (Figure 7a). In 2017, Wang et al. investigated the dependence of the SC spectral behavior on the waist diameter and the transition region length of the tapered MF (Figure 7b) [103]. They reported a 1.4–7.2-micrometer SC generation with an average power of 1.06 mW from a 12-cm-length As-S tapered MF pumped at 3.25 μm.

Tapering the fiber is also an important method to obtain a highly coherent SC by designing an all-normal dispersion of fibers to suppress soliton effects including soliton fission and Raman soliton self-frequency shift. In 2019, Li et al. reported a 1.5–8.3 μm SC generation with a high coherence property from As-S tapered MF pumped at 3.75 μm (Figure 7c) [104]. They found that the coherence properties gradually become worse when the waist core diameters of tapered MFs decrease and a flat and near-zero dispersion is beneficial for high-coherence SC generation.

To meet the demanding specifications including robust and alignment-free sources of broadband SC, the multimaterial all-fiber-based SC systems have drawn much attention. In 2015, Sun et al. fabricated multimaterial As_2Se_3-As_2S_3 ChG MFs with a zero-dispersion wavelength less than 2 μm [110]. The SC generation from 1.4 μm to longer than 4.8 μm was demonstrated by pumping the MF with 100 fs pulses at 3.4 μm. In 2017, Hudson et al. reported a 2.4 octave-SC generation by pumping a robust polymer-protected As_2Se_3-As_2S_3 MF with 230 fs, 4.2 kW peak power pulses at 3 μm (Figure 7d) [105]. An average power spectral density of 0.003 mW/nm was achieved. Furthermore, significant power existed at various octave points throughout the SC, making the system a good candidate for an *f*-2*f* interferometer.

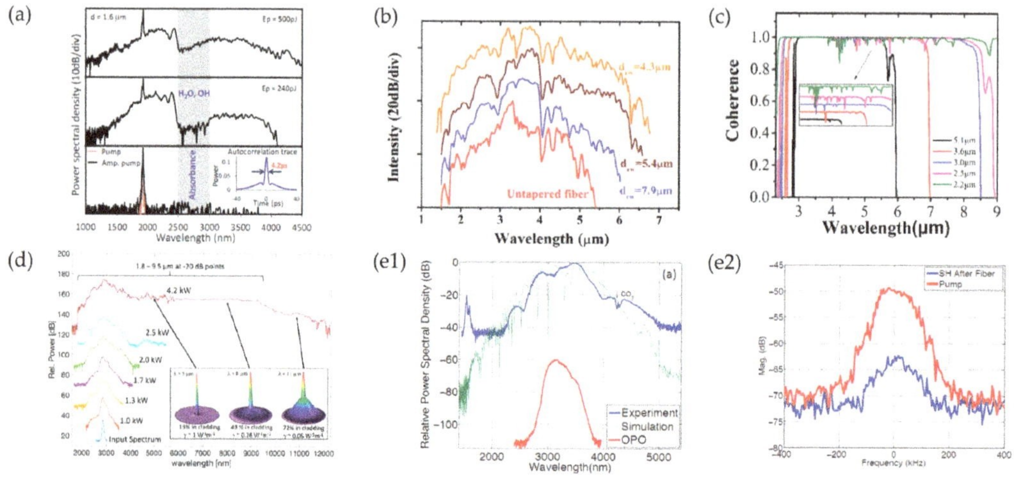

Figure 7. Supercontinuum generation and frequency comb generation in mid-IR MF. (**a**) SC spectra from amplified pump pulses using a 1.6-micrometer diameter As_2Se_3 MF for various amplified pump power levels. Inset: the autocorrelation trace of the amplified pump pulse. Reprinted with the permission from [102] © The Optical Society. (**b**) Measured SC spectra generated from 12-centimeter long As-S tapered MFs with diameters varying from 4.3 and 7.9 μm pumped at 3.25 μm. Reprinted with the permission from [103] © The Optical Society. (**c**) Simulated SC spectra coherence properties for different waist core diameters in As-S MFs. Reprinted with the permission from [104] © IEEE. (**d**) Spectral expansion at increasing peak power in a polymer-protected As_2Se_3/As_2S_3 MF pumping at 3 μm. The inset shows the mode profile in the MF section (3-μm core diameter) at various wavelengths. Reprinted with the permission from [105] © The Optical Society. (**e1**) Spectrum of SC generation around 3 μm from the As_2S_3 tapered MF compared with the OPO output and the simulation result (**e2**) RF spectrum of the interference of a c.w. laser at 1564 nm with the second harmonic (SH) of the tapered MF output. Reprinted with the permission from [106] © The Optical Society.

It is worth mentioning that another type of ChG MFs, tapered fibers with suspended-core geometry that operate in the anomalous dispersion regime, have been used to generate SC within a broad spectrum. In 2018, Anashkina et al. designed and developed $As_{39}Se_{61}$ suspended-core MFs for mid-IR SC generation. SC generation ranging from 1 to 10 μm was proposed with 150-fs 100-pJ pump pulses at 2 μm [111]. In 2020, Leonov et al. reported SC generation covering 1.4 to 4.2 μm in $As_{39}S_{61}$ suspended-core MFs pumped by a Cr:ZnSe laser. A maximum SC average power of ~35 mW was obtained with an optical efficiency of ~35% [112].

In addition, materials other than ChGs have also been investigated for SC generation. For example, compared to that of ChG, telluride glasses have shorter zero-dispersion wavelengths that are desirable for near-to-mid IR SC generation. In 2019, Saini et al. obtained an SC spectrum spanning 1.28–3.31 μm using a 3.2-centimeter-long tellurite tapered MF pumped with 200 fs laser pulse at 2.0 μm [26]. Compared to fluoride and chalcogenide fibers, fluorotellurite fibers based on TeO_2-BaF_2-Y_2O_3 have better chemical and thermal properties. In 2019, Li et al. demonstrated broadband SC generation from 0.6 to 5.4 μm in a fluorotellurite tapered MF pumped by a 2010 nm fs pulses [25]. Recently, polysilicon waveguides have emerged as an alternative platform as they are cheap and flexible to be incorporated into a wide range of geometries. In 2019, Ren et al. fabricated the tapered polycrystalline silicon core MFs (SCFs), and an SC spanning from 1.8 to 3.4 μm was obtained [33].

4.7. Frequency Combs

In the past years, frequency combs have been studied from the VIS and near-IR to the mid-IR regions. Mid-IR combs are of great use for molecular fingerprint spectroscopy and are typically produced by OPOs [113,114], difference frequency generation [115,116], microresonators [117,118] and fs laser generation [119]. Wide frequency combs spectra in mid-IR have also been demonstrated using PPLN-based OPOs pumped by fiber-laser-based near-IR combs [120]. The mid-IR MFs could further broaden the frequency combs by SC generation when the generated SC maintains coherence with the pump laser. In 2012, Marandi et al. used a subharmonic OPO and subsequent SC generation in the tapered MF to convert the pulses of a conventional 1.5-micrometer frequency comb source to mid-IR [106]. The spectrum of the MF output extended from 2.2 to 5 μm at 40 dB below the peak (Figure 7e1). The beat frequencies in Figure 7e2 verified that the coherence properties of the initial frequency combs presevered. It is worth mentioning that, to simplify the mid-IR frequency comb generation, in 2014, Lee et al. demonstrated mid-IR frequency comb generation using a hybrid ChG-silica nanospike waveguide [121]. The waveguide was pumped directly by a two-micrometer frequency comb and could minimize the unwanted dispersion of pump pulses. The waveguide was small and simple using free-space coupling into and out of the fiber without an intermediate OPO.

4.8. MF-Based Sensors

The optical fiber sensors operating in the mid-IR region where the vibrational and rotational states of molecules are located play an important role in biomedical [122], chemical [123] and environmental applications [124].

The principle of fiber evanescent wave spectroscopy (FEWS) is based on the interaction between evanescent fields and chemical or biological specimens to be investigated, showing desirable advantages of analyzing samples in situ in real time. In past decades, many FEWS experiments have been applied in different areas including the detection of contaminants in water [128], metabolic profiling of chronic diseases [129] and monitoring of chemical processes [130]. The characterization of human tissue based on mid-IR FEWS is an important method for early and rapid diagnosis of diseases including tumors or cancers [131]. Typically, the fiber diameter used for FEWS is much larger than the working wavelength, leaving much space for sensitivity enhancement.

Owing to the characteristic fingerprints of a wide range of gases (e.g., CO, CO_2, NO and CH_4) falling in the mid-IR region, MF-based gas sensors have been proposed based on mid-IR absorption spectroscopy in the last decade [132]. In 2019, Huang et al. demonstrated a nonlinear gas sensor based on third harmonic generation (THG) in cascaded ChG MFs for detecting CH_4 (Figure 8a1) [125]. Mid-IR fingerprint light was absorbed in the first section of MF and the rest transmission signal was converted into a near-IR signal after THG at the second section. The proposed mid-IR ChG MF CH_4 sensor had a theoretical detection limit as low as 7.4×10^{-8} (Figure 8a2). Another configuration combines graphene with MFs, which can provide high-sensitivity optical gas sensing from permittivity change of the graphene. In 2019, Wang et al. reported a mid-IR gas sensor based on the graphene Bragg grating integrated with the Si slot MF (Figure 8b1) [126]. The NO_2 concentration could be determined from the 3 dB bandwidth of the Bragg grating's reflection band (Figure 8b2). A detection limit below 1 ppm and a sensitivity of 1.02 nm/ppm were achieved.

The optimization of the tapered transition structure of the fiber sensor is essential to obtain high-efficiency excitation of the evanescent fields to further enhance the sensor sensitivity. In 2021, Wang et al. reported a geometrical optimization of Ge-Sb-Se tapered MF for concentration sensing of methanol (Figure 8c1,c2) [127]. The highest sensitivity of 0.0120 a.u./% was obtained from the MF with a down-taper transition length of 5.4 mm and an up-taper transition length of 7.9 mm. Meanwhile, Wang et al. from the same group investigated the dependence of the sensitivity (of aqueous ethanol solution) on the MF parameters (e.g., diameter and length) [133] and predicted that the sensitivity could be improved by decreasing the MF diameter and increasing the length of the taper waist.

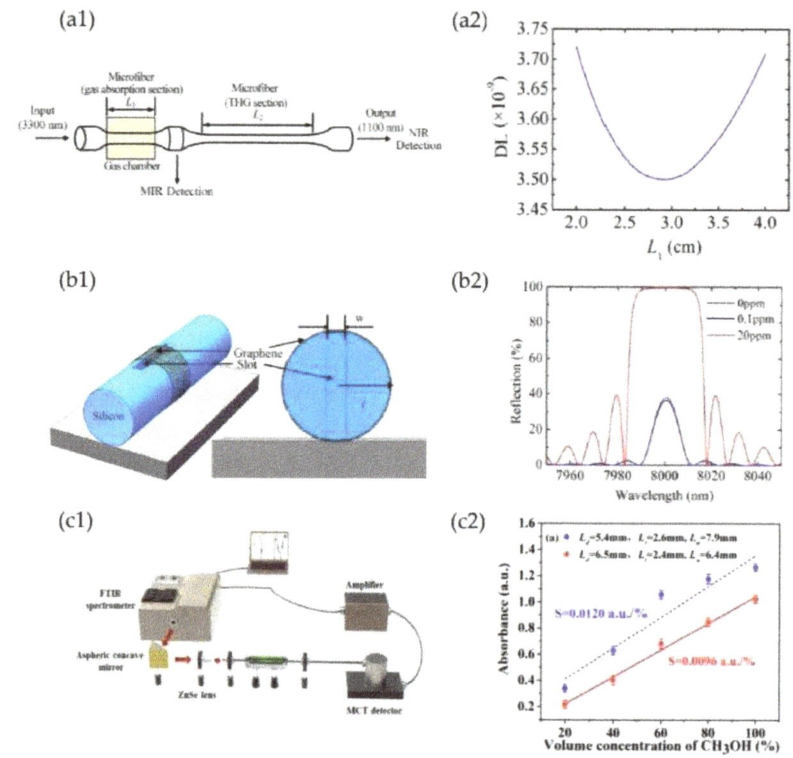

Figure 8. Mid-IR MF-based sensors. (**a1**) Schematic illustration of gas sensing based on cascaded ChG MFs. (**a2**) Detection limit of methane concentration for cascaded MFs sensing at different L_1. Hereon, L_2 = 9.5 cm and P_0 = 0.1 W. Reprinted with the permission from [125] © The Optical Society. (**b1**) Schematic illustration of graphene on-Si slot fiber structure. (**b2**) Reflection spectra of Bragg gratings based on graphene on-Si slot fiber with different NO_2 gas concentrations. Reprinted with the permission from [126] © IEEE. (**c1**) Experimental setup of $Ge_{15}Sb_{15}Se_{70}$ tapered MF sensor. (**c2**) Dependence of the absorbance at 9.78 μm on the methanol aqueous solution concentration for the asymmetric and symmetric structure tapered fiber sensors. Reprinted with the permission from [127] © IEEE.

5. Discussion

In this article, a comprehensive survey of recent works on ChG MFs for the mid-IR range is presented, where the fabrication methods and the optical waveguiding properties (i.e., n_{eff}, A_{eff} and η) of ChG MFs are discussed. So far, high-quality ChG MFs (e.g., As_2S_3 and As_2Se_3 MFs) have been fabricated by a high-temperature taper drawing process and have shown great promise for broadband low-loss optical waveguiding in the mid-IR region. Based on these MFs, a new category of compact and fiber-compatible photonic devices has been developed for passive, active and nonlinear applications in the mid-IR region. These initial results have unambiguously suggested the great opportunities of using the ChG MF as a versatile platform for a miniaturized fiber-based platform for mid-IR optics.

6. Conclusions

Looking into the future, there are obvious challenges and opportunities for ChG MFs. First, compared with silica glass, ChGs possess a much higher refractive index and optical nonlinearity that are advantageous for tighter optical confinement and lower threshold for nonlinear optical effects but exhibit much lower stability against optical, thermal, chemical and mechanical damages. Therefore, better package or protection techniques/protocols should be developed for real applications of ChG MFs, and more research efforts could be made on the material aspects. Secondly, compared to those for the VIS and near-IR spectral ranges, fiber-compatible devices for mid-IR, ranging from light sources, couplers, polarizers and filters to amplifiers, are typically not mature or optimized, leaving both challenges and opportunities in the future study of mid-IR MFs. Finally, other potentials, such as MF-incorporated plasmonic and atom optics that have been quite successful in the VIS/near-IR region, have not yet been equivalently investigated in mid-IR, which could be the new driving force in this field.

Author Contributions: Conceptualization, X.G., P.W. and L.T.; writing—original draft preparation, D.C. and Y.X.; writing—review and editing, D.C., Y.X., X.G., P.W. and L.T.; supervision, L.T.; project administration and funding acquisition, L.T. All authors have read and agreed to the published version of the manuscript.

Funding: This research was supported by the National Natural Science Foundation of China (11527901), Natural Science Foundation of Zhejiang Province (LR21F050002), and Fundamental Research Funds for the Central Universities.

Institutional Review Board Statement: Not applicable.

Informed Consent Statement: Not applicable.

Data Availability Statement: Not applicable.

Conflicts of Interest: The authors declare no conflict of interest.

References

1. Tong, L.M.; Gattass, R.R.; Ashcom, J.B.; He, S.L.; Lou, J.Y.; Shen, M.Y.; Maxwell, I.; Mazur, E. Subwavelength-diameter silica wires for low-loss optical wave guiding. *Nature* **2003**, *426*, 816–819. [CrossRef] [PubMed]
2. Brambilla, G.; Xu, F.; Horak, P.; Jung, Y.M.; Koizumi, F.; Sessions, N.P.; Koukharenko, E.; Feng, X.; Murugan, G.S.; Wilkinson, J.S.; et al. Optical fiber nanowires and microwires: Fabrication and applications. *Adv. Opt. Photonics* **2009**, *1*, 107–161. [CrossRef]
3. Tong, L.M.; Zi, F.; Guo, X.; Lou, J.Y. Optical microfibers and nanofibers: A tutorial. *Opt. Commun.* **2012**, *285*, 4641–4647. [CrossRef]
4. Wu, X.Q.; Tong, L.M. Optical microfibers and nanofibers. *Nanophotonics* **2013**, *2*, 407–428. [CrossRef]
5. Ismaeel, R.; Lee, T.; Ding, M.; Belal, M.; Brambilla, G. Optical microfiber passive components. *Laser Photonics Rev.* **2013**, *7*, 350–384. [CrossRef]
6. Ward, J.M.; O'Shea, D.G.; Shortt, B.J.; Morrissey, M.J.; Deasy, K.; Nic Chormaic, S.G. Heat-and-pull rig for fiber taper fabrication. *Rev. Sci. Instrum.* **2006**, *77*, 083105. [CrossRef]
7. Xu, Y.X.; Fang, W.; Tong, L.M. Real-time control of micro/nanofiber waist diameter with ultrahigh accuracy and precision. *Opt. Express* **2017**, *25*, 10434–10440. [CrossRef] [PubMed]
8. Tong, L.M.; Lou, J.Y.; Mazur, E. Single-mode guiding properties of subwavelength-diameter silica and silicon wire waveguides. *Opt. Express* **2004**, *12*, 1025–1035. [CrossRef] [PubMed]
9. Foster, M.A.; Moll, K.D.; Gaeta, A.L. Optimal waveguide dimensions for nonlinear interactions. *Opt. Express* **2004**, *12*, 2880–2887. [CrossRef]
10. Foster, M.A.; Turner, A.C.; Lipson, M.; Gaeta, A.L. Nonlinear optics in photonic nanowires. *Opt. Express* **2008**, *16*, 1300–1320. [CrossRef] [PubMed]
11. Leon-Saval, S.G.; Birks, T.A.; Wadsworth, W.J.; Russell, P.S.J.; Mason, M.W. Supercontinuum generation in submicron fibre waveguides. *Opt. Express* **2004**, *12*, 2864–2869. [CrossRef]
12. Balykin, V.I.; Hakuta, K.; Le Kien, F.; Liang, J.Q.; Morinaga, M. Atom trapping and guiding with a subwavelength-diameter optical fiber. *Phys. Rev. A* **2004**, *70*, 011401. [CrossRef]
13. Le Kien, F.; Balykin, V.I.; Hakuta, K. Atom trap and waveguide using a two-color evanescent light field around a subwavelength-diameter optical fiber. *Phys. Rev. A* **2004**, *70*, 063403. [CrossRef]
14. Sague, G.; Vetsch, E.; Alt, W.; Meschede, D.; Rauschenbeutel, A. Cold-atom physics using ultrathin optical fibers: Light-induced dipole forces and surface interactions. *Phys. Rev. Lett.* **2007**, *99*, 163602. [CrossRef] [PubMed]
15. Guo, X.; Ying, Y.B.; Tong, L.M. Photonic nanowires: From subwavelength waveguides to optical sensors. *Acc. Chem. Res.* **2014**, *47*, 656–666. [CrossRef] [PubMed]
16. Lou, J.Y.; Wang, Y.P.; Tong, L.M. Microfiber optical sensors: A review. *Sensors* **2014**, *14*, 5823–5844. [CrossRef] [PubMed]
17. Zhang, L.; Tang, Y.; Tong, L.M. Micro-/nanofiber optics: Merging photonics and material science on nanoscale for advanced sensing technology. *IScience* **2020**, *23*, 100810. [CrossRef] [PubMed]

18. Yang, Q.; Jiang, X.S.; Guo, X.; Chen, Y.; Tong, L.M. Hybrid structure laser based on semiconductor nanowires and a silica microfiber knot cavity. *Appl. Phys. Lett.* **2009**, *94*, 101108.
19. Jiang, X.S.; Song, Q.H.; Xu, L.; Fu, J.; Tong, L.M. Microfiber knot dye laser based on the evanescent-wave-coupled gain. *Appl. Phys. Lett.* **2007**, *90*, 233501. [CrossRef]
20. Jiang, X.S.; Yang, Q.; Vienne, G.; Li, Y.H.; Tong, L.M.; Zhang, J.J.; Hu, L.L. Demonstration of microfiber knot laser. *Appl. Phys. Lett.* **2006**, *89*, 143513. [CrossRef]
21. Nagel, S.R.; MacChesney, J.B.; Walker, K.L. *Optical Fiber Communications*, 1st ed.; Elsevier: New York, NY, USA, 1985; pp. 1–64.
22. Dai, S.X.; Wang, Y.Y.; Peng, X.F.; Zhang, P.Q.; Wang, X.S.; Xu, Y.S. A review of mid-Infrared supercontinuum generation in chalcogenide glass fibers. *Appl. Sci.* **2018**, *8*, 707. [CrossRef]
23. Gao, S.; Bao, X.Y. Chalcogenide taper and its nonlinear effects and sensing applications. *IScience* **2020**, *23*, 100802. [CrossRef] [PubMed]
24. Anashkina, E.A.; Andrianov, A.V.; Kim, A.V. Conversion of ultrashort laser pulses to wavelengths above 3 mm in tapered germanate fibres. *Quantum Electron.* **2015**, *45*, 437–442. [CrossRef]
25. Li, Z.R.; Yao, C.F.; Jia, Z.X.; Wang, F.; Qin, G.S.; Ohishi, Y.; Qin, W.P. Broadband supercontinuum generation from 600 to 5400 nm in a tapered fluorotellurite fiber pumped by a 2010 nm femtosecond fiber laser. *Appl. Phys. Lett.* **2019**, *115*, 091103. [CrossRef]
26. Saini, T.S.; Hoa, N.P.T.; Tuan, T.H.; Luo, X.; Suzuki, T.; Nish, Y. Tapered tellurite step-index optical fiber for coherent near-to-mid-IR supercontinuum generation: Experiment and modeling. *Appl. Opt.* **2019**, *58*, 415–421. [CrossRef] [PubMed]
27. Kubat, I.; Agger, C.S.; Moselund, P.M.; Bang, O. Mid-infrared supercontinuum generation to 45 μm in uniform and tapered ZBLAN step-index fibers by direct pumping at 1064 or 1550 nm. *J. Opt. Soc. Am. B* **2013**, *30*, 2743–2757. [CrossRef]
28. Chen, Z.G.; Taylor, A.J.; Efimov, A. Coherent mid-infrared broadband continuum generation in non-uniform ZBLAN fiber taper. *Opt. Express* **2009**, *17*, 5852–5860. [CrossRef] [PubMed]
29. Zakery, A.; Elliott, S.R. Optical properties and applications of chalcogenide glasses: A review. *J. Non-Cryst. Solids* **2003**, *330*, 1–12. [CrossRef]
30. Eggleton, B.J.; Luther-Davies, B.; Richardson, K. Chalcogenide photonics. *Nat. Photonics* **2011**, *5*, 141–148. [CrossRef]
31. Baker, C.; Rochette, M. Highly nonlinear hybrid AsSe-PMMA microtapers. *Opt. Express* **2010**, *18*, 12391–12398. [CrossRef]
32. Xin, C.G.; Wu, H.; Xie, Y.; Yu, S.L.; Zhou, N.; Shi, Z.X.; Guo, X.; Tong, L.M. CdTe microwires as mid-infrared optical waveguides. *Opt. Express* **2018**, *26*, 10944–10952. [CrossRef] [PubMed]
33. Ren, H.; Shen, L.; Wu, D.; Aktas, O.; Hawkins, T.; Ballato, J.; Gibson, U.J.; Peacock, A.C. Nonlinear optical properties of polycrystalline silicon core fibers from telecom wavelengths into the mid-infrared spectral region. *Opt. Mater. Express* **2019**, *9*, 1271–1279. [CrossRef]
34. Shiryaev, V.S.; Churbanov, M.F. Recent advances in preparation of high-purity chalcogenide glasses for mid-IR photonics. *J. Non-Cryst. Solids* **2017**, *475*, 1–9. [CrossRef]
35. Tao, G.M.; Ebendorff-Heidepriem, H.; Stolyarov, A.M.; Danto, S.; Badding, J.V.; Fink, Y.; Ballato, J.; Abouraddy, A.F. Infrared fibers. *Adv. Opt. Photonics* **2015**, *7*, 379–458. [CrossRef]
36. Churbanov, M.F.; Snopatin, G.E.; Shiryaev, V.S.; Plotnichenko, V.G.; Dianov, E.M. Recent advances in preparation of high-purity glasses based on arsenic chalcogenides for fiber optics. *J. Non-Cryst. Solids* **2011**, *357*, 2352–2357. [CrossRef]
37. Aktas, O. Chalcogenide microresonators tailored to distinct morphologies by the shaping of glasses on silica tapers. *Opt. Lett.* **2017**, *42*, 907–910. [CrossRef]
38. Zhang, Q.M.; Li, M.; Hao, Q.; Deng, D.H.; Zhou, H.; Zeng, H.P.; Zhan, L.; Wu, X.; Liu, L.Y.; Xu, L. Fabrication and characterization of on-chip optical nonlinear chalcogenide nanofiber devices. *Opt. Lett.* **2010**, *35*, 3829–3831. [CrossRef] [PubMed]
39. Ahmad, R.; Rochette, M. Photosensitivity at 1550 nm and Bragg grating inscription in As_2Se_3 chalcogenide microwires. *Appl. Phys. Lett.* **2011**, *99*, 061109. [CrossRef]
40. Ahmad, R.; Rochette, M.; Baker, C. Fabrication of Bragg gratings in subwavelength diameter As_2Se_3 chalcogenide wires. *Opt. Lett.* **2011**, *36*, 2886–2888. [CrossRef] [PubMed]
41. Florea, C.; Sanghera, J.S.; Shaw, B.; Aggarwal, I.D. Fiber Bragg gratings in As_2S_3 fibers obtained using a $0/-1$ phase mask. *Opt. Mater.* **2009**, *31*, 942–944. [CrossRef]
42. Gao, S.; Baker, C.; Cai, W.; Chen, L.; Bao, X.Y. 10 kHz-34 MHz ultrasound detection based on a dual-core hybrid taper. *APL Photonics* **2019**, *4*, 110805. [CrossRef]
43. Fan, H.B.; Chen, L.; Bao, X.Y. Chalcogenide microfiber-assisted silica microfiber for ultrasound detection. *Opt. Lett.* **2020**, *45*, 1128–1131. [CrossRef] [PubMed]
44. Yeom, D.I.; Magi, E.C.; Lamont, M.R.; Roelens, M.A.; Fu, L.B.; Eggleton, B.J. Low-threshold supercontinuum generation in highly nonlinear chalcogenide nanowires. *Opt. Lett.* **2008**, *33*, 660–662. [CrossRef]
45. Xie, Y.; Cai, D.W.; Wu, H.; Pan, J.; Zhou, N.; Xin, C.G.; Yu, S.L.; Wang, P.; Jiang, X.S.; Qiu, J.R.; et al. Mid-infrared chalcogenide microfiber knot resonators. *Photonics Res.* **2020**, *8*, 616–621. [CrossRef]
46. Cai, D.W.; Xie, Y.; Wang, P.; Zhang, L.; Guo, X.; Tong, L.M. Mid-infrared microfiber Bragg gratings. *Opt. Lett.* **2020**, *45*, 6114–6117. [CrossRef] [PubMed]
47. Tong, L.M.; Lou, J.Y.; Ye, Z.Z.; Svacha, G.T.; Mazur, E. Self-modulated taper drawing of silica nanowires. *Nanotechnology* **2005**, *16*, 1445–1448. [CrossRef]

48. Tong, L.M.; Hu, L.L.; Zhang, J.J.; Qiu, J.R.; Yang, Q.; Lou, J.Y.; Shen, Y.H.; He, J.L.; Ye, Z.Z. Photonic nanowires directly drawn from bulk glasses. *Opt. Express* **2006**, *14*, 82–87. [CrossRef]
49. Li, L.Z.; Al-Kadry, A.; Abdukerim, N.; Rochette, M. Design, fabrication and characterization of PC, COP and PMMA-cladded As_2Se_3 microwires. *Opt. Mater. Express* **2016**, *6*, 912–921. [CrossRef]
50. Baker, C.; Rochette, M. A generalized heat-brush approach for precise control of the waist profile in fiber tapers. *Opt. Mater. Express* **2011**, *1*, 1065–1076. [CrossRef]
51. Rodney, W.S.; Malitson, I.H.; King, T.A. Refractive index of arsenic trisulfide. *J. Opt. Soc. Am.* **1958**, *48*, 633–636. [CrossRef]
52. Van Popta, A.C.; DeCorby, R.G.; Haugen, C.J.; Robinson, T.; McMullin, J.N.; Tonchev, D.; Kasap, S.O. Photoinduced refractive index change in As_2Se_3 by 633 nm illumination. *Opt. Express* **2002**, *10*, 639–644. [CrossRef] [PubMed]
53. Huang, K.J.; Yang, S.Y.; Tong, L.M. Modeling of evanescent coupling between two parallel optical nanowires. *Appl. Opt.* **2007**, *46*, 1429–1434. [CrossRef]
54. Singh, V.; Lin, P.T.; Patel, N.; Lin, H.; Li, L.; Zou, Y.; Deng, F.; Ni, C.; Hu, J.; Giammarco, J.; et al. Mid-infrared materials and devices on a Si platform for optical sensing. *Sci. Technol. Adv. Mater.* **2014**, *15*, 014603. [CrossRef]
55. Hudson, D.D.; Dekker, S.A.; Maegi, E.C.; Judge, A.C.; Jackson, S.D.; Li, E.B.; Sanghera, J.S.; Shaw, L.B.; Aggarwal, I.D.; Eggleton, B.J. Octave spanning supercontinuum in an As_2S_3 taper using ultralow pump pulse energy. *Opt. Lett.* **2011**, *36*, 1122–1124. [CrossRef] [PubMed]
56. Guo, X.; Qiu, M.; Bao, J.M.; Wiley, B.J.; Yang, Q.; Zhang, X.N.; Ma, Y.G.; Yu, H.K.; Tong, L.M. Direct coupling of plasmonic and photonic nanowires for hybrid nanophotonic components and circuits. *Nano Lett.* **2009**, *9*, 4515–4519. [CrossRef] [PubMed]
57. Ding, Z.X.; Huang, Z.N.; Chen, Y.; Mou, C.B.; Lu, Y.Q.; Xu, F. All-fiber ultrafast laser generating gigahertz-rate pulses based on a hybrid plasmonic microfiber resonator. *Adv. Photon.* **2020**, *2*, 026002. [CrossRef]
58. Ma, Y.Q.; Farrell, G.; Semenova, Y.; Li, B.H.; Yuan, J.H.; Sang, X.Z.; Yan, B.B.; Yu, C.X.; Guo, T.; Wu, Q. Optical microfiber-loaded surface plasmonic TE-pass polarizer. *Opt. Laser Technol.* **2016**, *78*, 101–105. [CrossRef]
59. Vahala, K.J. Optical microcavities. *Nature* **2003**, *424*, 839–846. [CrossRef] [PubMed]
60. He, L.N.; Ozdemir, K.; Zhu, J.G.; Kim, W.; Yang, L. Detecting single viruses and nanoparticles using whispering gallery microlasers. *Nat. Nanotechnol.* **2011**, *6*, 428–432. [CrossRef]
61. Choi, J.; Noh, H. Enhanced absorption by coherent control in a photonic crystal resonator coupled with a microfiber. *Opt. Lett.* **2018**, *43*, 5532–5534. [CrossRef] [PubMed]
62. Kim, M.K.; Hwang, I.K.; Kim, S.H.; Chang, H.J.; Lee, Y.H. All-optical bistable switching in curved microfiber-coupled photonic crystal resonators. *Appl. Phys. Lett.* **2007**, *90*, 161118. [CrossRef]
63. Hwang, I.K.; Kim, S.K.; Yang, J.K.; Kim, S.H.; Lee, S.H.; Lee, Y.H. Curved-microfiber photon coupling for photonic crystal light emitter. *Appl. Phys. Lett.* **2005**, *87*, 131107. [CrossRef]
64. Lecaplain, C.; Javerzac-Galy, C.; Gorodetsky, M.L.; Kippenberg, T.J. Mid-infrared ultra-high-Q resonators based on fluoride crystalline materials. *Nat. Commun.* **2016**, *7*, 13383. [CrossRef] [PubMed]
65. Xie, Y.; Cai, D.W.; Pan, J.; Zhou, N.; Gao, Y.X.; Jin, Y.Y.; Jiang, X.S.; Qiu, J.R.; Wang, P.; Guo, X.; et al. Batch fabrication of high-quality infrared chalcogenide microsphere resonators. *Small* **2021**, *17*, 2100140. [CrossRef] [PubMed]
66. Heylman, K.D.; Knapper, K.A.; Horak, E.H.; Rea, M.T.; Vanga, S.K.; Goldsmith, R.H. Optical microresonators for sensing and transduction: A materials perspective. *Adv. Mater.* **2017**, *29*, 1700037. [CrossRef]
67. Sumetsky, M.; Dulashko, Y.; Fini, J.M.; Hale, A.; DiGiovanni, D.J. The microfiber loop resonator: Theory, experiment, and application. *J. Lightwave Technol.* **2006**, *24*, 242–250. [CrossRef]
68. Guo, X.; Li, Y.H.; Jiang, X.S.; Tong, L.M. Demonstration of critical coupling in microfiber loops wrapped around a copper rod. *Appl. Phys. Lett.* **2007**, *91*, 073512. [CrossRef]
69. Jiang, X.S.; Tong, L.M.; Vienne, G.; Guo, X.; Tsao, A.; Yang, Q.; Yang, D.R. Demonstration of optical microfiber knot resonators. *Appl. Phys. Lett.* **2006**, *88*, 223501. [CrossRef]
70. Wang, P.; Zhang, L.; Yang, Z.Y.; Gu, F.X.; Wang, S.S.; Yang, Q.; Tong, L.M. Fusion spliced microfiber closed-loop resonators. *IEEE Photonics Technol. Lett.* **2010**, *22*, 1075–1077. [CrossRef]
71. Li, W.; Wang, P.; Hu, Z.F.; Tong, L.M. Fusion splicing soft glass microfibers for photonic devices. *IEEE Photonics Technol. Lett.* **2011**, *23*, 831–833. [CrossRef]
72. Zhang, B.; Zeng, P.Y.; Yang, Z.L.; Xia, D.; Zhao, J.X.; Sun, Y.D.; Huang, Y.F.; Song, J.C.; Pan, J.S.; Cheng, H.J.; et al. On-chip chalcogenide microresonators with low-threshold parametric oscillation. *Photonics Res.* **2021**, *9*, 1272–1279. [CrossRef]
73. Sumetsky, M. Optical fiber microcoil resonators. *Opt. Express* **2004**, *12*, 2303–2316. [CrossRef]
74. Scheuer, J.; Sumetsky, M. Optical-fiber microcoil waveguides and resonators and their applications for interferometry and sensing. *Laser Photon. Rev.* **2011**, *5*, 465–478. [CrossRef]
75. Xu, Z.L.; Sun, Q.Z.; Li, B.R.; Luo, Y.Y.; Lu, W.G.; Liu, D.M.; Shum, P.P.; Zhang, L. Highly sensitive refractive index sensor based on cascaded microfiber knots with Vernier effect. *Opt. Express* **2015**, *23*, 6662–6672. [CrossRef] [PubMed]
76. Liu, M.; Tang, R.; Luo, A.P.; Xu, W.C.; Luo, Z.C. Graphene-decorated microfiber knot as a broadband resonator for ultrahigh-repetition-rate pulse fiber lasers. *Photonics Res.* **2018**, *6*, C1–C7. [CrossRef]
77. Luan, F.; Magi, E.; Gong, T.X.; Kabakova, I.; Eggleton, B.J. Photoinduced whispering gallery mode microcavity resonator in a chalcogenide microfiber. *Opt. Lett.* **2011**, *36*, 4761–4763. [CrossRef] [PubMed]

78. Ma, P.; Choi, D.Y.; Yu, Y.; Yang, Z.Y.; Vu, K.; Nguyen, T.; Mitchell, A.; Luther-Davies, B.; Madden, S. High Q factor chalcogenide ring resonators for cavity-enhanced MIR spectroscopic sensing. *Opt. Express* **2015**, *23*, 19969–19979. [CrossRef] [PubMed]
79. Grobnic, D.; Mihailov, S.J.; Ding, H.M.; Smelser, C.W. Bragg grating evanescent field sensor made in biconical tapered fiber with femtosecond IR radiation. *IEEE Photonics Technol. Lett.* **2006**, *18*, 160–162. [CrossRef]
80. Fang, X.; Liao, C.R.; Wang, D.N. Femtosecond laser fabricated fiber Bragg grating in microfiber for refractive index sensing. *Opt. Lett.* **2010**, *35*, 1007–1009. [CrossRef]
81. Ran, Y.; Jin, L.; Tan, Y.N.; Sun, L.P.; Li, J.; Guan, B.O. High-efficiency ultraviolet inscription of Bragg gratings in microfibers. *IEEE Photonics J.* **2012**, *4*, 181–186. [CrossRef]
82. Kou, J.L.; Qiu, S.J.; Xu, F.; Lu, Y.Q. Demonstration of a compact temperature sensor based on first-order Bragg grating in a tapered fiber probe. *Opt. Express* **2011**, *19*, 18452–18457. [CrossRef] [PubMed]
83. Liu, Y.X.; Meng, C.; Zhang, A.P.; Xiao, Y.; Yu, H.K.; Tong, L.M. Compact microfiber Bragg gratings with high-index contrast. *Opt. Lett.* **2011**, *36*, 3115–3117. [CrossRef] [PubMed]
84. Nayak, K.P.; Le Kien, F.; Kawai, Y.; Hakuta, K.; Nakajima, K.; Miyazaki, H.T.; Sugimoto, Y. Cavity formation on an optical nanofiber using focused ion beam milling technique. *Opt. Express* **2011**, *19*, 14040–14050. [CrossRef] [PubMed]
85. Guan, B.O.; Li, J.; Jin, L.; Ran, Y. Fiber Bragg gratings in optical microfibers. *Opt. Fiber Technol.* **2013**, *19*, 793–801. [CrossRef]
86. Gao, S.; Jin, L.; Ran, Y.; Sun, L.P.; Li, J.; Guan, B.O. Temperature compensated microfiber Bragg gratings. *Opt. Express* **2012**, *20*, 18281–18286. [CrossRef] [PubMed]
87. Luo, W.; Kou, J.L.; Chen, Y.; Xu, F.; Lu, Y.Q. Ultra-highly sensitive surface-corrugated microfiber Bragg grating force sensor. *Appl. Phys. Lett.* **2012**, *101*, 133502. [CrossRef]
88. Bernier, M.; El-Amraoui, M.; Couillard, J.F.; Messaddeq, Y.; Vallee, R. Writing of Bragg gratings through the polymer jacket of low-loss As_2S_3 fibers using femtosecond pulses at 800 nm. *Opt. Lett.* **2012**, *37*, 3900–3902. [CrossRef] [PubMed]
89. Baker, C.; Gao, S.; Chen, L.; Bao, X.Y. Self-inscribed antisymmetric long-period grating in a dual-core As_2Se_3-PMMA fiber. *Opt. Express* **2017**, *25*, 12409–12414. [CrossRef]
90. Wang, L.L.; Ma, W.Q.; Zhang, P.Q.; Yang, D.D.; Zhu, L.; Wang, X.S.; Dai, S.X. Design and analysis of long-period fiber gratings in tapered multimode chalcogenide glass fiber for temperature measurement. *J. Opt. Soc. Am. B* **2019**, *36*, 1792–1798. [CrossRef]
91. Supradeepa, V.R.; Feng, Y.; Nicholson, J.W. Raman fiber lasers. *J. Opt.* **2017**, *19*, 023001. [CrossRef]
92. Pierce, M.C.; Jackson, S.D.; Dickinson, M.R.; King, T.A.; Sloan, P. Laser-tissue interaction with a continuous wave 3-μm fibre laser: Preliminary studies with soft tissue. *Lasers Surg. Med.* **2000**, *26*, 491–495. [CrossRef]
93. Jackson, S.D. Towards high-power mid-infrared emission from a fibre laser. *Nat. Photonics* **2012**, *6*, 423–431. [CrossRef]
94. White, R.T.; Monro, T.M. Cascaded Raman shifting of high-peak-power nanosecond pulses in As_2S_3 and As_2Se_3 optical fibers. *Opt. Lett.* **2011**, *36*, 2351–2353. [CrossRef] [PubMed]
95. Abdukerim, N.; Li, L.Z.; El Amraoui, M.; Messaddeq, Y.; Rochette, M. 2 μm Raman fiber laser based on a multimaterial chalcogenide microwire. *Appl. Phys. Lett.* **2017**, *110*, 161103. [CrossRef]
96. Guo, X.H.; Li, Z.R.; Jiao, Y.D.; Zhao, Z.P.; Yao, C.F.; Jia, Z.X.; Dong, K.G.; Zhang, H.Y.; Lin, H.H.; Qin, W.P.; et al. A shower of mid-infrared Raman solitons at designed wavelength of similar to 3 μm from a tapered fluorotellurite fiber. *Laser Phys.* **2021**, *31*, 095103. [CrossRef]
97. Alamgir, I.; Abdukerim, N.; Rochette, M. In situ fabrication of far-detuned optical fiber wavelength converters. *Opt. Lett.* **2019**, *44*, 4467–4470. [CrossRef] [PubMed]
98. Godin, T.; Combes, Y.; Ahmad, R.; Rochette, M.; Sylvestre, T.; Dudley, J.M. Far-detuned mid-infrared frequency conversion via normal dispersion modulation instability in chalcogenide microwires. *Opt. Lett.* **2014**, *39*, 1885–1888. [CrossRef] [PubMed]
99. Li, L.Z.; Abdukerim, N.; Rochette, M. Mid-infrared wavelength conversion from As_2Se_3 microwires. *Opt. Lett.* **2017**, *42*, 639–642. [CrossRef]
100. Liang, F.; Kang, L.; Lin, Z.S.; Wu, Y.C. Mid-infrared nonlinear optical materials based on metal chalcogenides: Structure-property relationship. *Cryst. Growth Des.* **2017**, *17*, 2254–2289. [CrossRef]
101. Abdukerim, N.; Li, L.Z.; Rochette, M. Chalcogenide-based optical parametric oscillator at 2 μm. *Opt. Lett.* **2016**, *41*, 4364–4367. [CrossRef]
102. Al-Kadry, A.; El Amraoui, M.; Messaddeq, Y.; Rochette, M. Two octaves mid-infrared supercontinuum generation in As_2Se_3 microwires. *Opt. Express* **2014**, *22*, 31131–31137. [CrossRef] [PubMed]
103. Wang, Y.Y.; Dai, S.X.; Li, G.T.; Xu, D.; You, C.Y.; Han, X.; Zhang, P.Q.; Wang, X.S.; Xu, P.P. 1.4–7.2 μm broadband supercontinuum generation in an As-S chalcogenide tapered fiber pumped in the normal dispersion regime. *Opt. Lett.* **2017**, *42*, 3458–3461. [CrossRef] [PubMed]
104. Li, G.T.; Peng, X.F.; Dai, S.X.; Wang, Y.Y.; Xie, M.; Yang, L.; Yang, C.F.; Wei, W.Y.; Zhang, P.Q. Highly coherent 1.5–8.3 μm broadband supercontinuum generation in tapered As-S chalcogenide fibers. *J. Lightwave Technol.* **2019**, *37*, 1847–1852. [CrossRef]
105. Hudson, D.D.; Antipov, S.; Li, L.Z.; Alamgir, I.; Hu, T.; Amraoui, M.; Messaddeq, Y.; Rochette, M.; Jackson, S.D.; Fuerbach, A. Toward all-fiber supercontinuum spanning the mid-infrared. *Optica* **2017**, *4*, 1163–1166. [CrossRef]
106. Marandi, A.; Rudy, C.W.; Plotnichenko, V.G.; Dianov, E.M.; Vodopyanov, K.L.; Byer, R.L. Mid-infrared supercontinuum generation in tapered chalcogenide fiber for producing octave-spanning frequency comb around 3 μm. *Opt. Express* **2012**, *20*, 24218–24225. [CrossRef] [PubMed]

107. Dudley, J.M.; Genty, G.; Coen, S. Supercontinuum generation in photonic crystal fiber. *Rev. Mod. Phys.* **2006**, *78*, 1135–1184. [CrossRef]
108. Al-kadry, A.; Baker, C.; El Amraoui, M.; Messaddeq, Y.; Rochette, M. Broadband supercontinuum generation in As_2Se_3 chalcogenide wires by avoiding the two-photon absorption effects. *Opt. Lett.* **2013**, *38*, 1185–1187. [CrossRef]
109. Rudy, C.W.; Marandi, A.; Vodopyanov, K.L.; Byer, R.L. Octave-spanning supercontinuum generation in in situ tapered As_2S_3 fiber pumped by a thulium-doped fiber laser. *Opt. Lett.* **2013**, *38*, 2865–2868. [CrossRef]
110. Sun, Y.A.; Dai, S.X.; Zhang, P.Q.; Wang, X.S.; Xu, Y.S.; Liu, Z.J.; Chen, F.F.; Wu, Y.A.; Zhang, Y.J.; Wang, R.P.; et al. Fabrication and characterization of multimaterial chalcogenide glass fiber tapers with high numerical apertures. *Opt. Express* **2015**, *23*, 23472–23483. [CrossRef]
111. Anashkina, E.A.; Shiryaev, V.S.; Koptev, M.Y.; Stepanov, B.S.; Muravyev, S.V. Development of of As-Se tapered suspended-core fibers for ultra-broadband mid-IR wavelength conversion. *J. Non-Cryst. Solids* **2018**, *480*, 43–50. [CrossRef]
112. Leonov, S.O.; Wang, Y.C.; Shiryaev, V.S.; Snopatin, G.E.; Stepanov, B.S.; Plotnichenko, V.G.; Vicentini, E.; Gambetta, A.; Coluccelli, N.; Svelto, C.; et al. Coherent mid-infrared supercontinuum generation in tapered suspended-core $As_{39}Se_{61}$ fibers pumped by a few-optical-cycle Cr:ZnSe laser. *Opt. Lett.* **2020**, *45*, 1346–1349. [CrossRef]
113. Leindecker, N.; Marandi, A.; Byer, R.L.; Vodopyanov, K.L. Broadband degenerate OPO for mid-infrared frequency comb generation. *Opt. Express* **2011**, *19*, 6304–6310. [CrossRef] [PubMed]
114. Reid, D.T.; Gale, B.J.S.; Sun, J. Frequency comb generation and carrier-envelope phase control in femtosecond optical parametric oscillators. *Laser Phys.* **2008**, *18*, 87–103. [CrossRef]
115. Keilmann, F.; Gohle, C.; Holzwarth, R. Time-domain mid-infrared frequency-comb spectrometer. *Opt. Lett.* **2004**, *29*, 1542–1544. [CrossRef]
116. Erny, C.; Moutzouris, K.; Biegert, J.; Kuhlke, D.; Adler, F.; Leitenstorfer, A.; Keller, U. Mid-infrared difference-frequency generation of ultrashort pulses tunable between 3.2 and 4.8 μm from a compact fiber source. *Opt. Lett.* **2007**, *32*, 1138–1140. [CrossRef]
117. Del'Haye, P.; Herr, T.; Gavartin, E.; Gorodetsky, M.L.; Holzwarth, R.; Kippenberg, T.J. Octave spanning tunable frequency comb from a microresonator. *Phys. Rev. Lett.* **2011**, *107*, 063901. [CrossRef] [PubMed]
118. Del'Haye, P.; Schliesser, A.; Arcizet, O.; Wilken, T.; Holzwarth, R.; Kippenberg, T.J. Optical frequency comb generation from a monolithic microresonator. *Nature* **2007**, *450*, 1214–1217. [CrossRef]
119. Bernhardt, B.; Sorokin, E.; Jacquet, P.; Thon, R.; Becker, T.; Sorokina, I.T.; Picque, N.; Hansch, T.W. Mid-infrared dual-comb spectroscopy with 2.4 μm Cr^{2+}:ZnSe femtosecond lasers. *Appl. Phys. B* **2010**, *100*, 3–8. [CrossRef]
120. Adler, F.; Cossel, K.C.; Thorpe, M.J.; Hartl, I.; Fermann, M.E.; Ye, J. Phase-stabilized, 1.5 W frequency comb at 2.8–4.8 μm. *Opt. Lett.* **2009**, *34*, 1330–1332. [CrossRef] [PubMed]
121. Lee, K.F.; Granzow, N.; Schmidt, M.A.; Chang, W.; Wang, L.; Coulombier, Q.; Troles, J.; Leindecker, N.; Vodopyanov, K.L.; Schunemann, P.G.; et al. Midinfrared frequency combs from coherent supercontinuum in chalcogenide and optical parametric oscillation. *Opt. Lett.* **2014**, *39*, 2056–2059. [CrossRef] [PubMed]
122. Sanghera, J.S.; Shaw, L.B.; Busse, L.E.; Talley, D.; Aggarwal, I.D. Infrared transmitting fiber optics for biomedical applications. In Proceedings of the BiOS '99 International Biomedical Optics Symposium, San Jose, CA, USA, 23–29 January 1999; Volume 3596, pp. 178–187.
123. Sanghera, J.S.; Kung, F.H.; Busse, L.E.; Pureza, P.C.; Aggarwal, I.D. Infrared evanescent absorption-spectroscopy of toxic-chemicals using chalcogenide class fibers. *J. Am. Ceram. Soc.* **1995**, *78*, 2198–2202. [CrossRef]
124. Michel, K.; Bureau, B.; Pouvreau, C.; Sangleboeuf, J.C.; Boussard-Pledel, C.; Jouan, T.; Rouxel, T.; Adam, J.L.; Staubmann, K.; Steinner, H.; et al. Development of a chalcogenide glass fiber device for in situ pollutant detection. *J. Non-Cryst. Solids* **2003**, *326*, 434–438. [CrossRef]
125. Huang, P.; Huang, T.Y.; Zeng, S.W.; Pan, J.X.; Wu, X.; Zhao, X.; Wu, Y.H.; Ping, P.S.; Brambilla, G. Nonlinear gas sensing based on third-harmonic generation in cascaded chalcogenide microfibers. *J. Opt. Soc. Am. B* **2019**, *36*, 300–305. [CrossRef]
126. Wang, J.Q.; Chen, Y.Z.; Geng, Y.F.; Hong, X.M.; Li, X.J. Theoretical design of mid-infrared graphene optical gas sensor based on slot Si core fiber. *IEEE Photonics Technol. Lett.* **2019**, *31*, 1096–1099. [CrossRef]
127. Wang, M.; Yang, F.; Dai, S.X.; Cao, Z.F.; Su, J.X.; Ding, S.J.; Zhang, P.Q. Effect of the geometries of Ge-Sb-Se chalcogenide glass tapered fiber on the sensitivity of evanescent wave sensors. *J. Lightwave Technol.* **2021**, *39*, 4828–4836. [CrossRef]
128. Michel, K.; Bureau, B.; Boussard-Pledel, C.; Jouan, T.; Adam, J.L.; Staubmann, K.; Baumann, T. Monitoring of pollutant in waste water by infrared spectroscopy using chalcogenide glass optical fibers. *Sens. Actuators B Chem.* **2004**, *101*, 252–259. [CrossRef]
129. Keirsse, J.; Boussard-Pledel, C.; Loreal, O.; Sire, O.; Bureau, B.; Turlin, B.; Leroyer, P.; Lucas, J. Chalcogenide glass fibers used as biosensors. *J. Non-Cryst. Solids* **2003**, *326*, 430–433. [CrossRef]
130. Le Coq, D.; Michel, K.; Keirsse, J.; Boussard-Pledel, C.; Fonteneaua, G.; Bureau, B.; Le Quere, J.M.; Sire, O.; Lucas, J. Infrared glass fibers for in-situ sensing, chemical and biochemical reactions. *Comptes Rendus Chim.* **2002**, *5*, 907–913. [CrossRef]
131. Hocde, S.; Loreal, O.; Sire, O.; Boussard-Pledel, C.; Bureau, B.; Turlin, B.; Keirsse, J.; Leroyer, P.; Lucas, J. Metabolic imaging of tissues by infrared fiber-optic spectroscopy: An efficient tool for medical diagnosis. *J. Biomed. Opt.* **2004**, *9*, 404–407. [CrossRef] [PubMed]

132. Gutierrez-Arroyo, A.; Baudet, E.; Bodiou, L.; Nazabal, V.; Rinnert, E.; Michel, K.; Bureau, B.; Colas, F.; Charrier, J. Theoretical study of an evanescent optical integrated sensor for multipurpose detection of gases and liquids in the mid-infrared. *Sens. Actuators B Chem.* **2017**, *242*, 842–848. [CrossRef]
133. Wang, X.M.; Su, J.X.; Wang, Y.Y.; Yang, C.F.; Dai, S.X.; Zhang, P.Q. High-sensitivity sensing in bare Ge-Sb-Se chalcogenide tapered fiber with optimal structure parameters. *J. Non-Cryst. Solids* **2021**, *559*, 120686. [CrossRef]

photonics

Letter

Operation of a Single-Frequency Bismuth-Doped Fiber Power Amplifier near 1.65 μm

Grzegorz Gomółka [1], Monika Krajewska [1], Małgorzata Kaleta [1], Aleksandr M. Khegai [2], Sergey V. Alyshev [2], Aleksey S. Lobanov [3], Sergei V. Firstov [2] and Michał Nikodem [1,*]

1 Department of Optics and Photonics, Wroclaw University of Science and Technology, Wybrzeze Wyspianskiego 27, 50-370 Wroclaw, Poland; grzegorz.gomolka@pwr.edu.pl (G.G.); 236704@student.pwr.edu.pl (M.K.); 245004@student.pwr.edu.pl (M.K.)
2 Prokhorov General Physics Institute of the Russian Academy of Sciences, Dianov Fiber Optics Research Center, 38 Vavilov str., 119333 Moscow, Russia; khegai@fo.gpi.ru (A.M.K.); alyshe@fo.gpi.ru (S.V.A.); fir@fo.gpi.ru (S.V.F.)
3 Institute of Chemistry of High Purity Substances, Russian Academy of Sciences, 29 Tropinin str., 603600 Nizhny Novgorod, Russia; lobanov@ihps.nnov.ru
* Correspondence: michal.nikodem@pwr.edu.pl

Received: 6 November 2020; Accepted: 8 December 2020; Published: 9 December 2020

Abstract: The spectral range between 1650 and 1700 nm is an interesting region due to its potential applications in optical telecommunication and optical-based methane sensing. Unfortunately, the availability of compact and simple optical amplifiers with output powers exceeding tens of milliwatts in this spectral region is still limited. In this paper, a single-frequency continuous-wave bismuth-doped fiber amplifier (BDFA) operating at 1651 and 1687 nm is presented. With the improved signal/pump coupling and modified pump source design, the output powers of 163 mW (at 1651 nm) and 197 mW (at 1687 nm) were obtained. Application of the BDFA to the optical spectroscopy of methane near 1651 nm is also described. We demonstrate that the BDFA can be effectively used for signal amplitude enhancement in photothermal interferometry.

Keywords: optical amplifiers; doped fiber amplifiers; near-infrared; molecular gas spectroscopy; photothermal spectroscopy; heterodyne detection

1. Introduction

Over the last few decades, fiber-based optical amplifiers have proven to be effective power amplifiers with a number of advantages, such as high gain, a high output power, a low noise figure, polarization independence, and broad amplification spectral bands. Fiber-based amplifiers are capable of providing gain in numerous spectral regions, from the visible light to near-infrared region. However, in some spectral regions, the availability of proper gain medium, fiber chemical composition, and manufacturing issues are still critical problems. One of the research gaps is the spectral range beyond 1650 nm, which can potentially be used not only to extend existing telecommunication windows, but also in applications such as laser eye surgery [1,2] and the chemical sensing of hydrocarbons such as methane [3–5] and ethane [6]. Thulium-doped fiber amplifiers (TDFAs) can be used in this spectral region. However, because these wavelengths are at the edge of the thulium gain band, the highest demonstrated saturation power near 1650 nm was less than 50 mW [7,8]. Raman amplifiers are capable of providing much higher output powers, but require long (about 1 km or more) pieces of specialty fiber [9] and very high (watt level) pumping powers [10]. In this respect, bismuth-doped fiber amplifiers are interesting alternatives that may be used to obtain decent output powers (beyond 100 mW) in a simple and compact configuration, similar to rare earth-doped amplifiers.

Bismuth-doped fibers (BDFs) are an interesting group of active fibers that are capable of covering spectral regions unavailable for rare-earth-doped fibers [11]. As Bi-doped fibers have luminescence properties dependent on the host glass, they are a quite flexible medium for photonic system development [12]. For example, broadband emission covering a range from 1150 up to 1500 nm has been demonstrated using Bi-doped aluminosilicate and phosphosilicate glass fibers [13–16], while Bi-doped germanosilicate fibers can also be used as gain media beyond 1600 nm [17–20].

This paper presents a continuation of our earlier experiments, in which the performance of a bismuth-doped fiber amplifier (BDFA) at 1651 [21] and 1687 nm [22] was analyzed. Previously, the output powers of ~87 and ~107 mW were obtained at 1651 and 1687 nm, respectively. These results were limited by two issues. The first was stimulated Brillouin scattering (SBS) in BDF, which turned out to affect the BDFA gain when high pumping powers were used [21]. The other issue was the efficiency of coupling the signal (at 1651 or 1687 nm) and the pump (at 1550 nm). In this paper, a modified configuration of the amplifier is demonstrated. Improved signal/pump coupling and a new pump source enabled output powers of 163 mW (at 1651 nm) and 197 mW (at 1687 nm) to be obtained, which are almost twice as high as those achieved with the previous amplifier design.

The paper is organized as follows. The second section describes the experimental setup. We present a new pump source (Section 2.1) and demonstrate its impact on the reduction of backscattered light due to SBS (Section 3.1). The performance of the BDFA (including the output power dependence on the pump power and the input power at two wavelengths) is presented in Section 3.2. In Section 3.3, we show the application of BDFA to photothermal spectroscopy of methane near 1651 nm. We demonstrate that this technique (as well as fundamentally similar photoacoustic spectroscopy) can greatly benefit from using optical power amplifiers for signal enhancement. Section 4 contains a discussion and conclusions.

2. Materials and Methods (Experimental Setup)

A schematic of the BDFA is shown in Figure 1. A 90 m-long germanosilicate BDF was used as an active fiber (the same fiber was also used in our previous works [21,22]). The GeO_2 content of the fiber was ~50 mol.%. Although its core diameter was only 2.2 μm, it could be spliced to the standard single-mode fiber with relatively low losses (below 1 dB). Bismuth doping combined with the presence of Ge in the fiber core results in luminescence in the spectral region from 1600 to 1800 nm when the fiber is pumped near 1.55 μm [11]. Therefore, the pump source was constructed using a continuous-wave erbium-doped fiber laser (EDFL) amplified with a commercially available erbium/ytterbium-doped fiber amplifier (EYDFA; model GOA-S320 from BKtel, with a maximum output power of 32 dBm). A wavelength division multiplexing (WDM) coupler (from Haphit) was used to couple the pump light into the BDF and out-couple the amplified signals from single-frequency laser diodes: A distributed feed-back (DFB) laser diode was employed at 1651 nm (from NTT Electronics) and a discrete mode (DM) laser diode was applied at 1687 nm (from Eblana Photonics). A polarization-insensitive fiber-based optical isolator (from Opto-link) was placed between the seed source and the BDF to block any backscattered radiation or unabsorbed pump light. It was designed to operate in the C and L band (from 1530 to 1610 nm), but its transmission at longer wavelengths was still relatively high (more than 85% at 1651 nm and ~75% at 1687 nm, including connectors).

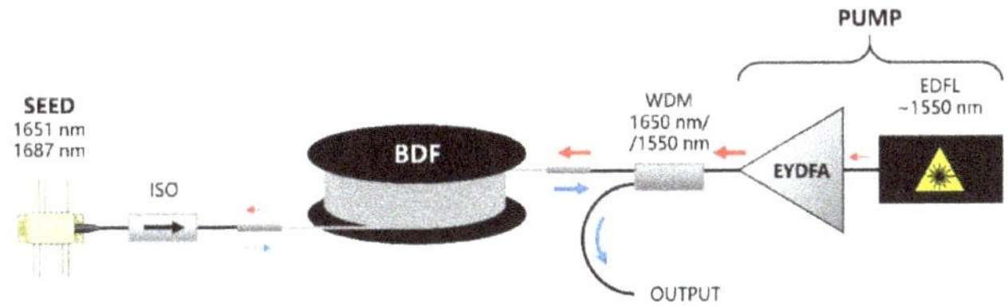

Figure 1. Bismuth-doped fiber amplifier (BDFA) setup: SEED—seed laser diode; ISO—fiber isolator' BDF—bismuth-doped fiber; WDM—wavelength division multiplexing coupler; EYDFA—erbium-ytterbium-doped fiber amplifier; and EDFL—erbium-doped fiber laser (see Section 2.1).

2.1. 1.55 μm Pump Source

As mentioned in the introduction, the main limitation of our previous BDFA configuration was SBS in the bismuth-doped fiber. This phenomenon was affecting the efficiency of BDFA. We believe that, due to backscattering that was observed for high pumping powers, part of the BDF was not pumped sufficiently, so the overall gain was reduced. As a result, the achievable output power was limited. The SBS was observed because a narrowband distributed feed-back (DFB) laser diode (LD) was used as a seed for EYDFA. Therefore, the simplest way to increase the SBS threshold was to use a spectrally broad source which was broader than the gain bandwidth associated with the SBS [23,24].

The schematic diagram of an erbium-doped fiber laser (EDFL) used as a BDFA's pump source is shown in Figure 2. A 10 m-long erbium-doped fiber (EDF; model M5-980-125 from Thorlabs) was pumped using a 974 nm laser diode (from JDSU). A fiber Bragg grating (FBG) was placed inside the ring cavity to make sure that EDFL operated near 1550 nm. The spectrum of the FBG is shown in Figure 3a and the 3 dB bandwidth was ~36 GHz. An optical circulator (from Opto-link) was used not only to insert the FBG into the laser cavity, but also to force a unidirectional operation. An optical coupler (50:50) was used as an output of the laser. Approximately 10 mW of optical power at 1550 nm was obtained with a pump power (at 974 nm) of ~60 mW. Figure 3b shows the optical spectrum emitted from the EDFL with a 3 dB bandwidth of ~4 GHz.

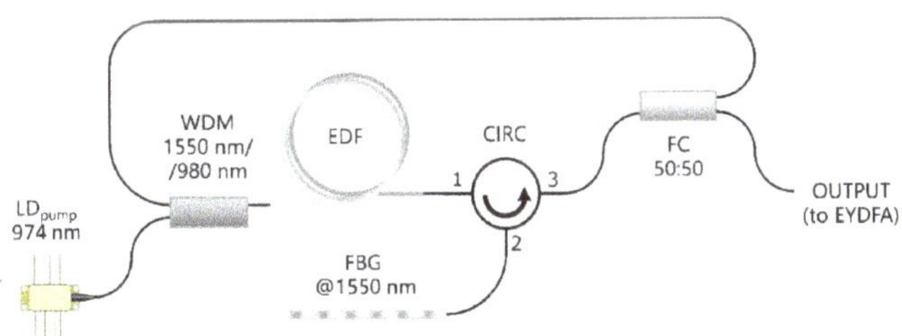

Figure 2. Erbium-doped fiber laser used as a new pump source for BDFA: EDF—erbium-doped fiber; CIRC—optical fiber circulator; FBG—fiber Bragg grating; and FC—fiber coupler.

For comparison, a linewidth of the DFB laser diode used in our previous configurations was also measured. A self-heterodyne method [25] was used, in which light emitted from a narrow-linewidth laser is divided into two beams. One is frequency-shifted by Ω and delayed using an optical fiber (the fiber must be longer than the coherence length of the source under study). This beam is then coupled with the second beam (emitted directly from the source) and the beatnote signal at Ω is analyzed using a radio frequency (RF) spectrum analyzer. Figure 4c shows the RF spectrum recorded using a 1550 nm DFB laser diode as a source and a 600 m-long delay line. Based on this measurement, the linewidth (full width at half maximum) of the pump source used in our previous configurations was estimated to be approximately 1.1 MHz. This is much lower than typical values of the SBS gain bandwidth [26] and more than three orders of magnitude less than the linewidth of the fiber laser used in this work.

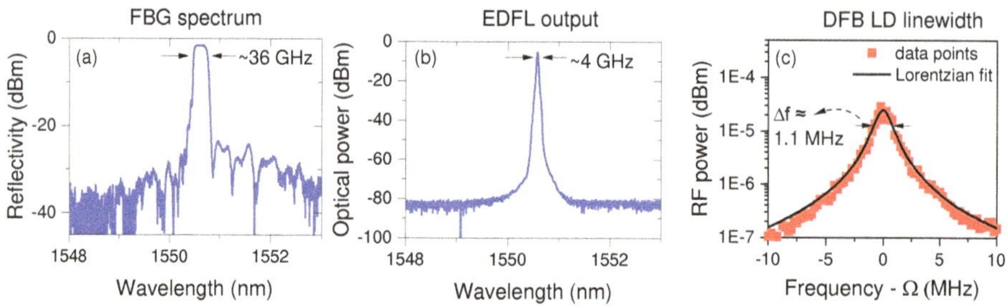

Figure 3. (**a**) Fiber Bragg grating reflectivity near 1550 nm (3 dB bandwidth is ~36 GHz); (**b**) output spectrum of the EDFL emission (3 dB bandwidth of ~4 GHz); (**c**) the result of the self-heterodyne linewidth measurement [25] shows the RF spectrum of the beatnote signal around Ω = 200 MHz. Based on this measurement, the linewidth Δf of the distributed feed-back (DFB) diode was estimated to be ~1.1 MHz.

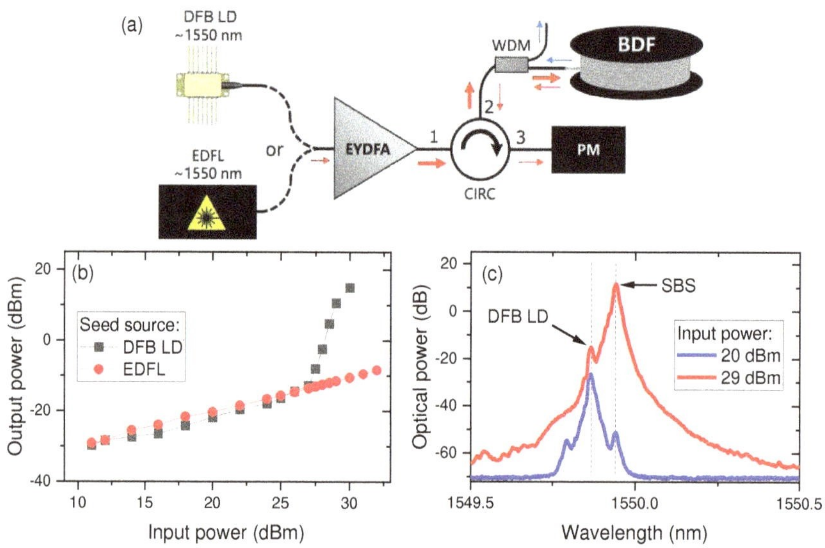

Figure 4. (**a**) Experimental setup used to measure the optical power of light backscattered in the BDF: PM—power meter. A WDM coupler was used to block backscattered amplified spontaneous emission (ASE), as indicated with blue arrows; (**b**) dependence of the backscattered output power on the input power for two light sources; (**c**) spectra of backscattered light recorded with a 0.1 nm resolution for two different input power levels. The peak frequency shifted by ~9.5 GHz that appears for a higher power is clear evidence of stimulated Brillouin scattering (SBS).

3. Results

3.1. SBS Suppression with the New Pump Source

To verify that with the new source, the SBS threshold is reduced, we built the setup shown in Figure 4a. An optical circulator was used to couple the amplified light into a BDF. The third port of the circulator was used to measure the power of the backscattered radiation. Additionally, a WDM coupler was employed between the circulator and BDF to block backward amplified spontaneous emission (ASE). First, the light from a narrow-linewidth DFB laser diode was used as a seed source for EYDFA. Subsequently, the DFB laser diode was replaced with a new EDFL source. Figure 4b shows how the backscattered optical power depends on the input power (i.e., at the output of the EYDFA) for both seed sources.

In both cases, we observed some linear dependence between the input power and the output (backscattered) power. This might be due to Rayleigh scattering, residual reflection from the splices or connectors, or cross-talk between the 1st and 3rd port of the circulator. However, when a DFB laser diode was used as a seed and the EYDFA power exceeded approximately 27 dBm (~500 mW), SBS became visible. It manifested itself through a rapid, nonlinear increase of the backscattered light power level. This phenomenon was not observed when EDFL was used as a source, even for much higher input powers (up to 32 dBm). The presence of the SBS was additionally confirmed by measuring the spectrum of the backscattered radiation. For a higher input power, a signal frequency shift of ~9.5 GHz was observed, which is expected for Brillouin scattering in a silica fiber doped with Ge [27].

3.2. BDFA Performance

In Figure 5, we present the performance of the BDFA. Figure 5a shows how the output power depends on the pump power for two seed wavelengths: 1651 and 1687 nm. In both cases, an almost linear dependence was obtained for a pump power up to 32 dBm (~1.58 W). The BDFA output power dependence on the seed laser power is shown in Figure 5b. For the highest pump power and the highest available seed power at 1651 nm (15.54 mW), the output power of 163 mW (or 22.1 dBm) was obtained. For a longer wavelength (i.e., 1687 nm), the measured power was even higher: 197 mW (or 22.94 dBm) when a seed power of 5.48 mW was used. These results are almost two times better than those demonstrated previously (87 mW at 1651 nm in [21] and 107 mW at 1687 nm in [22]). The measured gain (shown in Figure 5c) varied from 26 dB (for a low input power at 1687 nm) to approximately 10 dB (for a high input power at 1651 nm).

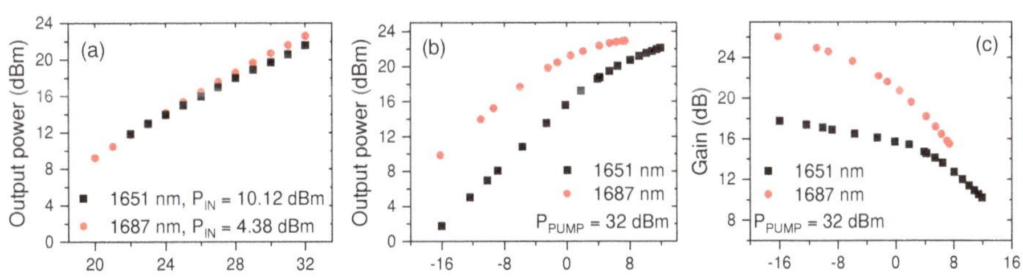

Figure 5. (**a**) BDFA output power versus pump power at 1651 and 1687 nm. (**b**) and (**c**) BDFA output power and gain versus seed power at both wavelengths for a pump power of 32 dBm.

The optical spectra recorded when the BDFA was seeded with a 1651 and 1687 nm laser diode are shown in Figure 6. In both cases, the maximum pump power (32 dBm) was used. A very significant power level difference (more than 40 dB) between the signal and the amplified spontaneous emission (ASE) was obtained. For comparison, the spectra of the two seed sources are also shown in Figure 6a.

Figure 7 presents the stability of the output power of BDFA. It was measured using a power meter (PM100D from Thorlabs with S132C sensor) and a 1651 nm seed laser diode. The input power and the pump power were set to 10.1 and 30 dBm, respectively. The output power was recorded for more than 30 min. The peak-to-peak fluctuations were ~0.32% of the mean value (95.5 mW). Similar output power fluctuations were obtained when lower input powers were used (Figure 7b,c).

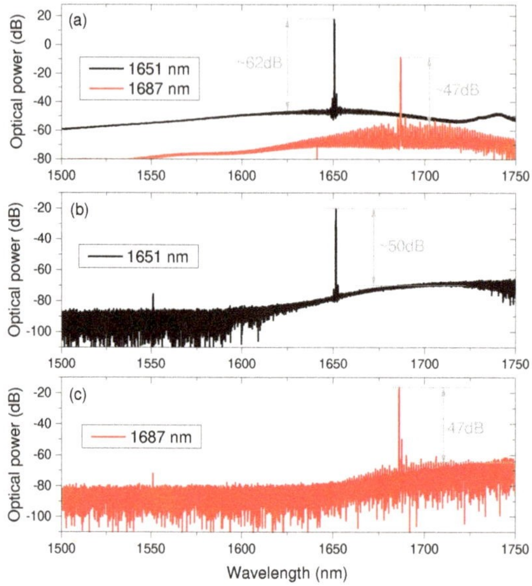

Figure 6. Spectra of the seed sources (**a**) and the output of the amplifier, with a seed wavelength of 1651 (**b**) and 1687 nm (**c**). Pump power was 32 dBm.

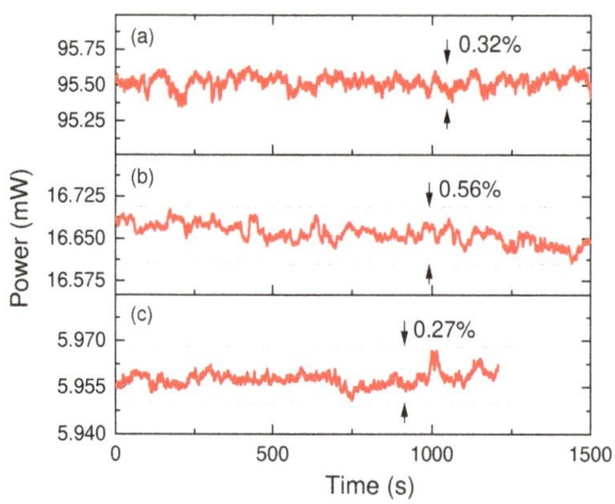

Figure 7. BDFA output powers measured with a 1651 nm seed and 30 dBm pump power. The input powers were ~10 (**a**), ~0.5 (**b**), and ~0.2 mW (**c**). All data were acquired after warm-up (~1 h).

3.3. Photothermal Spectroscopy Using a BDFA-Amplified Source

In order to demonstrate how BDFA can be used in laser spectroscopy, a setup that enables photothermal spectroscopy (PTS) to be performed was built. PTS can be used to detect and quantify gas samples by observing indirect effects of light–matter interaction. Energy absorbed by gas is turned to heat, leading to a local rise of the gas temperature and change of the gas refractive index. The refractive index change can be measured using, e.g., an interferometric setup [28,29]. Because the photothermal signal is proportional to the absorbed energy, an optical amplifier can be used to improve the performance of PTS. This was previously demonstrated using photoacoustic spectroscopy (a technique which is fundamentally similar to PTS) performed in a more convenient telecom wavelength region (below 1.6 μm, where erbium-doped fiber amplifiers can be used) [30–34]. Here, we show BDFA-enhanced PTS of methane using the setup schematically shown in Figure 8.

The light emitted from a 1651 nm DFB laser diode was amplified with the BDFA and entered the gas sample (enclosed in a glass cell) through a 90:10 fiber coupler. The DFB laser diode was current modulated with a sinewave (f_m = 0.5 kHz) to produce periodic changes of the refractive index inside the gas cell. These changes were measured using an interferometer formed between two fiber couplers (90:10 and 50:50). Additional saw-tooth modulation (1 Hz) was added to provide a wavelength scan across the target transition. The light from the 1550 nm DFB laser diode was used as a probe beam. An acousto-optic frequency shifter/modulator (AOM) driven at f_a = 200 MHz was placed in one arm of the interferometer, enabling heterodyne detection of the photothermal signal: Changes of the optical length in one arm were detected as changes of the phase of the heterodyne signal. The main advantage of heterodyne-based detection [35] compared to homodyne-based sensing is that a heterodyne-based setup does not require the arms of the interferometer to be stabilized [36,37]. A high-speed lock-in amplifier (UHFLI from Zurich Instruments) was used to perform phase demodulation of the beat note at 200 MHz, as well as subsequent signal demodulation (filtering) at 2 × f_m (this helps with removing background signals due to, e.g., non-resonant absorption in the setup).

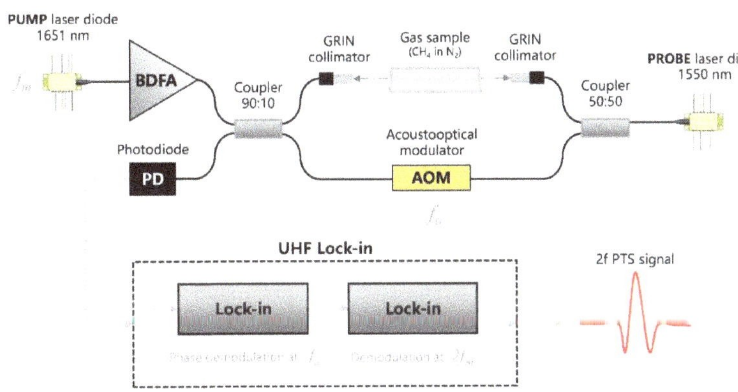

Figure 8. Heterodyne interferometric setup for photothermal gas detection.

Figure 9a shows the PTS spectra measured for a 100 mm-long sample containing 4.29% methane (balanced with nitrogen, total pressure of 740 torr; the concentrations was determined based on direct absorption measurement and spectral fitting with data from the HITRAN database [38]). Four different powers at the output of BDFA were used (the optical power was measured after the collimator and before the gas cell). 2*f* PTS spectra are baseline-free and their amplitude clearly depends on the optical power.

A linear dependence between the signal to noise ratio (SNR) and the optical power level was also confirmed, as shown in Figure 9b. For this measurement, a shorter gas cell (25 mm) was used (with a methane concentration of approximately 5.35%). The signal was measured as the 2*f* PTS amplitude at the transition center and the noise was calculated as the standard deviation of the signal recorded when the sample was removed from the setup. Based on this measurement (i.e., SNR ≈ 52 for the incident power of ~120 mW), we could calculate the noise equivalent concentration, which was shown to be approximately 19.3 ppm × m × $Hz^{-1/2}$, corresponding to the noise equivalent absorbance of ~7.2×10^{-4} $Hz^{-1/2}$. This result is comparable to the sensitivity demonstrated previously with a similar method [35]. We expect that a significantly better detection limit could be obtained if photothermal spectroscopy was performed inside a hollow core fiber [36,37].

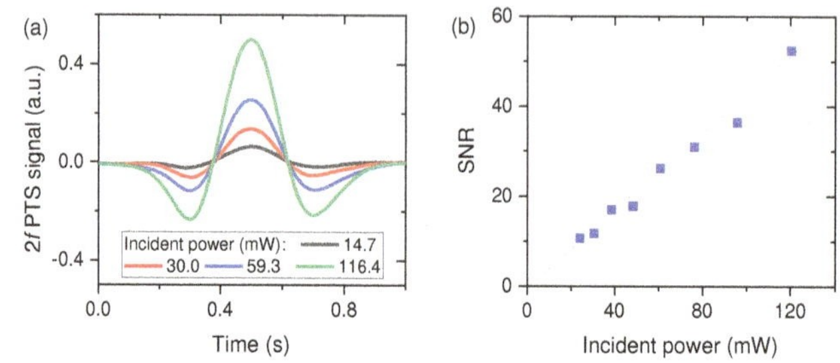

Figure 9. (**a**) 2*f* photothermal spectroscopy (PTS) signals for various incident power levels for a 100 mm-long cell with a methane concentration of 4.29%. A 1 Hz saw-tooth modulation was used to scan the wavelength across the methane transition near 1651 nm. (**b**) The dependence of the 2*f* PTS signal/noise ratio on the incident power at 1651 nm; the signal was measured using a 25 mm-long cell with a methane concentration of 5.35%.

4. Discussion

In this paper, the bismuth-doped fiber amplifier was presented. The new fiber-based pump source helped to suppress the unwanted SBS (observed in the previous setup) and enabled the delivery of a much higher pumping power (up to almost 1.6 W) to the BDF compared to the configuration presented in [21,22]. The performance of the amplifier was analyzed at two wavelengths: 1651 and 1687 nm. The output powers of 163 mW (1651 nm) and 197 mW (1687 nm) were obtained. These power levels are almost two times higher than those demonstrated earlier with the previous BDFA configuration [21,22]. More importantly, the linear dependence between the output power and the pump power suggests that a further increase of the output power should be possible when a higher pump power level is used.

In addition to the high power at the output, the BDFA presented in this paper also offers a significant small signal gain: 18 dB at 1651 nm and 26 dB at 1687 nm. These values are higher than previously published for BDFAs [17] and also higher then reported for the short-wavelength Al/Tm-doped fiber amplifier presented in [7]. A similar small signal gain was obtained in [8], where a Ge/Tm-doped fiber was used.

A comparison of the presented BDFA with some previously demonstrated amplifiers is presented in Table 1. The BDFA reported in this work uses a longer fiber than Tm-doped fiber amplifiers, but provides a higher output power while requiring a smaller pump power.

Table 1. Summary of the parameters of various previously reported fiber amplifiers operating near 1650 nm and the results obtained in this work.

Ref.	Fiber Type	Fiber Length	Pump Power	Small Signal Gain @1650 nm	Output Power @1650 nm
[7]	Al/Tm-doped	5 m	37 dBm	8 dB	8 dBm
[8]	Ge/Tm-doped	4.5 m	36.9 dBm	18.7 dB	15 dBm
[17]	Ge/Bi-doped	50 m	24.77 dBm	~5 dB	10 dBm (@1680 nm)
This work	Ge/Bi-doped	90 m	32 dBm	18 dB	22.1 dBm

We also analyzed the noise at the output of the amplifier. Output power fluctuations below 1% were obtained not only for high, but also for low, input powers. These features (high gain and output power stability for low input powers) make the presented BDFA an attractive booster amplifier in the wavelength region that cannot be easily covered by semiconductor amplifiers.

As a potential application of BDFA, we have presented the photothermal spectroscopy of methane at 1651 nm. We have demonstrated that the amplifier can be used to improve the signal amplitude and signal/noise ratio. Similar sensitivity enhancement can be expected when more common photoacoustic spectroscopy is used [39,40].

Author Contributions: Bismuth-doped fiber design, fabrication, and characterization: A.M.K., S.V.A., A.S.L., and S.V.F.; BDFA design and characterization: G.G., M.K. (Małgorzata Kaleta), and M.N.; Photothermal spectroscopy experiments: M.K. (Monika Krajewska), G.G., and M.N.; manuscript writing: M.N., G.G., and S.V.F. All authors have read and agreed to the published version of the manuscript.

Funding: This research was funded by The National Science Centre, Poland, grant no. UMO-2018/29/B/ST7/01730. M.N. acknowledges funding within the National Laboratory for Photonics and Quantum Technologies (NLPQT) project (POIR.04.02.00-00-B003/18), co-financed by the European Regional Development Fund (ERDF), and the Russian Science Foundation, grant no. 19-72-10003 (fabrication and characterization of Bi-doped fiber).

Conflicts of Interest: The authors declare no conflict of interest.

References

1. Crotti, C.; Deloison, F.; Alahyane, F.; Aptel, F.; Kowalczuk, L.; Legeais, J.-M.; Peyrot, D.A.; Savoldelli, M.; Plamann, K. Wavelength optimization in femtosecond laser corneal surgery. *Investig. Ophthalmol. Vis. Sci.* **2013**, *54*, 3340–3349. [CrossRef]
2. Qin, Y.; Batjargal, O.; Cromey, B.; Kieu, K. All-fiber high-power 1700 nm femtosecond laser based on optical parametric chirped-pulse amplification. *Opt. Express* **2020**, *28*, 2317–2325. [CrossRef]
3. Zhang, E.J.; Teng, C.C.; van Kessel, T.G.; Klein, L.; Muralidhar, R.; Wysocki, G.; Green, W.M.J. Field deployment of a portable optical spectrometer for methane fugitive emissions monitoring on oil and gas well pads. *Sensors* **2019**, *19*, 2707. [CrossRef]
4. Tian, X.; Cao, Y.; Chen, J.; Liu, K.; Wang, G.; Tan, T.; Mei, J.; Chen, W.; Gao, X. Dual-gas sensor of CH4/C2H6 based on wavelength modulation spectroscopy coupled to a home-made compact dense-pattern multipass cell. *Sensors* **2019**, *19*, 820. [CrossRef]
5. Liu, K.; Wang, L.; Tan, T.; Wang, G.; Zhang, W.; Chen, W.; Gao, X. Highly sensitive detection of methane by near-infrared laser absorption spectroscopy using a compact dense-pattern multipass cell. *Sens. Actuators B Chem.* **2015**, *220*, 1000–1005. [CrossRef]
6. Cheng, G.; Cao, Y.; Liu, K.; Zhu, G.; Wang, G.; Gao, X. Photoacoustic measurement of ethane with near-infrared DFB diode laser. *J. Spectrosc.* **2018**, *2018*, 9765806. [CrossRef]
7. Li, Z.; Jung, Y.; Daniel, J.M.O.; Simakov, N.; Tokurakawa, M.; Shardlow, P.C.; Jain, D.; Sahu, J.K.; Heidt, A.M.; Clarkson, W.A.; et al. Exploiting the short wavelength gain of silica-based thulium-doped fiber amplifiers. *Opt. Lett.* **2016**, *41*, 2197–2200. [CrossRef]
8. Chen, S.; Jung, Y.; Alam, S.; Richardson, D.J.; Sidharthan, R.; Ho, D.; Yoo, S.; Daniel, J.M.O. Ultra-short wavelength operation of thulium-doped fiber amplifiers and lasers. *Opt. Express* **2019**, *27*, 36699–36707. [CrossRef]
9. Tsuzaki, T.; Kakui, M.; Hirano, M.; Onishi, M.; Nakai, Y.; Nishimura, M. Broadband discrete fiber Raman amplifier with high differential gain operating over 1.65 μm-band. In Proceedings of the OFC 2001 Optical Fiber Communication Conference and Exhibit. Technical Digest Postconference Edition (IEEE Cat. 01CH37171), Anaheim, CA, USA, 17 March 2001; Volume 1, p. MA3.
10. Bauer, R.; Legg, T.; Mitchell, D.; Flockhart, G.M.H.; Stewart, G.; Johnstone, W.; Lengden, M. Miniaturized photoacoustic trace gas sensing using a raman fiber amplifier. *J. Light. Technol.* **2015**, *33*, 3773–3780. [CrossRef]
11. Dianov, E.M. Bismuth-doped optical fibers: A challenging active medium for near-IR lasers and optical amplifiers. *Light Sci. Appl.* **2012**, *1*, e12. [CrossRef]
12. Thipparapu, N.K.; Wang, Y.; Wang, S.; Umnikov, A.A.; Barua, P.; Sahu, J.K. Bi-doped fiber amplifiers and lasers. *Opt. Mater. Express* **2019**, *9*, 2446–2465. [CrossRef]
13. Bufetov, I.A.; Melkumov, M.A.; Firstov, S.V.; Riumkin, K.E.; Shubin, A.V.; Khopin, V.F.; Guryanov, A.N.; Dianov, E.M. Bi-doped optical fibers and fiber lasers. *IEEE J. Sel. Top. Quantum Electron.* **2014**, *20*, 111–125. [CrossRef]
14. Thipparapu, N.K.; Wang, Y.; Umnikov, A.A.; Barua, P.; Richardson, D.J.; Sahu, J.K. 40 dB gain all fiber bismuth-doped amplifier operating in the O-band. *Opt. Lett.* **2019**, *44*, 2248. [CrossRef] [PubMed]
15. Thipparapu, N.K.; Umnikov, A.A.; Barua, P.; Sahu, J.K. Bi-doped fiber amplifier with a flat gain of 25 dB operating in the wavelength band 1320–1360 nm. *Opt. Lett.* **2016**, *41*, 1518–1521. [CrossRef]

16. Dianov, E.M.; Mel'kumov, M.A.; Shubin, A.V.; Firstov, S.V.; Khopin, V.F.; Gur'yanov, A.N.; Bufetov, I.A. Bismuth-doped fibre amplifier for the range 1300–1340 nm. *Quantum Electron.* **2009**, *39*, 1099–1101. [CrossRef]
17. Firstov, S.V.; Alyshev, S.V.; Riumkin, K.E.; Khopin, V.F.; Guryanov, A.N.; Melkumov, M.A.; Dianov, E.M. A 23-dB bismuth-doped optical fiber amplifier for a 1700-nm band. *Sci. Rep.* **2016**, *6*, 28939. [CrossRef]
18. Kharakhordin, A.V.; Alyshev, S.V.; Firstova, E.G.; Lobanov, A.S.; Khopin, V.F.; Khegai, A.M.; Melkumov, M.A.; Guryanov, A.N.; Firstov, S.V. Lasing properties of thermally treated GeO2-SiO2 glass fibers doped with bismuth. *Appl. Phys. B* **2020**, *126*, 87. [CrossRef]
19. Alyshev, S.; Kharakhordin, A.; Firstova, E.; Khegai, A.; Melkumov, M.; Khopin, V.; Lobanov, A.; Guryanov, A.; Firstov, S. Photostability of laser-active centers in bismuth-doped GeO2-SiO2 glass fibers under pumping at 1550 nm. *Opt. Express* **2019**, *27*, 31542–31552. [CrossRef]
20. Firstov, S.V.; Alyshev, S.V.; Riumkin, K.E.; Khegai, A.M.; Kharakhordin, A.V.; Melkumov, M.A.; Dianov, E.M. Laser-active fibers doped with bismuth for a wavelength region of 1.6–1.8 μm. *IEEE J. Sel. Top. Quantum Electron.* **2018**, *24*, 1–15. [CrossRef]
21. Nikodem, M.; Khegai, A.; Firstov, S.V. Single-frequency bismuth-doped fiber power amplifier at 1651 nm. *Laser Phys. Lett.* **2019**, *16*, 115102. [CrossRef]
22. Gomolka, G.; Khegai, A.M.; Alyshev, S.V.; Lobanov, A.S.; Firstov, S.V.; Nikodem, M. Characterization of a single-frequency bismuth-doped fiber power amplifier with a continuous wave and modulated seed source at 1687 nm. *Appl. Opt.* **2020**, *59*, 1558–1563. [CrossRef] [PubMed]
23. Zeringue, C.; Dajani, I.; Naderi, S.; Moore, G.T.; Robin, C. A theoretical study of transient stimulated Brillouin scattering in optical fibers seeded with phase-modulated light. *Opt. Express* **2012**, *20*, 21196–21213. [CrossRef] [PubMed]
24. Garmire, E. Stimulated brillouin review: Invented 50 years ago and applied today. *Int. J. Opt.* **2018**, *2018*, 2459501. [CrossRef]
25. Okoshi, T.; Kikuchi, K.; Nakayama, A. Novel method for high resolution measurement of laser output spectrum. *Electron. Lett.* **1980**, *16*, 630–631. [CrossRef]
26. Preussler, S.; Schneider, T. Stimulated brillouin scattering gain bandwidth reduction and applications in microwave photonics and optical signal processing. *Opt. Eng.* **2015**, *55*, 1–11. [CrossRef]
27. Yeniay, A.; Delavaux, J.-M.; Toulouse, J. Spontaneous and stimulated brillouin scattering gain spectra in optical fibers. *J. Light. Technol.* **2002**, *20*, 1425. [CrossRef]
28. Owens, M.A.; Davis, C.C.; Dickerson, R.R. A photothermal interferometer for gas-phase ammonia detection. *Anal. Chem.* **1999**, *71*, 1391–1399. [CrossRef] [PubMed]
29. Waclawek, J.P.; Kristament, C.; Moser, H.; Lendl, B. Balanced-detection interferometric cavity-assisted photothermal spectroscopy. *Opt. Express* **2019**, *27*, 12183–12195. [CrossRef]
30. Webber, M.E.; Pushkarsky, M.; Patel, C.K.N. Fiber-amplifier-enhanced photoacoustic spectroscopy with near-infrared tunable diode lasers. *Appl. Opt.* **2003**, *42*, 2119–2126. [CrossRef]
31. He, Y.; Ma, Y.; Tong, Y.; Yu, X.; Tittel, F.K. HCN ppt-level detection based on a QEPAS sensor with amplified laser and a miniaturized 3D-printed photoacoustic detection channel. *Opt. Express* **2018**, *26*, 9666–9675. [CrossRef]
32. Wu, H.; Sampaolo, A.; Dong, L.; Patimisco, P.; Liu, X.; Zheng, H.; Yin, X.; Ma, W.; Zhang, L.; Yin, W.; et al. Quartz enhanced photoacoustic H_2S gas sensor based on a fiber-amplifier source and a custom tuning fork with large prong spacing. *Appl. Phys. Lett.* **2015**, *107*, 111104. [CrossRef]
33. Wu, H.; Dong, L.; Liu, X.; Zheng, H.; Yin, X.; Ma, W.; Zhang, L.; Yin, W.; Jia, S. Fiber-amplifier-enhanced QEPAS sensor for simultaneous trace gas detection of NH3 and H2S. *Sensors* **2015**, *15*, 26743–26755. [CrossRef]
34. Ma, Y.; He, Y.; Tong, Y.; Yu, X.; Tittel, F.K. Ppb-level detection of ammonia based on QEPAS using a power amplified laser and a low resonance frequency quartz tuning fork. *Opt. Express* **2017**, *25*, 29356–29364. [CrossRef]
35. Krzempek, K.; Dudzik, G.; Abramski, K.; Wysocki, G.; Jaworski, P.; Nikodem, M. Heterodyne interferometric signal retrieval in photoacoustic spectroscopy. *Opt. Express* **2018**, *26*, 1125–1132. [CrossRef]
36. Jin, W.; Cao, Y.; Yang, F.; Ho, H.L. Ultra-sensitive all-fibre photothermal spectroscopy with large dynamic range. *Nat. Commun.* **2015**, *6*, 6767. [CrossRef]
37. Yao, C.; Wang, Q.; Lin, Y.; Jin, W.; Xiao, L.; Gao, S.; Wang, Y.; Wang, P.; Ren, W. Photothermal CO detection in a hollow-core negative curvature fiber. *Opt. Lett.* **2019**, *44*, 4048–4051. [CrossRef]

38. Rothman, L.S.; Gordon, I.E.; Barbe, A.; Benner, D.C.; Bernath, P.F.; Birk, M.; Boudon, V.; Brown, L.R.; Campargue, A.; Champion, J.-P.; et al. The HITRAN 2008 molecular spectroscopic database. *J. Quant. Spectrosc. Radiat. Transf.* **2009**, *110*, 533–572. [CrossRef]
39. Hu, L.; Zheng, C.; Zhang, M.; Yao, D.; Zheng, J.; Zhang, Y.; Wang, Y.; Tittel, F.K. Quartz-enhanced photoacoustic spectroscopic methane sensor system using a quartz tuning fork-embedded, double-pass and off-beam configuration. *Photoacoustics* **2020**, *18*, 100174. [CrossRef]
40. Milde, T.; Hoppe, M.; Tatenguem, H.; Assmann, C.; Schade, W.; Sacher, J. Comparison of the spectral excitation behavior of methane according to InP, GaSb, IC, and QC lasers as excitation source by sensor applications. *Appl. Opt.* **2019**, *58*, C84–C91. [CrossRef]

Publisher's Note: MDPI stays neutral with regard to jurisdictional claims in published maps and institutional affiliations.

photonics

Communication

All-Fiber Gas Raman Laser by D_2-Filled Hollow-Core Photonic Crystal Fibers

Wenxi Pei [1,2], Hao Li [1,3], Wei Huang [1,3], Meng Wang [1,2,3] and Zefeng Wang [1,2,3,*]

1 College of Advanced Interdisciplinary Studies, National University of Defense Technology, Changsha 410073, China; peiwenxi@nudt.edu.cn (W.P.); lihao18c@nudt.edu.cn (H.L.); huangw2684@nudt.edu.cn (W.H.); wangmeng@nudt.edu.cn (M.W.)
2 Hunan Provincial Key Laboratory of High Energy Laser Technology, Changsha 410073, China
3 State Key Laboratory of Pulsed Power Laser Technology, Changsha 410073, China
* Correspondence: zefengwang@nudt.edu.cn

Abstract: We report here an all-fiber structure tunable gas Raman laser based on deuterium-filled hollow-core photonic crystal fibers (HC-PCFs). An all-fiber gas cavity is fabricated by fusion splicing a 49 m high-pressure deuterium-filled HC-PCF with two solid-core single-mode fibers at both ends. When pumped with a pulsed fiber amplifier seeded by a tunable laser diode at 1.5 μm, Raman lasers ranging from 1643 nm to 1656 nm are generated. The maximum output power is ~1.2 W with a Raman conversion efficiency of ~45.6% inside the cavity. This work offers an alternative choice for all-fiber lasers operating at 1.6–1.7 μm band.

Keywords: photonic crystal fibers; hollow-core fibers; fiber lasers; stimulated Raman scattering; gas Raman lasers

Citation: Pei, W.; Li, H.; Huang, W.; Wang, M.; Wang, Z. All-Fiber Gas Raman Laser by D_2-Filled Hollow-Core Photonic Crystal Fibers. *Photonics* **2021**, *8*, 382. https://doi.org/10.3390/photonics8090382

Received: 18 August 2021
Accepted: 6 September 2021
Published: 9 September 2021

Publisher's Note: MDPI stays neutral with regard to jurisdictional claims in published maps and institutional affiliations.

1. Introduction

Stimulated Raman scattering (SRS) of gas medium has received great attention since it was first reported [1]. It provides a novel way to generate lasers with unobtainable wavelength for traditional schemes, especially in the ultraviolet and infrared spectral range. However, due to the short reaction length and weak reaction intensity, gas Raman lasers based on the traditional gas cell usually require high pump power, and the corresponding conversion is always inefficient. Moreover, there is often more than one Raman line at the output except the desired line, which exacerbates the inefficiency. The advent of the hollow-core fibers (HCFs) provides new opportunities for efficient gas SRS [2,3]. This kind of innovative gas cell based on HCFs can provide a much longer reaction distance, and the pump laser and gas molecules are confined in the micron-level air core, which enhances the reaction intensity greatly. More importantly, the HCF makes it easy to control the transmission spectrum by designing fiber structure, so the conversion of unwanted Raman lines can be well suppressed, which further improves the conversion efficiency. Therefore, fiber gas Raman lasers (FGRLs) have drawn great attention with the fast development of the HCFs [4–7].

Fiber lasers operating at 1.6~1.7 μm band have been studied extensively, owing to their significant applications [8]. They play a great role in medical treatment [9,10], material processing [11], the generation of mid-infrared lasers [12] and so on. Several approaches for generating the laser at this band have been reported, which can be divided into two categories. One is based on the rare-earth doped fibers, such as thulium-doped fibers (TDFs) and bismuth-doped fibers (BDFs), which is mainly limited by the low gain of the fiber in this band or immature manufacturing technology. The other is to take advantage of nonlinear effects based on SRS in the solid-core fibers, while the complex cascade structure is always necessary. The advent of fiber gas lasers based on SRS provides a novel method for the generation of laser emission at some wavelengths not easily achievable with traditional ways, including 1.6–1.7 μm band [13–17]. By using proper HCFs and proper active gas medium, laser emissions can be realized flexibly and effectively. Moreover, due to the higher damage threshold of the HCF, it is more possible for FGRLs to realize higher output power. Recently, we have reported some works on 1.6–1.7 μm gas-filled FGRLs [13–18]. The common gas media used in our work are hydrogen and deuterium. Compared with hydrogen, D_2 has a larger molecular weight, making it easier to confine it inside the HCF. Besides, D_2 has a much smaller coefficient of Raman shift, which generates and enriches the unobtainable laser wavelength in this band. In our previous work, we demonstrated a subwatt 1.65 μm FGRL based on D_2-filled hollow-core photonic crystal fibers (HC-PCFs) [16], and then further improved its output power [17]. However, both of them sealed the output end in a customized gas chamber, making the system bulky and poorly stabilized, which limits its application in the future.

Here, we report a 1.65 μm all-fiber tunable pulsed FGRL based on D_2-filled HC-PCFs. The all-fiber gas cavity is fabricated by fusion splicing a 49 m high-pressure deuterium-filled HC-PCF with two solid-core single-mode fibers at both ends. When pumped by a homemade fiber amplifier at 1.5 μm band, the pure rotational SRS occurs in the gas cavity, generating first-order Raman lasers ranging from 1643 nm to 1656 nm. The maximum output Raman power of ~1.2 W at 1645 nm is obtained with the repetition frequency of 2.2 MHz when pump power is at the maximum. The Raman conversion efficiency inside the cavity is ~45.6%. This work enriches the research on FGRLs and provides a convenient choice for generation of fiber lasers operating at the 1.6–1.7 μm band.

2. Experimental Setup

Figure 1a shows the schematic diagram of experimental setup, which is similar to our previous work [15], but a longer HC-PCF is introduced. A homemade narrow-linewidth pulsed Erbium-doped fiber amplifier (EDFA) seeded by a tunable laser diode at 1.5 μm works as the pump source. The width of the output pulse is 20 ns and it remains the same when the repetition frequency or output power changes. The repetition frequency is adjustable, and the power characteristics are basically the same, with the pulse width of 20 ns in different repetition frequency. The pump source can be tuned in the range of 1538 nm to 1545 nm, which retains the similar characteristics when the wavelength changes. Figure 1b–d show the output spectrum of the EDFA with different operating wavelengths when the output power is the maximum of ~7 W. It can be seen that although the amplified spontaneous emission is generated during the amplification, it can always be well suppressed. An ultra-low-loss fiber coupler with the measured coupling ratio of 99.06:0.94 is spliced to the pigtail of the EDFA to monitor the real output power. Its signal end is spliced to the port1 of a circulator, which is introduced to protect the pump source and measure backward light at port3, as marked in the figure. The loss of the circulator from port1 to port2 is ~0.67 dB. An all-fiber gas cavity is fabricated by a 49 m long HC-PCF (HC-1550-02, NKT Photonics) and two solid-core single-mode fibers (SMF-28e, Corning), as shown within the dashed box in Figure 1a. One of the solid-core fibers is employed to be the input end of the cavity, which is spliced with the circulator port2 and one end of the HC-PCFs, respectively. After being vacuumized, the HC-PCF is inflated with high-pressure D_2 and stands until the gas pressure is balanced. Then the other end that was sealed in the gas chamber is spliced to another solid-core fiber, which works as the output end of the gas cavity. The final gas pressure in the cavity is estimated to be 25.5 bar by calculating the time and gas leakage ratio. The core diameter of the HC-PCFs is 10 μm and its cross section by optical microscope is shown in Figure 1e. Due to the similar mode field diameter (~9 μm) of SMF-28e and HC-1550-02, the loss of fusion splicing directly is acceptable. The losses of the fusion splicing points marked as splice1 and splice2 are 1.4 dB and 2 dB, respectively. Two lenses with the focal length of 15 mm are placed at the output end of the gas cavity and the circulator port3 separately to collimate the output beam. Two long-pass filters (transmission ~95% > 1600 nm) play the role of separating the pump light and Raman light.

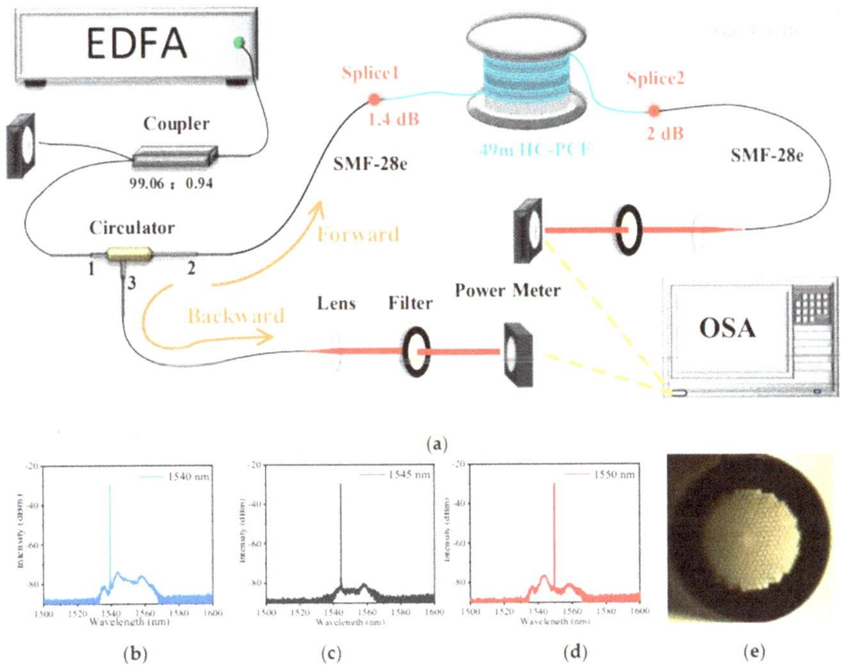

Figure 1. (**a**) The schematic diagram of experimental setup. OSA: optical spectrum analyzer. EDFA: Erbium-doped fiber amplifier; (**b**–**d**) the output spectrum of the EDFA at the maximum output power; (**e**) the cross section of the used HC-PCFs by optical microscope.

3. Results and Discussion

3.1. Spectrum Characteristics

Figure 2a shows the measured output spectrum with different pump wavelengths when the pump source works at the maximum output power. The repetition frequency (RF) is set to 4 MHz. When pumped by the lines at 1538 nm to 1550 nm, the pure rotational SRS of D_2 occurs in the cavity, generating Raman lines ranging from 1643 nm to 1656 nm with the Raman shift of ~414 cm^{-1}. Each pump line corresponds to only one Raman line. Figure 2b shows the fine spectrum of the Raman line at 1645 nm (in red line) and the pump line at 1540 nm (in blue dotted line), respectively. It can be seen that the central wavelength of the pump line is at 1539.9 nm. For ease of expression and readability of the paper, we express it as 1540 nm.

Figure 3a presents the measured transmission spectrum of the HC-PCF, covering 1530 nm to 1760 nm. When the pump wavelength is set to 1540 nm, we measured the output spectrum with different RF at the maximum output power, as shown in Figure 3b–f. It can be seen in Figure 3b that when the RF is 1 MHz, due to the high peak power, more than one Raman line occurs in the gas cavity. In addition to the pump line at 1540 nm and the first-order Raman line at 1645 nm, there are 3 s order Raman lines. The lines at 1695 nm, 1730 nm and 1765 nm correspond to the frequency shift of ~179 cm^{-1}, ~298 cm^{-1} and ~414 cm^{-1}, respectively. High peak power of the pump pulse is more likely to exceed the high-order Raman threshold, leading to the cascaded Raman conversion. The output power of the first-order Raman laser and its conversion efficiency may be reduced in this case. Compared with H_2, D_2 has a much smaller coefficient of Raman shift, permitting the existence of some high-order Raman lines in the transmission band of the HC-PCF. It not only enriches the output wavelength in 1.7 μm band, but also makes it possible to realize the cascade Raman laser output if the fiber Bragg granting is introduced.

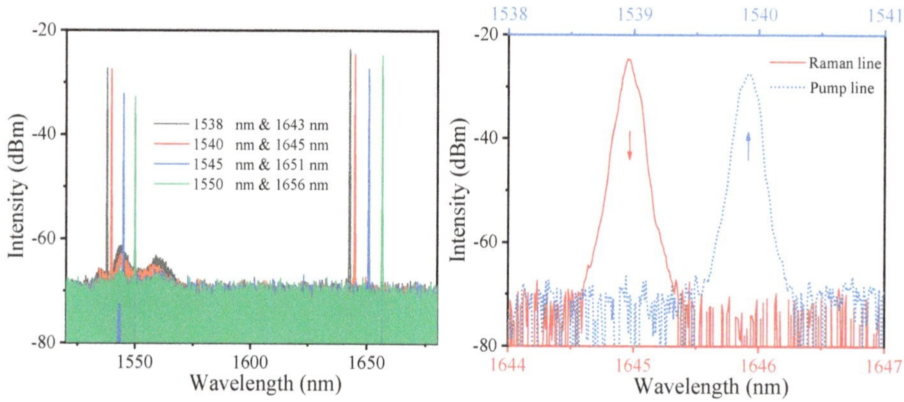

(a) (b)

Figure 2. (**a**) The output spectrum of the fiber gas Raman laser pumped by different pump wavelengths at the maximum output power. (**b**) The fine spectrum of the Raman lines at 1645 nm (in red line) and the pump lines at 1540 nm (in blue dotted line).

When the RF is 2 MHz in Figure 3c, besides the pump lines and the first-order Raman lines, there is only a weak second-order Raman line at 1765 nm due to the reduction of the pulse peak power. It also indicates that for D_2 inside the gas cavity, the rotational SRS whose Raman shift is ~414 cm^{-1} corresponds to the stronger Raman gain (both the frequency shifts of these two Raman lines are ~414 cm^{-1}).

Interestingly, when the RF is set to 3 MHz in Figure 3d, there are two second-order Raman lines at 1695 nm and 1730 nm again. This can be explained by the following theory. For the gas cavity, we regard it as a special resonant cavity formed by the 49 m HC-PCFs and two fusion splicing points (splice1 and splice2). Its average refractive index is ~1, so the resonant frequency can be calculated through the formula: $\upsilon_p = qc/2L$, where υ_p is the resonant frequency, q is a positive integer (the number of the pulse round trip through the cavity), c is the speed of light (2.997*10^8 m/s) and L is the length of resonant cavity. In our work, L is 49 m; then the υ_p can be calculated to be ~3 MHz when the q is 1. For the pump pulse RF of 3 MHz, it is consistent with the resonant frequency. While the pump pulses oscillate in the cavity, they can coincide with previous pulses, which means that the resonant cavity can provide positive feedback to reduce the Raman threshold and benefit the conversion. Therefore, there are still high-order Raman lines in this case. The absence of the line at 1765 nm is mainly due to the high transmission loss and the lower peak power. Furthermore, the specific Raman threshold power has been measured, as discussed in the following Section 3.2. Power Characteristics, which also supports this theory.

As we can see in Figure 3e,f, there are no high-order lines when the RFs are 4 MHz and 5 MHz, which is due to the continuous decrease of the peak power. It should be noted that when the RF is 5 MHz, the intensity of the first-order Raman line at 1645 nm is 3.065 dB weaker than that of the pump line at 1540 nm, which indicates the insufficient conversion in the gas cavity.

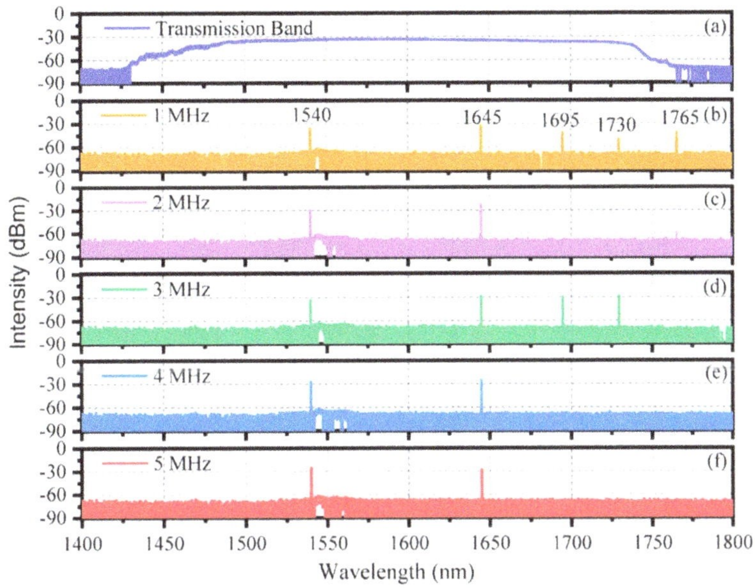

Figure 3. (**a**) The measured transmission spectrum of the HC-PCF. (**b–f**) The output spectrum of the fiber gas Raman laser pumped by the pulse with different RF at the maximum output power.

3.2. Power Characteristics

To explore the maximum output power and the corresponding conversion efficiency of the first-order Raman laser at 1645 nm, we measured the power characteristics with the RF of 2 MHz, 4 MHz and 5 MHz, as shown in Figure 4. All the powers in the figure are the average powers of the pulse train. The power characteristics with the RF of 1 MHz or 3 MHz are not measured due to their complex spectrum characteristics (as shown in Figure 3), which have multi-wavelength output at the maximum pump power and cannot separate the unwanted lines easily. The incident pump power is the output power of the circulator port2. Figure 4a,b show the characteristics of forward Raman power and pump power. It can be seen that the Raman threshold average power with the RF of 2 MHz is much lower than that of the RF in 4 MHz and 5 MHz, as the peak power of pump pulse in 2 MHz is much higher. When the incident pump power exceeds threshold power, all the curves of the pump power drop and Raman power increases gradually. The maximum forward Raman power of ~1.13 W is achieved at the maximum pump power when the RF is 2 MHz. Figure 4c presents the backward Raman power characteristics. Obviously, backward Raman power is much lower than forward Raman power, as it is reflected by the splice2. We calculated the ratio of the maximum backward Raman power to the maximum forward Raman power; the value is ~3.28%, ~15.94% and ~16.92% when the corresponding RF is 2 MHz, 4 MHz and 5MHz, respectively. This indicates that the high RF is conductive to the amplification of the backward Raman power. The curves of backward pump power in Figure 4d are basically the same, and the reason is that they are incident pump power that is reflected by the splice1. All the pump powers in Figure 4d keep increasing as the incident pump power increases.

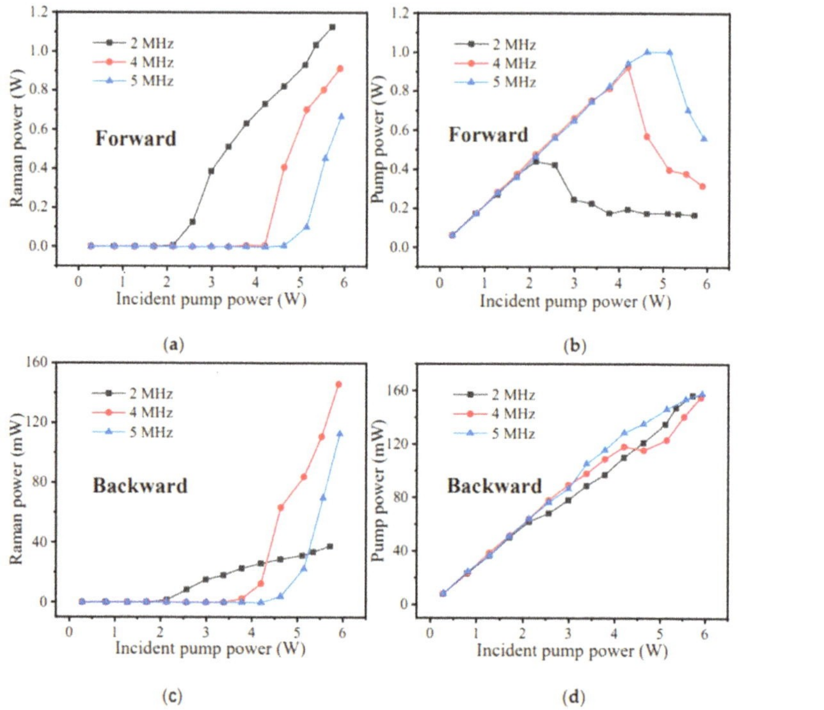

Figure 4. The output power characteristics of the fiber gas Raman laser with the repetition frequency of 2 MHz, 4 MHz and 5 MHz: (**a**) forward Raman power; (**b**) forward pump power; (**c**) backward Raman power; (**d**) backward pump power.

Figure 5a shows the evolution of the Raman threshold average power and peak power with different RF of the pump pulse. Here we define the Raman threshold power as the minimum pump power coupled in the HC-PCF when the first-order Raman laser can be detected by the OSA at the output end of the cavity. We measured the Raman threshold power with the RF from 1 MHz to 5 MHz; it can be seen that except for the case with the RF of 3 MHz, all the peak power of the Raman threshold is basically the same and the Raman threshold average power increases with the increasing of the RF. This is not unexpected, because the peak power of the Raman threshold is constant for a certain cavity, which is mainly determined by the gas pressure inside the cavity or the length of the HC-PCF. However, when the RF of 3 MHz is consistent with the resonant frequency mentioned above, the positive feedback caused by the cavity and the coincidence between pulses greatly reduce the peak power of the Raman threshold. Therefore, its threshold characteristics differ from those of the other RF, meaning that there may be high-order Raman laser when the pump power increases in this case, as discussed above. Furthermore, we also measured the threshold power with the RF of 6 MHz, which is another resonant frequency of the cavity (the corresponding q is 2), as shown in Figure 5a. It can be seen that its peak power of the Raman threshold is the same as that of 3 MHz, but the Raman threshold average power is double, which is because the RF of 6 MHz is the double of 3 MHz. This confirms the conclusion and provides a new way for realizing multi-wavelength FGRL with all-fiber structure. Moreover, we calculated the total Raman power and internal Raman power (Raman power inside the gas cavity) with different RF when the pump power is at the maximum, as shown in Figure 5b. The Raman power at 1645 nm with RF of 1 MHz or 3 MHz is not obtained because of the existence of the high-order Raman power with low-threshold characteristic. Obviously, the maximum Raman power can be achieved when the RF is 2 MHz. The total Raman power is ~1.16 W, and the internal Raman power is calculated to be 1.85 W, which is much higher than that with other RF.

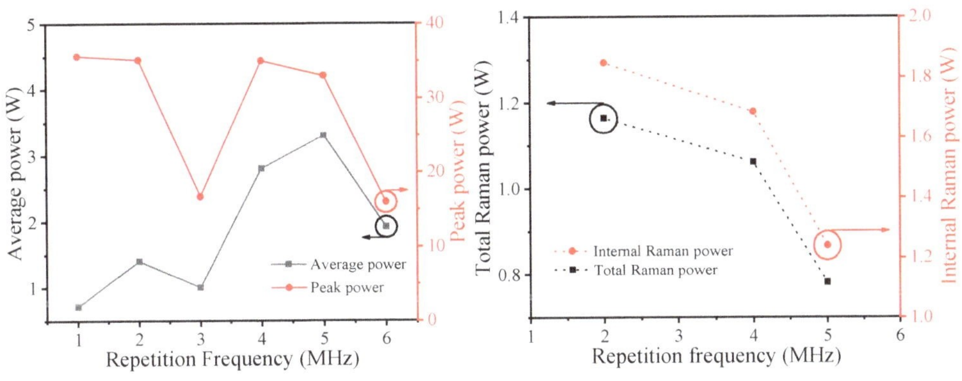

Figure 5. (**a**) The evolution of the Raman threshold average power and the corresponding peak power with repetition frequency ranging from 1 MHz to 6 MHz. (**b**) The variation of total Raman power and the corresponding internal Raman power inside the HC-PCF with the repetition frequency when the pump power is the maximum.

Then, we changed the RF in steps of 200 kHz and measured their Raman power characteristics, as shown in Figure 6. As shown in Figure 6a,b, both the maximums of forward and backward Raman power are achieved at the maximum pump power when the RF is set to 2.2 MHz. When the RF is 1.6 MHz and 1.8 MHz, the corresponding Raman powers increase at first, but decrease later when the incident pump power is high enough. This is mainly because the peak power of the pump pulse with these RF is higher than that of the other RF, and the first-order Raman laser is converted into high-order Raman laser when its power is high enough. However, due to the high transmission loss, the power of the high-order Raman laser is exhausted and cannot be measured. Figure 6c shows the evolution of the total and internal Raman power with different RF when the pump power is at the maximum, respectively. The maximum total Raman power of ~1.2 W is obtained when the RF is 2.2 MHz, and the corresponding internal Raman power is ~1.9 W. Its optical–optical conversion efficiency is ~21.2% and the corresponding conversion efficiency inside the cavity is ~45.6%. By contrast, the power with the RF of 1.6 MHz and 1.8 MHz is much lower than that with other RF, which is mainly due to the high-order Raman conversion. Figure 6d presents the maximum optical–optical conversion efficiency with different RF. The efficiency inside the gas cavity without the loss caused by splice1 and splice2 is also calculated. This shows that the maximum of total conversion efficiency is above 20% with different RF, but when the splice loss is considered, the efficiency is more than 40%. It enlightens us that if the splice loss of the gas cavity can be optimized, the conversion efficiency can be further improved.

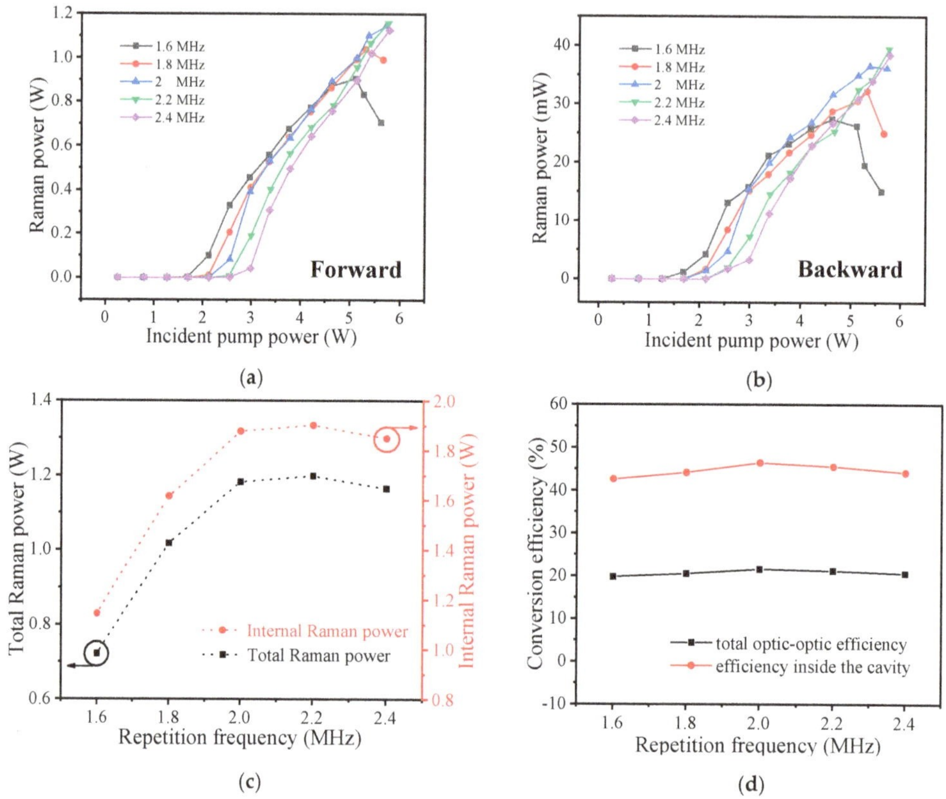

Figure 6. The output power characteristics of the fiber gas Raman laser with the repetition frequency of 1.6 MHz to 2.4 MHz: (**a**) forward Raman power; (**b**) backward Raman power; (**c**) the evolution of the total and internal Raman power with different repetition frequency when the pump power is at the maximum; (**d**) the evolution of the maximum conversion efficiency with different repetition frequency.

3.3. Pulse Shapes

Figure 7 shows the shapes of output pulse when the maximum is obtained with the pulse RF of 2.2 MHz. A bandpass filter (1530–1570 nm) and a long-pass filter (transmission ~95% > 1600 nm) are used for separating pump pulse and Raman pulse. The pulse can be detected by a fast photodetector (EOT ET5000) and a broadband oscilloscope (Tektronix MDO3104). It can be seen from Figure 7a that the shape of the residual forward pump pulse has a dip in the middle. The reason is that for a pump pulse in the cavity, only the part with higher energy can exceed the Raman threshold power and be converted into the Raman pulse, which results in the dip at its middle part. Thereby, the width of the Raman pulse is close to that of the dip, which is narrower than the pump pulse. Backward pulse shapes in Figure 7b differ from those of the forward pulse. It can be seen that pump pulse and Raman pulse are not caught at the same time, and there is a ~330 ns time delay. As analyzed above, the backward Raman pulse is reflected by the splice2, which has quite a low power, leading to the weak intensity of the measured pulse shape. The transmission length of the backward Raman pulse is double the length of the gas cavity (~100 m), which causes the time delay of ~330 ns.

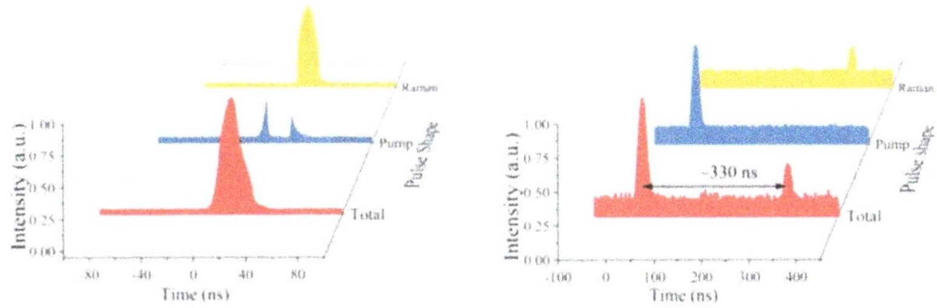

(a) (b)

Figure 7. The output pulse shapes of the fiber gas Raman laser with the repetition frequency of 2.2 MHz at the maximum pump power; (**a**) forward output pulse; (**b**) backward output pulse.

4. Conclusions

We have demonstrated a 1.65 μm all-fiber tunable FGRL based on the pure rotational SRS in the D_2.-filled HC-PCFs. An all-fiber gas cavity is realized by a 49 m long HC-PCF and two solid-core fibers. Pumped by a 1.5 μm homemade pulsed EDFA, 1643 nm to 1656 nm Raman laser is generated in the gas cavity. The maximum output power of ~1.2 W at 1645 nm is obtained when the pump laser is at 1540 nm. The efficiency inside the cavity is ~45.6%; however, the corresponding total conversion efficiency is only 21.2% due to the relative high splicing loss at present, which could be greatly reduced by more precise mode field matching methods [19,20]. This work provides a convenient alternative choice for generation of fiber lasers operating at 1.6–1.7 μm band. By optimizing the length of HC-PCF and reducing the loss of the splicing point, the conversion efficiency and output laser power could be greatly improved.

Author Contributions: Conceptualization, Z.W. and W.P.; methodology, W.P.; software, W.H. and H.L.; validation, Z.W., H.L., W.H. and W.P.; formal analysis, Z.W. and H.L.; investigation, W.P.; resources, M.W. and W.P.; data curation, H.L.; writing—original draft preparation, W.P.; writing—review and editing, Z.W.; visualization, W.P. All authors have read and agreed to the published version of the manuscript.

Funding: This work was supported by the Outstanding Youth Science Fund Project of Hunan Province Natural Science Foundation (2019JJ20023), National Natural Science Foundation of China (NSFC) (11974427, 12004431), and State Key Laboratory of Pulsed Power Laser Technology (SKL2020ZR05, SKL2021ZR01).

Data Availability Statement: The data presented in this study are available on request from the corresponding author. The data are not publicly available due to privacy.

Conflicts of Interest: The authors declare no conflict of interest.

References

1. Minck, R.W.; Terhune, R.W.; Rado, W.G. Laser-stimulated Raman Effect and Resonant Four-photon Interactions in gases H_2, D_2, and CH_4. *Appl. Phys. Lett.* **1963**, *3*, 181–184. [CrossRef]
2. Cregan, R.F. Single-Mode Photonic Band Gap Guidance of Light in Air. *Science* **1999**, *285*, 1537–1539. [CrossRef] [PubMed]
3. Benabid, F.; Knight, J.C.; Antonopoulos, G.; Russell, P.S.J. Stimulated Raman Scattering in Hydrogen-Filled Hollow-Core Photonic Crystal Fiber. *Science* **2002**, *298*, 399–402. [CrossRef] [PubMed]
4. Benabid, F.; Bouwmans, G.; Knight, J.C.; Russell, P.S.J.; Couny, F. Ultrahigh Efficiency Laser Wavelength Conversion in a Gas-Filled Hollow Core Photonic Crystal Fiber by Pure Stimulated Rotational Raman Scattering in Molecular Hydrogen. *Phys. Rev. Lett.* **2004**, *93*, 123903. [CrossRef] [PubMed]
5. Couny, F.; Benabid, F.; Light, P.S. Subwatt Threshold Cw Raman Fiber-Gas Laser Based on H_2 -Filled Hollow-Core Photonic Crystal Fiber. *Phys. Rev. Lett.* **2007**, *99*, 143903. [CrossRef] [PubMed]
6. Wang, Z.; Yu, F.; Wadsworth, W.J.; Knight, J.C. Efficient 1.9 μm Emission in H_2-Filled Hollow Core Fiber by Pure Stimulated Vibrational Raman Scattering. *Laser Phys. Lett.* **2014**, *11*, 105807. [CrossRef]

7. Li, Z.; Huang, W.; Cui, Y.; Wang, Z. Efficient Mid-Infrared Cascade Raman Source in Methane-Filled Hollow-Core Fibers Operating at 2.8 μm. *Opt. Lett.* **2018**, *43*, 4671. [CrossRef] [PubMed]
8. Zhang, Y.; Zhang, P.; Liu, P.; Han, K.; Du, Q.; Wang, T.; Zhang, L.; Tong, S.; Jiang, H. Fiber Light Source at 1.7 μm Waveband and Its Applications. *Laser Optoelectron. Prog.* **2016**, *53*, 090002. [CrossRef]
9. Piao, Z.; Ma, T.; Li, J.; Wiedmann, M.T.; Huang, S.; Yu, M.; Kirk Shung, K.; Zhou, Q.; Kim, C.-S.; Chen, Z. High Speed Intravascular Photoacoustic Imaging with Fast Optical Parametric Oscillator Laser at 1.7 Mm. *Appl. Phys. Lett.* **2015**, *107*, 083701. [CrossRef] [PubMed]
10. Alexander, V.V.; Ke, K.; Xu, Z.; Islam, M.N.; Freeman, M.J.; Pitt, B.; Welsh, M.J.; Orringer, J.S. Photothermolysis of Sebaceous Glands in Human Skin Ex Vivo with a 1708 nm Raman Fiber Laser and Contact Cooling: PHOTOTHERMOLYSIS OF SEBACEOUS GLANDS. *Lasers Surg. Med.* **2011**, *43*, 470–480. [CrossRef] [PubMed]
11. Mingareev, I.; Weirauch, F.; Olowinsky, A.; Shah, L.; Kadwani, P.; Richardson, M. Welding of Polymers Using a 2 μm Thulium Fiber Laser. *Opt. Laser Technol.* **2012**, *44*, 2095–2099. [CrossRef]
12. Quimby, R.S.; Shaw, L.B.; Sanghera, J.S.; Aggarwal, I.D. Modeling of Cascade Lasing in Dy: Chalcogenide Glass Fiber Laser with Efficient Output at 4.5 μm. IEEE Photon. *Technol. Lett.* **2008**, *20*, 123–125. [CrossRef]
13. Huang, W.; Li, Z.; Cui, Y.; Zhou, Z.; Wang, Z. Efficient, Watt-Level, Tunable 1.7 μm Fiber Raman Laser in H_2-Filled Hollow-Core Fibers. *Opt. Lett.* **2020**, *45*, 475–478. [CrossRef]
14. Li, H.; Pei, W.; Huang, W.; Wang, M.; Wang, Z. Highly Efficient Nanosecond 1.7 μm Fiber Gas Raman Laser by H_2-Filled Hollow-Core Photonic Crystal Fibers. *Crystals* **2020**, *11*, 32. [CrossRef]
15. Pei, W.; Li, H.; Huang, W.; Wang, M.; Wang, Z. All-Fiber Tunable Pulsed 1.7 μm Fiber Lasers Based on Stimulated Raman Scattering of Hydrogen Molecules in Hollow-Core Fibers. *Molecules* **2021**, *26*, 4561. [CrossRef] [PubMed]
16. Cui, Y.; Huang, W.; Li, Z.; Zhou, Z.; Wang, Z. High-Efficiency Laser Wavelength Conversion in Deuterium-Filled Hollow-Core Photonic Crystal Fiber by Rotational Stimulated Raman Scattering. *Opt. Express* **2019**, *27*, 30396. [CrossRef] [PubMed]
17. Li, H.; Huang, W.; Cui, Y.; Zhou, Z.; Wang, Z. 3 W Tunable 1.65 μm Fiber Gas Raman Laser in D_2-Filled Hollow-Core Photonic Crystal Fibers. *Opt. Laser Technol.* **2020**, *132*, 106474. [CrossRef]
18. Li, H.; Huang, W.; Cui, Y.; Zhou, Z.; Wang, Z. Pure rotational stimulated Raman scattering in H_2-filled hollow-core photonic crystal fibers. *Opt. Express* **2020**, *28*, 23881. [CrossRef] [PubMed]
19. Yu, R.; Wang, C.; Benabid, F.; Chiang, K.; Xiao, L. Robust mode matching between structurally dissimilar optical fiber waveguides. *ACS Photonics* **2021**, *8*, 857. [CrossRef]
20. Wang, C.; Yu, R.; Debord, B.; Gerome, F.; Benabid, F.; Chiang, K.; Xiao, L. Ultralow-loss fusion splicing between negative curvature hollow-core fibers and conventional SMFs with a reverse-tapering method. *Opt. Express* **2021**, *29*, 22470. [CrossRef] [PubMed]

photonics

MDPI

Communication

All-Fiber Gas Cavity Based on Anti-Resonant Hollow-Core Fibers Fabricated by Splicing with End Caps

Jing Shi [1], Xinyu Ye [1], Yulong Cui [1,2,*], Wei Huang [1,2], Hao Li [1,2], Zhiyue Zhou [1,2], Meng Wang [1,2,3], Zilun Chen [1,2,3] and Zefeng Wang [1,2,3,*]

1 College of Advanced Interdisciplinary Studies, National University of Defense Technology, Changsha 410073, China; shijing18@nudt.edu.cn (J.S.); yexinyu17@nudt.edu.cn (X.Y.); huangw2684@nudt.edu.cn (W.H.); lihao18c@nudt.edu.cn (H.L.); zhouzhiyue12@nudt.edu.cn (Z.Z.); wangmeng@nudt.edu.cn (M.W.); chenzilun@nudt.edu.cn (Z.C.)
2 State Key Laboratory of Pulsed Power Laser Technology, Changsha 410073, China
3 Hunan Provincial Key Laboratory of High Energy Laser Technology, Changsha 410073, China
* Correspondence: cuiyulong@nudt.edu.cn (Y.C.); zefengwang_nudt@163.com or zefengwang@nudt.edu.cn (Z.W.)

Abstract: In recent years, fiber gas lasers have obtained a rapid development, however, efficient and stable pump coupling is a key limitation for their applications in the future. Here, we report an all-fiber gas cavity based on anti-resonant hollow-core fibers which have the beneficial properties of adjustable broad transmission bands and potential low transmission attenuation, especially in the mid-infrared. This kind of all-fiber gas cavity is fabricated by directly splicing with end caps at both ends for the first time. The high-power laser transmission characteristics were studied, and the experimental results show that the all-fiber gas cavities have a very stable performance. The maximum input laser power at 1080 nm is about 260 W, and the output power is 203 W, giving a total transmission efficiency of 78.1%. This work opens a new opportunity for the development of high-power all-fiber structured fiber gas lasers.

Keywords: hollow-core fibers; anti-resonant hollow-core fibers; tapered fibers; fiber gas lasers; fiber end cap

Citation: Shi, J.; Ye, X.; Cui, Y.; Huang, W.; Li, H.; Zhou, Z.; Wang, M.; Chen, Z.; Wang, Z. All-Fiber Gas Cavity Based on Anti-Resonant Hollow-Core Fibers Fabricated by Splicing with End Caps. *Photonics* **2021**, *8*, 371. https://doi.org/10.3390/photonics8090371

Received: 4 August 2021
Accepted: 2 September 2021
Published: 3 September 2021

1. Introduction

Anti-resonant hollow-core fibers (AR-HCFs) are a new kind of HCF with very simple structures, mainly including revolver type [1] and ice-cream type [2]. In the past decade, AR-HCFs have attracted rapid development due to the beneficial characteristics of broad and easily adjustable transmission bands, potential low transmission-loss, particularly in the mid-infrared range, etc. [3,4]. AR-HCFs provide an ideal environment for the interaction of laser beams and gases, making a novel kind of laser source, namely fiber gas lasers (FGLs) [5,6]. In the past few years, FGLs have obtained enormous attention because they have been demonstrated to be an effective new method for generating mid-infrared laser emission [5–19] in addition to traditional ways, such as solid-state lasers [20], gas lasers [21], quantum-cascade lasers [22] and rare-earth-doped fiber lasers [23–25]. Due to the perfect combination of the advantages of both gas lasers and fiber lasers, FGLs have the potential to achieve high-power mid-infrared lasers with abundant wavelengths and portable structures [6]. To date, the maximum output power for CW mid-infrared FGLs is 1 W at 3.1 μm [16], while for wavelengths above 4 μm it is only dozens of milliwatt [18], which is mainly limited by the maximum input power of the pump. Therefore, highly efficient and stable coupling of high-power pump light is a key problem that requires solving. Up to now, almost all the FGLs have been based on free-space coupling for the pump light through gas cells. In this structure, both ends of the HCFs are sealed in the gas cells, but when high power is injected into the gas cell, heat easily accumulates at the junction of the rubber plug and the HCF, resulting in a higher temperature of the rubber plug and deformation which changes the coupling position of the HCF, thereby affecting the coupling efficiency. So, all-fiber structure FGLs are needed for most applications, especially at a high-power level, and all-fiber cavities are the key components.

High-power delivery in HCFs has been researched in recent years. In 2016, the long-term stable transmission of kW-level average power through a hollow capillary and a Kagome-type photonic crystal fiber was demonstrated. It was the highest-power delivery in a HCF with a cooling system [26]. In 2020, a 300 W delivery in AR-HCF was reported in an uncooled system and a near-diffraction-limited beam was measured [27].

In this paper, we reported an AR-HCF based all-fiber gas cavity using fiber end caps for the first time. The cavity is composed of a 1-meter-long ice-cream type of AR-HCFs and two silica end caps, which are spliced with the AR-HCFs at both ends using a large diameter splicing system. The main characteristics of the cavity were measured, including transmission efficiency, high-power capacities, heat treatment, and so on. Results show that the all-fiber gas cavity has a very stable performance. The maximum injecting laser power at 1080 nm is about 260 W without active cooling, and the transmitted laser power at the output end is 203 W, corresponding to a total transmission efficiency of 78.1%. Compared to the previous work, although we don't transmit power that is as high as before, a new coupling method using HCF end cap is suggested and the stability is confirmed. It could be further improved by designing a cooling system for higher power transmission. This work is very important for the development of high-power all-fiber FGLs.

2. Fabrication

The HCF used in this paper is a kind of ice-cream, anti-resonant HCF and the structure is shown in the inset picture of Figure 1. The core is surrounded by eight ice-cream capillaries forming a negative curvature boundary, which defines the transmission band of the HCF. It has a core diameter of around 46 μm and a cladding diameter of around 280 μm. From 1000 to 1100 nm the HCF has lower transmission loss, and for the wavelength we used at 1080 nm it has a loss of around 0.12 dB/m.

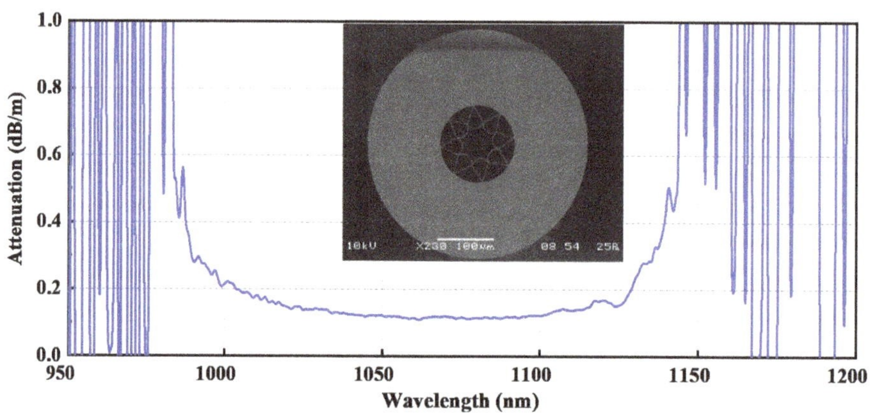

Figure 1. The transmission loss of the ice-cream type AR-HCF used for the splicing. Inset: the cross-section electronic micrograph.

The schematic of the splicing process between the HCF and the end cap is shown in Figure 2a. Figure 2b is the picture of the end cap used for splicing. During splicing, the AR-HCF is fixed by the fiber clamp at the left side, and at the right side the end cap is fixed by a self-made clamp which is composed of three glass tubes. The splicing end of the end cap is designed as a 3 mm-long cone. The diameter of the splicing end of the cone is 1 mm, which is easy for splicing at a lower temperature, and the diameter of the larger end is 2.5 mm. The right cylinder of the end cap has a diameter of 8.2 mm and a length of 17 mm, as shown in

Figure 2b. The output end is coated with an antireflection coating in the 1080 nm band. To obtain the results of the splicing, we needed to make the surfaces of the HCF and the end cap smooth. After the cutting treatment, the angle of the surface for splicing is less than 0.5°. Before splicing, the center of the HCF needed to be aligned with the center of the end cap. Then, the HCF is inserted into the end cap when the front end of the conical is heated into a molten state by the electrodes.

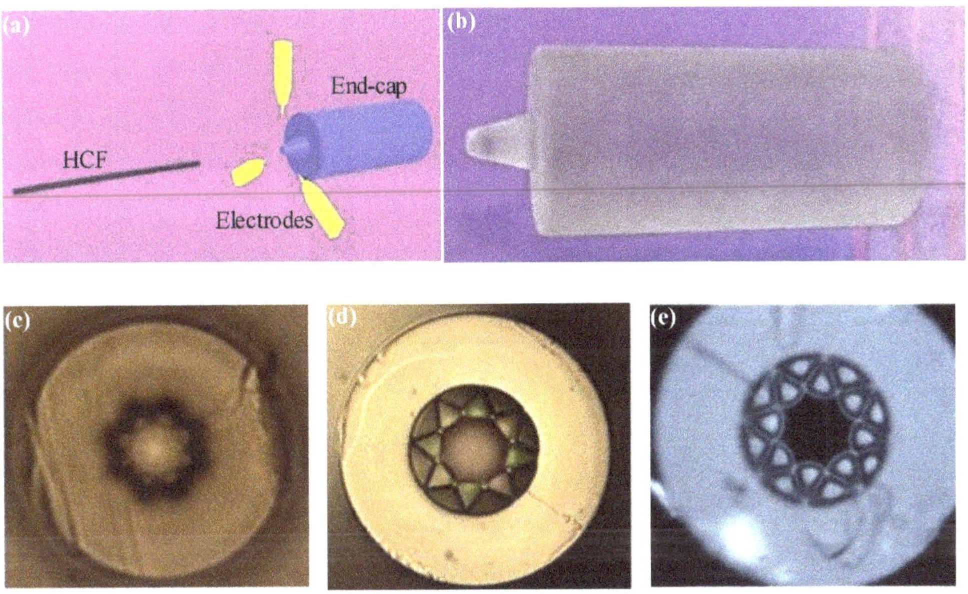

Figure 2. (**a**) Schematic of the splicing process between the HCF and the end cap; (**b**) the picture of the end cap used for splicing; (**c**) the cross section view of the HCF after splicing at a higher temperature; (**d**) the cross section view of the HCF after splicing at a suitable temperature; (**e**) the cross section view of the HCF after splicing at a lower temperature.

Due to the hollow-core structure of the HCFs, through the splicing process, the microstructure of the HCFs will be damaged which will lead the collapse of the cladding. To control the damage of the cladding, the heating temperature of the splicing system should be controlled, and waste heat should be used to maintain the molten state through moving the heating electrode some distance away from the HCF. The temperature when the end cap is heated to the molten state is a key parameter in fabrication. Figure 2c–e show the cross section view of the HCF with splicing at different temperature conditions. Figure 2c shows the cross section after splicing when the temperature is too high, and the microstructure of the HCF is almost completely destroyed. Figure 2d shows that only the negative curvature is destroyed and the microstructure still exists when the temperature is suitable. Figure 2e shows that when the temperature is too low, the HCF and the end cap are only bonded to each other but not strongly enough. In order to achieve a better splicing effect, controlling the temperature makes sure that the cross section appears as in Figure 2d, and the deformation distance is as short as possible. After obtaining great splicing, we fabricated a nice all-fiber cavity with both ends of HCF splicing with end caps, as shown in Figure 3.

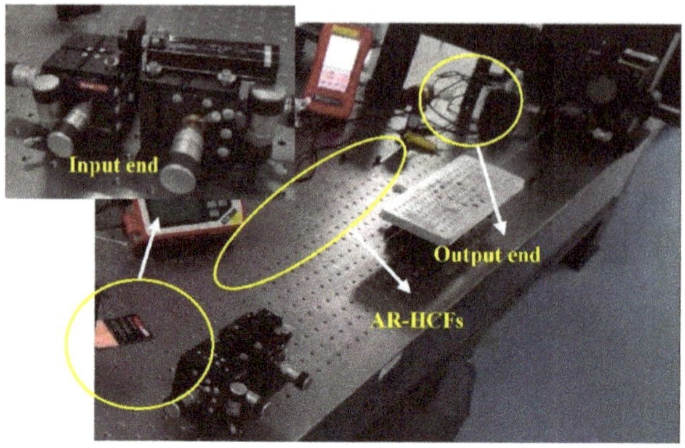

Figure 3. Picture of the fabricated all-fiber gas cavity based on AR-HCFs and end caps in the test system.

3. Experiments

The test system of the all-fiber gas cavity is shown in Figure 4. The light source for the test is a kW-level fiber laser oscillator with a 20/400 fiber end cap. The fiber laser source has a good beam quality, as shown in Figure 4a. At the left side is the beam profile at different positions, which are both fundamental modes, and M^2 is 1.13 for a and 1.21 for b when the output power is 1500 W. Two plano-convex lenses with focal lengths both of 25 mm, and two reflectors with high reflectivity at near-infrared bands are used to couple the incident light into the HCF splicing with the end cap. The core diameter of the HCF we used is 46 µm, according to the principle of mode field-matching, these two lenses are suitable for high coupling efficiency. The HCF end cap is fixed by a fixture designed by us, as shown in Figure 3. Because the anti-resonant HCF is a type of leaky fiber, an area where the cladding of the HCF and the fixture of the end cap make contact has leakage of light, which leads to heat accumulation as per the inset picture in Figure 4. This temperature is measured by a thermal imager from Fluke. The matching paste is applied evenly near the heating area, causing the temperature at this area to be significantly reduced. In order to avoid the influence of fiber bending on the measurement, the whole HCF is straightened during the experiments.

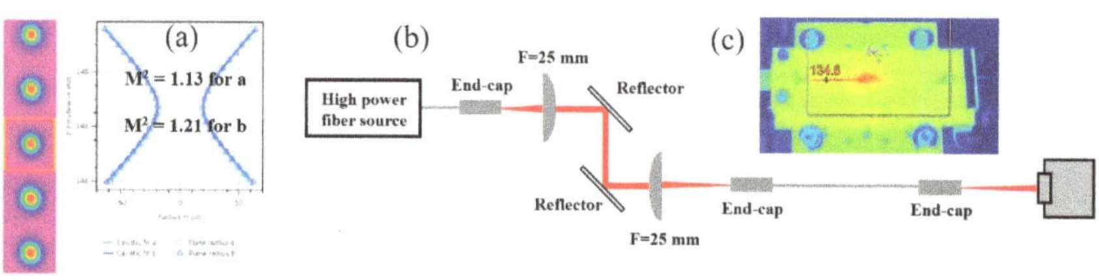

Figure 4. (**a**) Measured beam profile and quality factor M^2 of the fiber laser source; (**b**) the experimental setup of the system used to test the all-fiber gas cavity based on AR-HCFs; (**c**) thermal image of the HCF end cap at the input end.

Based on the testing system above, the total efficiency from the fiber source to the output end is tested and the result is shown in Figure 5a. As the power increases, the heat accumulation of the lenses affects the efficiency, so a manual adjustment is needed to make sure the coupling efficiency is at the maximum as the power increases. After the adjustment, the efficiency stays at around 80% when the input power increases under 200 W. The total loss comes from the lenses, the reflectors, the coupling efficiency, the Fresnel reflection and the HCF itself. For the lenses, the transmission is 97.5% and the reflectivity of the reflectors is 99.5%. The transmission loss of the 1-meter-long HCF is 0.12 dB. Additionally, the Fresnel reflection is around 3.9%. According to the loss above, the coupling efficiency is estimated as 88.8%, which dominates the main loss. The output power and the monitored temperature as the input power increased is also shown. The total incident power is 260 W and the output power is 203 W. Figure 5b shows the temperature of the cladding and the coating— it is relatively moderate when the input power is below 200 W. As the input power increases above 200 W, the temperature has a sharp rise which is due to the efficiency starting to decease corresponding to the drop of the efficiency in Figure 5a. In Figure 5b, there are also two drop points in the temperature of the coating, which also correspond to the rising point in Figure 5a, which are the results of manual adjustment of the coupling efficiency. The intensity distribution in Figure 5c,d is consistent with the Gaussian distribution, which means the main energy is still the fundamental mode, and the beam profile is shown in the inset picture of Figure 5c. The beam profile varies between Figure 5c,d as the input power increases to 200 W power level. In the beam profiles, the beam from the fiber source is shown. Comparing the profiles, the edge of the beam from the end cap is not as good as that of the pump source, thus the collapse of air holes of the negative curvature of the HCF when splicing with the end cap has a slight influence on the beam quality of the output light, but the main energy is still concentrated on the center of the beam.

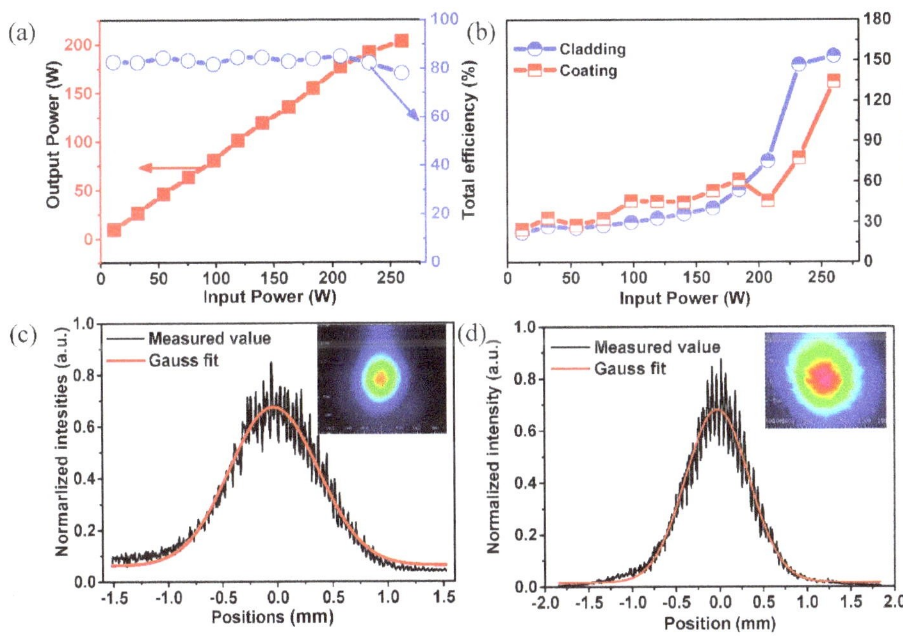

Figure 5. (**a**) The total efficiency and the output power vary with the input power; (**b**) Measured results of the temperature of the all-fiber cavity; (**c**) Intensity distribution of the output beam at lower power; (**d**) Intensity distribution of the output beam at higher power. Inset: Beam profile of the output.

The measurements of power stability of the fiber laser source and the output power, and the temperature of the cladding and coating are carried out when the output power is 100 W, and the results are shown in Figure 6. The output power shows cyclical fluctuations near 100 W with a fluctuation the same as the power fluctuation of the fiber laser source, as Figure 6a shows. This is caused by the intermittent operation of the cooling system of the fiber laser source, during its intermittent working the power will decrease and then increase slowly as before. There are two major hot points, one is mentioned above as Figure 4c shows, the other is the front end of the coating. As Figure 6b shows, at first the temperatures of the two positions slightly fluctuate, and then the temperature of the cladding stabilizes at around 38 °C and the temperature of the coating stabilizes at around 44 °C as the time gets longer. The ice-cream AR-HCF is a type of leaky fiber, so the change in temperature is mainly due to the leakage of the input light. If the coupling efficiency does not change, the temperature will not have an obvious change. The higher temperature is caused by the leakage of light and it may be the surface temperature of the HCF, and although the temperature is higher than that of an all-fiber structure solid-core fiber laser, it is still safe to work stably.

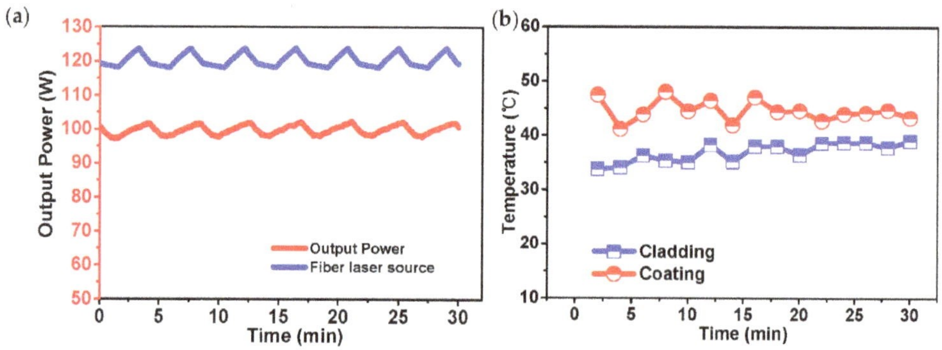

Figure 6. Test results of (**a**) power stability and (**b**) temperature stability of the all-fiber gas cavity at about 100 W.

4. Conclusions

We have demonstrated the first AR-HCFs-based all-fiber gas cavity based on fiber end caps. The cavity is fabricated by splicing a 1-meter-long ice-cream type of AR-HCFs with two silica end caps at both ends using a large diameter splicing system. With a high-power fiber source at 1080 nm, the characteristics of transmission efficiency, high-power capacity and stability, heat treatment, etc., were tested without active cooling. Experimental results show good stability of the all-fiber cavity at a high-power level. The maximum injecting laser power is about 260 W with a total transmission efficiency of 78.1%. Under 200 W incident power, the system can maintain a normal temperature. If a proper cooling system is designed and adopted in the future, the power capacity can be greatly enhanced. This work opens a new way for the development of high-power all-fiber FGLs.

Author Contributions: Conceptualization, Z.W. and Y.C.; methodology, W.H. and Z.Z.; validation, Z.W., M.W. and Z.C.; formal analysis, Z.W.; investigation, J.S., X.Y. and Y.C.; resources, M.W.; data curation, X.Y. and J.S.; writing—original draft preparation, J.S. and X.Y.; writing—review and editing, Z.W., Y.C. and H.L. All authors have read and agreed to the published version of the manuscript.

Funding: This research was funded by the Outstanding Youth Science Fund Project of Hunan Province Natural Science Foundation (2019JJ20023), National Natural Science Foundation of China (NSFC) (11974427, 12004431) and Postgraduate Scientific Research Innovation Project of Hunan Province (CX20200017).

Institutional Review Board Statement: Not applicable.

Informed Consent Statement: Not applicable.

Data Availability Statement: The data presented in this study are available on request from the corresponding author. The data are not publicly available due to privacy restrictions.

Acknowledgments: We thanks Jonathan C. Knight from University of Bath in UK for providing the AR-HCFs for our experiments.

Conflicts of Interest: The authors declare no conflict of interest.

References

1. Pryamikov, A.D.; Biriukov, A.S.; Kosolapov, A.F.; Plotnichenko, V.G.; Semjonov, S.L.; Dianov, E.M. Demonstration of a waveguide regime for a silica hollow-core microstructured optical fiber with a negative curvature of the core boundary in the spectral region >3.5 μm. *Opt. Express* **2011**, *19*, 1441–1448. [CrossRef]
2. Yu, F.; Wadsworth, W.J.; Knight, J.C. Low loss silica hollow core fibers for 3–4 μm spectral region. *Opt. Express* **2012**, *20*, 11153. [CrossRef] [PubMed]
3. Kolyadin, A.N.; Kosolapov, A.F.; Pryamikov, A.D.; Biriukov, A.S.; Plotnichenko, V.G.; Dianov, E.M. Light transmission in negative curvature hollow core fiber in extremely high material loss region. *Opt. Express* **2013**, *21*, 9514–9519. [CrossRef] [PubMed]
4. Ding, W.; Wang, Y.Y.; Gao, S.F.; Wang, M.L.; Wang, P. Recent Progress in Low-Loss Hollow-Core Anti-Resonant Fibers and Their Applications. *IEEE J. Sel. Top. Quantum Electron.* **2020**, *264*, 4400312. [CrossRef]
5. Jones, A.M.; Nampoothiri, A.V.V.; Ratanavis, A.; Fiedler, T.; Rudolph, W. Mid-infrared gas filled photonic crystal fiber laser based on population inversion. *Opt. Express* **2011**, *19*, 2309–2316. [CrossRef]
6. Nampoothiri, A.V.V.; Jones, A.M.; Fourcade-Dutin, C.; Mao, C.; Dadashzadeh, N.; Baumgart, B.; Wang, Y.Y.; Alharbi, M.; Bradley, T.; Campbell, N. Hollow-core Optical Fiber Gas Lasers HOFGLAS: A review. *Opt. Mater. Express* **2012**, *2*, 948. [CrossRef]
7. Benabid, F. Stimulated Raman Scattering in Hydrogen-Filled Hollow-Core Photonic Crystal Fiber. *Science* **2002**, *298*, 399–402. [CrossRef]
8. Gladyshev, A.V.; Kosolapov, A.F.; Khudyakov, M.M.; Yatsenko, Y.P.; Kolyadin, A.N.; Krylov, A.A.; Pryamikov, A.D.; Biriukov, A.S.; Likhachev, M.E.; Bufetov, I.A. 4.4 μm Raman laser based on hollow-core silica fibre. *Quantum Electron.* **2017**, *47*, 491–494. [CrossRef]
9. Gladyshev, A.V.; Kosolapov, A.F.; Khudyakov, M.M.; Yatsenko, Y.; Kolyadin, A.N.; Krylov, A.A.; Pryamikov, A.; Biriukov, A.S.; Likhachev, M.E.; Bufetov, I.A. 2.9, 3.3 and 3.5 μm Raman Lasers Based on Revolver Hollow-Core Silica Fiber Filled by H2/D2 Gas Mixture. *IEEE J. Sel. Top. Quantum Electron.* **2018**, *24*, 0903008. [CrossRef]
10. Li, Z.; Huang, W.; Cui, Y.; Wang, Z. Efficient mid-infrared cascade Raman source in methane-filled hollow-core fibers operating at 2.8 μm. *Opt. Lett.* **2018**, *43*, 4671–4674. [CrossRef]
11. Ling, C.; Gao, S.F.; Peng, Z.G.; Wang, X.C.; Pu, W. High peak power 2.8 μm Raman laser in a methane-filled negative-curvature fiber. *Opt. Express* **2018**, *26*, 5609.
12. Wei, H.; Cui, Y.; Zhixian, L.; Zhiyue, Z.; Zefeng, W. 1.56 μm and 2.86 μm Raman lasers based on gas-filled anti-resonance hollow-core fiber. *Chin. Opt. Lett.* **2019**, *7*, 71406.
13. Astapovich, M.S.; Gladyshev, A.V.; Khudyakov, M.M.; Kosolapov, A.F.; Likhachev, M.E.; Bufetov, I.A. Watt-level Nanosecond 4.42-μm Raman Laser Based on Silica Fiber. *IEEE Photonic Technol. Lett.* **2019**, *1*, 78–81. [CrossRef]
14. Wang, Z.; Belardi, W.; Yu, F.; Wadsworth, W.J.; Knight, J.C. Efficient diode-pumped mid-infrared emission from acetylene-filled hollow-core fiber. *Opt. Express* **2014**, *22*, 21872. [CrossRef]
15. Hassan, M.R.A.; Fei, Y.; Wadsworth, W.J.; Knight, J.C. Cavity-based mid-IR fiber gas laser pumped by a diode laser. *Optica* **2016**, *3*, 218. [CrossRef]
16. Xu, M.; Fei, Y.; Jonathan, K. Mid-infrared 1 W hollow-core fiber gas laser source. *Opt. Lett.* **2017**, *42*, 4055–4058. [CrossRef]
17. Zhou, Z.; Ni, T.; Zhixian, L.; Wei, H.; Zefeng, W.; Wuming, W.; Weihong, H. High-power tunable mid-infrared fiber gas laser source by acetylene-filled hollow-core fibers. *Opt. Express* **2018**, *26*, 19144. [CrossRef]
18. Cui, Y.; Huang, W.; Wang, Z.; Wang, M.; Zhou, Z.; Li, Z.; Gao, S.; Wang, Y.; Pu, W. 4.3 μm fiber laser in CO_2 fibers. *Optica* **2019**, *6*, 951–954. [CrossRef]
19. Aghbolagh, F.B.A.; Nampoothiri, V.; Debord, B.; Gerome, F.; Vincetti, L.; Benabid, F.; Rudolph, W. Mid IR hollow core fiber gas laser emitting at 4.6 um. *Opt. Lett.* **2019**, *44*, 383–386. [CrossRef] [PubMed]
20. Fedorov, V.V.; Mirov, S.B.; Gallian, A.; Badikov, D.V.; Voronov, A.A. 3.77–5.05-μm tunable solid-state lasers based on Fe^{2+}-doped ZnSe crystals operating at low and room temperatures. *IEEE J. Quantum Electron.* **2006**, *42*, 907–917. [CrossRef]
21. Chang, T.; Wood, O. An optically pumped CO_2 laser. *IEEE J. Quantum Electron.* **1972**, *8*, 598. [CrossRef]
22. Vurgaftman, I.; Meyer, J.R. Analysis of limitations to wallplug efficiency and output power for quantum cascade lasers. *J. Appl. Phys.* **2006**, *99*, 123101–123108. [CrossRef]
23. Maes, F.; Fortin, V.; Bernier, M.; Vallée, R. 5.6 W monolithic fiber laser at 3.55 μm. *Opt. Lett.* **2017**, *42*, 2054. [CrossRef] [PubMed]
24. Maes, F.; Fortin, V.; Poulain, S.; Poulain, M.; Vallée, R. Room-temperature fiber laser at 3.92 μm. *Optica* **2018**, *5*, 761. [CrossRef]
25. Fortin, V.; Jobin, F.; Larose, M.; Bernier, M.; Vallée, R. 10-W-level monolithic dysprosium-doped fiber laser at 3.24 μm. *Opt. Lett.* **2019**, *44*, 491–494. [CrossRef]
26. Drich, S.H.; Rothhardt, J.; Demmler, S.; Tschernajew, M.; Hoffmann, A.; Krebs, M.; Liem, A.; Vries, O.D.; Pl Tner, M.; Fabian, S. Scalability of components for kW-level average power few-cycle lasers. *Appl. Opt.* **2016**, *55*, 1636–1640.
27. Yu, F.; Wu, D.; Wang, Y.; Li, Q.; Chen, S. Delivery of CW laser power up to 300 watts at 1 μm by uncooled low-loss antiresonant hollow-core fiber. *Opt. Express* **2020**, *29*, 1492–1501.

photonics

Review

Review on All-Fiber Online Raman Sensor with Hollow Core Microstructured Optical Fiber

Haonan Ding [1], Dora Juan Juan Hu [2], Xingtao Yu [1], Xiaoxian Liu [1], Yifan Zhu [1] and Guanghui Wang [1,*]

[1] College of Engineering and Applied Sciences, Nanjing University, Nanjing 210009, China; mf21340003@smail.nju.edu.cn (H.D.); mf1834023@smail.nju.edu.cn (X.Y.); dg21340008@smail.nju.edu.cn (X.L.); mf20340022@smail.nju.edu.cn (Y.Z.)

[2] Institute for Infocomm Research, Agency for Science, Technology and Research, Singapore 138632, Singapore; jjhu@i2r.a-star.edu.sg

* Correspondence: wangguanghui@nju.edu.cn

Abstract: Raman spectroscopy is widely used for qualitative and quantitative analysis of trace components in scientific fields such as food safety monitoring, drug testing, environmental monitoring, etc. In addition to its demonstrated advantages of fast response, non-destructive, and non-polluting characteristics, fast online Raman detection is drawing growing attention for development. To achieve this desirable capability, hollow core optical fibers are employed as a common transmission channel for light and fluid in the Raman sensor. By enhancing the interaction process between light and matter, the detection sensitivity is improved. At the same time, the Raman spectroscopy signal light collection efficiency is significantly improved. This article summarizes enhancement techniques reported for Raman sensors, followed by a detailed review on fiber-based Raman sensor techniques including theoretical analyses, fabrication, and application based on hollow core photonic crystal fibers and capillary-based hollow core fibers. The prospects of using these fibers for Raman spectroscopy are discussed.

Keywords: Raman spectroscopy; capillary-based hollow core fibers; hollow core photonic crystal fiber; detection

Citation: Ding, H.; Hu, D.J.J.; Yu, X.; Liu, X.; Zhu, Y.; Wang, G. Review on All-Fiber Online Raman Sensor with Hollow Core Microstructured Optical Fiber. *Photonics* **2022**, *9*, 134. https://doi.org/10.3390/photonics9030134

Received: 25 January 2022
Accepted: 23 February 2022
Published: 25 February 2022

1. Introduction

Rapid detection and identification of chemical substances are of utmost importance in many fields, such as food safety, drug monitoring, industrial wastewater discharging, etc. [1,2]. There are many detection methods which can achieve very accurate qualitative and quantitative detection for their respective application scenarios [3,4]. However, due to the large size of the instrument, the complicated detection process, and the long detection time, these detection methods cannot meet the requirements of fast online detection. Raman spectroscopy is considered as the 'fingerprint spectrum' of the matter, which can be used as non-destructive testing and is very suitable for rapid liquid sample testing [5]. However, the Raman signal itself is very weak [6]. This shortcoming limits the application of Raman detection. Much effort has been made to enhance the Raman signal, such as surface-enhanced Raman spectroscopy, tip-enhanced Raman spectroscopy, etc., but these methods are not suitable for fast detection. Moreover, multiple tests cannot be performed, which leads to cross-contamination of the sample, and any impurities can seriously affect the Raman scattering signal. The detection range is concentrated on the focal position of the excitation light, and the interaction volume between the light and the sample is small.

The hollow core fiber (HCF) can confine the excitation light and the Raman signal in the cavity to propagate at the same time. Through internal multiple reflections, the optical path is lengthened, thereby increasing the volume of interaction between the light and the sample and increasing the intensity of the Raman signal. At the same time, within the fiber-enhanced Raman spectroscopy (FERS) structure, the diameter of the HCF used can reach hundreds of microns, which can perform liquid exchange faster, so as to achieve rapid switching of test samples, which is more suitable for rapid test scenarios.

This paper presents a review of all-fiber online Raman sensors with hollow core microstructured optical fibers. Firstly, the principle of Raman spectroscopy and the methods for Raman signal enhancement are introduced in Section 2. Secondly, based on the overview of Raman sensors with field enhancements, FERS is introduced with different types of commonly used hollow core microstructured optical fibers in Section 3. The hollow core can be used to guide light and sample flow simultaneously. Based on its enhancement of interaction length, in Section 4, we theoretically analyze its enhancement mechanism. Moreover, several in-fiber optical path modification methods are proposed to further enhance the interaction length. Then, in Sections 5 and 6, we present a comprehensive review of their applications in detail, focusing on hollow core photonic crystal fiber and capillary-based hollow core fiber, respectively. In total, four kinds of optical fibers are discussed and their applications are summarized. In Section 7, from the perspective of testing samples, we also summarize the application cases of fiber-enhanced Raman sensors. Finally, the prospects of using these hollow core microstructured optical fibers in Raman spectroscopy are summarized and concluded.

2. Principle of Raman Spectroscopy and Overview of Raman Sensor with Enhancement

2.1. Principle of Raman Spectroscopy

In 1928, Indian scientist C. V. Raman discovered a new scattering effect, known as the Raman effect, when studying the liquid scattering of monochromatic light by a mercury lamp [7]. The effect explains the wavelength shift of scattered light from a small fraction of the molecules, which has a frequency different from that of the incident light, depending on the chemical structure of the corresponding molecule. Raman spectroscopy uses scattered light to obtain information about molecular vibration, which can provide specific information about structure, symmetry, molecular bonding, electronic environment, etc. [8]. Therefore, Raman spectroscopy can be applied to qualitative and quantitative analysis of various compounds.

When light with a frequency interacts with matter, inelastic scattering will occur. The wavelength of the scattered light of a small part of the molecule is shifted, and the photon obtains vibration energy from the molecule. The energy of the scattered light is increased and is called anti-Stokes Raman scattering, and on the contrary, the energy of the scattered light is reduced to Stokes Raman scattering [9]. The energy level diagram is shown in Figure 1. Since most of the molecules are in the ground state at room temperature, it can be seen from the Boltzmann distribution that the Stokes scattering is much stronger than the anti-Stokes scattering. Therefore, the Stokes scattering spectrum is usually used when studying Raman scattering spectra. Raman spectroscopy uses scattered light to obtain information about molecular vibration. As the vibrations of material molecules are different, the Raman spectra are also different, so Raman spectroscopy is also called the 'fingerprint spectrum' of the object [5]. Using this characteristic of Raman spectroscopy, the composition and concentration of the sample can be distinguished, and the qualitative and quantitative analysis of the sample can be realized.

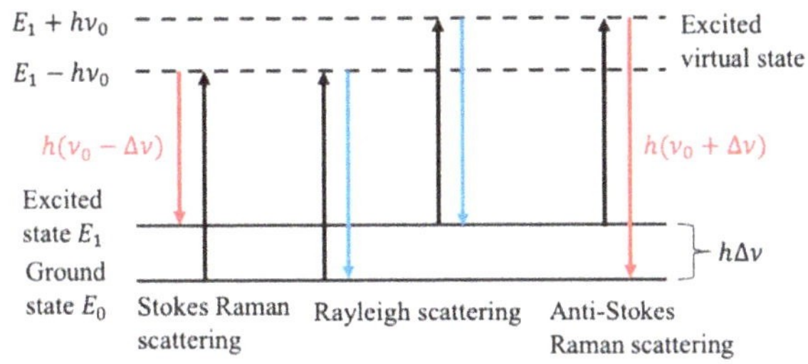

Figure 1. Rayleigh scattering and Raman scattering energy level diagram. Reproduced with permission [10].

2.2. Overview of Raman Sensor with Enhancement

As the Raman scattering signal is very weak, in recent years, many researchers have devoted themselves to researching methods to enhance the Raman signal with great progress. They proposed enhancement methods include resonance Raman spectroscopy (RRS), coherent anti-Stokes Raman spectroscopy (CARS) [11], cavity-enhanced Raman spectroscopy (CERS) [12], surface-enhanced Raman spectroscopy (SERS) [13], fiber-enhanced Raman spectroscopy (FERS), etc.

Resonance Raman spectroscopy (RRS): In the 1950s, Shorygin et al. proposed that when the excitation photon energy roughly matches the energy required for the electron transition in the sample, the measured Raman spectrum intensity will increase significantly. This method is called RRS, which can detect the Raman spectra of samples with concentrations as low as 10^{-8} mol/L. RRS has been widely used in the analysis of various luminescent biological samples, such as enzymes. The excitation light can be adjusted to the specific absorption band of the active part of the enzyme. Therefore, RRS can selectively determine its structure [14]. In addition, using different wavelengths of excitation light, the vibrational spectra of various parts of biomolecules can be obtained [15].

Coherent anti-Stokes Raman spectroscopy (CARS): In 1965, Maker et al. [16] first reported the 'three-wave mixing experiment'. In 1974, Begley et al. [11] called it CARS. This technology realizes vibration imaging with high sensitivity, high spectral resolution, and a three-dimensional slice function. CARS is a non-linear detection technology based on inducing coherent Raman on the target molecule. Two lasers are detected by the third laser, and a coherent signal with a blue shift of frequency is generated in the direction of phase matching. CARS is an effective method to measure the temperature and main component concentration in the reaction fluid [17], and it can detect different molecular bonds in various biological systems. CARS based on nanosecond, picosecond, and femtosecond lasers also has a wide range of applications in gas-phase reaction streams [18].

Cavity-enhanced Raman spectroscopy (CERS): Since the Raman signal and laser power are linear, the problem of low sensitivity of spontaneous Raman scattering can be solved by using a higher power laser. One method is to place the Raman cavity and the gain medium together in the laser resonator, so as to utilize the efficient laser power in the laser resonator (in-cavity laser Raman spectrometer) [12]. Another method is to use the laser power accumulated in the external optical cavity. Its simplest form is composed of two highly reflective mirrors with two parallel sides. If the laser frequency is stabilized at the resonant frequency of the cavity, then light can be efficiently coupled into the cavity, and the laser power can be enhanced by several orders of magnitude. This kind of power accumulation in an optical cavity can be used to improve the efficiency of spontaneous Raman scattering, which is called CERS [19]. In 2001, Taylor et al. [20] realized the specific application of CERS for the first time. Through the electronic feedback mechanism, the 1 W, 532 nm laser was stabilized in a short linear cavity, and the Raman cavity was placed in the optical cavity. The Raman signal of hydrogen is detected from the 90° direction, but the Raman cavity window causes a lot of optical loss, resulting in the enhancement factor being limited to about 50 times.

Surface-enhanced Raman spectroscopy (SERS): In 1974, Fleischmann et al. [13] discovered the phenomenon of SERS. SERS can be used to detect and identify very low-concentration molecular species. This method is based on adsorbing the analyte to the surface of a metal structure, usually gold, silver, or copper, and mainly uses the local electromagnetic field generated by the assembly of nanostructured materials and the strong field enhancement generated by the electromagnetic coupling between the nanoparticles, which overcomes the obstacle of small scattering cross-section [21]. Compared with normal Raman scattering, the Raman scattering cross-section of the analyte in the SERS detection method is increased by up to 15 orders of magnitude, and the Raman signal of the target molecule is also increased by several orders of magnitude, so that SERS can even detect the Raman signal of a single molecule. In recent years, in order to improve the stability and reproducibility of SERS substrates, and to further increase the enhancement rate of SERS, metal nanoresonators with more controllable shapes and reproducible shapes have been introduced on solid substrates, such as a metal film on a nanosphere substrate and metal-coated alumina nanoparticles.

Tip-enhanced Raman spectroscopy (TERS): Like other pure optical techniques, the spatial resolution of standard SERS measurements is limited by the Abbe diffraction limit $\lambda/2$, where λ is the wavelength of the incident radiation. However, this diffraction limit may be overcome. In 2000, Zenobi et al. [22] showed that it is possible to couple Raman spectroscopy with scanning tunneling microscopes or atomic force microscopes to form a new spectral microscope tool, the so-called tip-enhanced Raman spectroscopy. In TERS detection, the sharp AFM (or STM) tip made of SERS active metal is moved toward the sample to be tested and irradiated with a laser [23], and the Raman scattering generated by the sample under the tip is enhanced. Neacsu et al. [24] achieved the tip-induced Raman scattering enhancement in the experiment, reaching 5×109. In this way, TERS can detect Raman spectra with high spatial resolution. The TERS detection method is relatively complicated. The current research of TERS aims to improve the sensitivity of detection, such as detecting the Raman signal of a single molecule.

Fiber-enhanced Raman spectroscopy (FERS): Compared with normal Raman scattering, the Raman signal enhancement multiples obtained by SERS, TERS, CERS, and other methods are as high as ten orders of magnitude, but in most cases, the volume of light–object interaction is very small, and the volume is proportional to the intensity of the Raman signal. In addition, the molecules adsorbed on the metal surface are difficult to clean, and any tiny impurities or pollution on the metal surface will significantly affect the Raman signal [25,26], so these methods are not suitable for online detection that needs to be reused. Therefore, this article mainly introduces the method of FERS. The Raman scattering spectrum of the sample is detected by injecting the sample into the HCF. This method is simple to use and is suitable for online rapid Raman detection. FERS is discussed in detail in subsequent sections.

3. Fiber-Enhanced Raman Spectroscopy (FERS)

FERS is based on confining light and substances in the hollow core channel of the HCF. On one hand, the sample and the excitation light can be fully used, and on the other hand, the fiber can enhance the Raman light collection effect, enhancing the Raman signal [27]. Traditional waveguide transmission is based on total internal reflection, requiring the core layer to have a higher refractive index than the cladding layer. However, for HCF, the core layer is air, and the refractive index is lower than that of the cladding layer, which cannot meet the condition of total internal reflection. Therefore, other methods are needed to realize the hollow core waveguide, such as leveraging the photonic band gap effect or anti-resonance to achieve hollow core guidance, or forming a liquid core with a higher refractive index than cladding to achieve total internal reflection, coating a highly reflective metal layer on the inner wall or hollow core photonic crystal fiber, etc., so that light can be transmitted in the hollow core. According to specific applications, the current studies on FERS are mainly divided into hollow core photonic crystal fibers (HCPCFs) and capillary-based HCFs.

3.1. Hollow Core Photonic Crystal Fiber (HCPCF)

HCPCF is a special type of microstructured optical fiber. The current applications in Raman detection mainly include two types of HCPCF, hollow core photonic band gap fiber (HCPBGF) and hollow core anti-resonant photonic crystal fiber (HCARF). HCPCF has the characteristics of hollow structure, small dispersion, small non-linear coefficient, and low transmission loss. It has important application value and broad application prospects in the interaction of light and gas, high-energy laser transmission, pulse compression, optical fiber sensing, and next-generation optical communications [28].

For HCPBGF, several capillaries are removed in the center to form the hollow core, and the surrounding capillaries are arranged in a honeycomb structure to form a cladding, as shown in Figure 2a. For the HCARF, the cladding can be composed of several non-contact thin-walled capillaries, and the beam is confined in the core by enhancing the reflection of the incident light when it meets the thin wall of the cladding, as shown in Figure 2b.

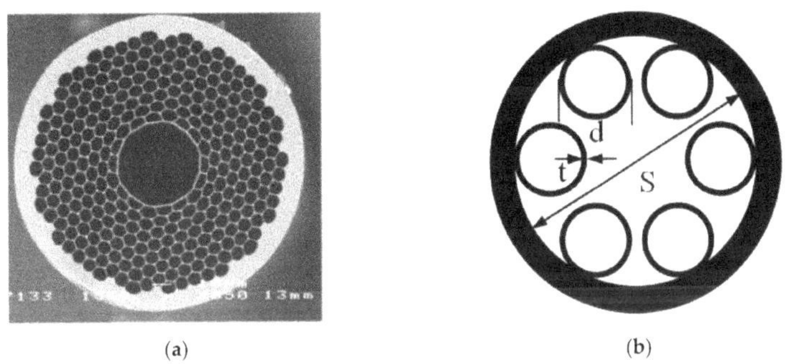

Figure 2. (**a**) Cross-section of HCPBGF. Reprinted with permission from [29]; © The Optical Society. (**b**) Cross-section of HCARF. Reprinted from [30], copyright (2019), with permission from Elsevier.

3.2. Capillary-Based Hollow Core Fiber (HCF)

Compared with HCPCF, capillary-based HCF has a large hollow core aperture, which can increase the volume of light and confine the light while realizing rapid switching of the sample, without requiring a large pressure to inject the sample. Capillary-based HCFs have two main categories, namely metal-lined hollow core fiber (MLHCF) and liquid core fiber (LOF).

For MLHCF, the inner wall of the glass capillary tube is plated with metal to increase the reflectivity to ensure the transmission of light in the core. Due to the high reflectivity of the metal, the Raman signal can be bound in the capillary core to improve the signal collection efficiency. In addition, the inner diameter of the inner aluminized capillary tube is large, liquid can be injected directly under normal pressure, and cross-contamination is small when changing the sample, which is suitable for fast and convenient online detection. The hollow core aperture of the MLHCF is generally several hundred microns. The sample is injected into the fiber, the excitation light emitted by the laser is coupled into the hollow core from one end of the inner MLHCF and is bound in the capillary under the action of the metal layer [31], and the light interacts with the sample to generate Raman light. A part of the Raman scattered light continues to propagate forward, and a part is reflected back to the incident end [32,33]. Therefore, the receiving end of the Raman signal can be divided into transmission Raman detection and reflection Raman detection.

For LOF, its discovery originated in the mid-19th century, when people first conducted scientific research on light guidance in liquids. It was not until 1970 that scientists began to use LOFs as potential communication media [34,35]. The liquid core fiber is composed of a small-diameter glass tube filled with high refractive index liquid, and has low loss in the near-infrared band [34,36], which has obvious practical significance. Compared with traditional experimental equipment, LOF has great advantages in analog communication [37]. In spontaneous Raman spectroscopy, by limiting the excitation radiation and effectively collecting Raman scattered light with a long interaction length, the liquid core fiber can produce an enhancement factor that is 1000–3000 times higher than that of the traditional fiber [38]. There are also huge advantages in the application of absorption spectroscopy [39] and fluorescence spectroscopy [40].

In addition to the enhancement based on HCPCF and HCF, further enhancement can be achieved by introducing optical path configurations, such as a Sagnac loop, capillary with an inserted reflector, and capillary with an inserted FP cavity. The theoretical part of the enhancement by the optical path configuration will be explained in Section 4.2, and the experiment will be shown in Section 6.1.2.

4. Theoretical Analysis

4.1. Theoretical Analysis of FERS

The signal enhancement is mainly caused by the increased interaction length between confined light and samples in the HCF, which provides high-efficiency Raman signal collection. The Raman scattering signal can be described by the equation [41]:

$$P \propto m\sigma V I_0 (\nu_0 - \nu_r)^4. \tag{1}$$

In Equation (1), m is the number of molecules per unit volume, σ is the Raman scattering cross-section of each molecule, V is the sample volume that is irradiated by the pump light, and the scattered light can be received. I_0, ν_0, and ν_r are the intensity of the excitation light, Raman light frequency, and excitation light frequency, respectively. The enhancement of the Raman signal can be obtained by increasing the intensity of excitation light including selecting a high-power laser and optical focusing. However, the former easily causes optical loss and thermal effect. Additionally, the latter reduces the volume of the light object while increasing the intensity of light, which is only applicable to the case of limited sample volume [42].

When the laser power is constant P_p, if the power distribution of the excitation light on the sample is uniform, the light intensity can be expressed as P_p/S, where S is the cross-sectional area of the sample, L is the optical path length, and the volume can be expressed as SL. Hence, the increase in optical path length could enhance the Raman light.

Another key consideration in increasing the Raman signal is the collection efficiency of the signal. As shown in Figure 3a, in the case of beam focusing, only the Raman light at the focal point in the lens numerical aperture (NA) can be accepted, and the collection efficiency is $\eta = S_{foucus} \cdot h \cdot \Omega_{obj}/4\pi$. For the case of introducing the HCF, as shown in Figure 3b, the HCF can collect the Raman signal in the entire hollow core, which is a large-volume signal. This enhancement mechanism is called volume-enhanced Raman scattering. At this time, the collection efficiency is $\eta = 1/2 \cdot SL \cdot \min\left(\Omega_{obj}/\Omega_{HCF}, 1\right)$. Ω_{obj} and Ω_{HCF} are the solid angles corresponding to the NA of the lens and the HCF, respectively. For the focusing structure, only the samples in the excitation light focus volume $S_{focus} \cdot h$ can be excited by the excitation laser in the optical path to generate Raman signal (S_{focus} is the focal area and h is the depth of focus). In addition, the Raman scattered light diverges uniformly in all directions, and only the Raman light within the solid angle $\Omega_{obj}/4\pi$ corresponding to the NA can be received by the lens. Compared with the lens, HCF is more advantageous because of the large NA and the capacity for confining the Raman light within the entire fiber. At the output end, if the NA of the lens is smaller than that of the HCF, then the Raman optical coupling efficiency is $\Omega_{obj}/\Omega_{HCF}$.

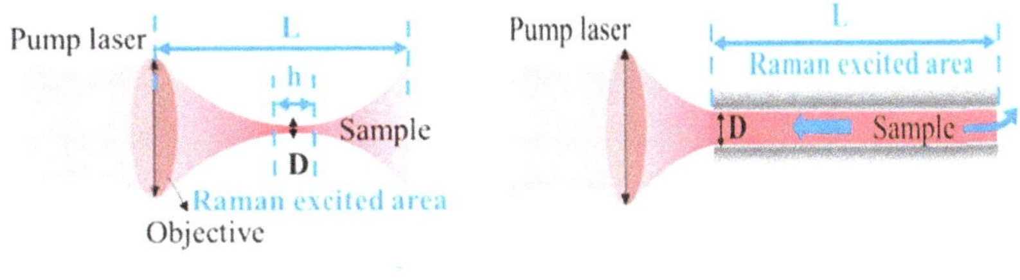

(a) (b)

Figure 3. (**a**) The excitation light is focused on the object. (**b**) The excitation light is coupled into the HCF and has a long acting distance. Reprinted with permission from [43] © The Optical Society.

4.2. Theoretical Calculation of Raman Signal Enhancement

Through theoretical calculations, the Raman signal strength can be estimated. Here, we mainly discuss the calculation of Raman signal intensities in direct detection and the structure with MLHCF.

In the case of direct detection by the Raman probe, only the Raman light within the focal volume of the probe lens can be received by the probe. The focus volumes of different lenses are different, but the focus volume is very small compared to the hollow core volume of the MLHCF. The Raman signal intensity P_{probe} in the direct detection structure can be expressed by Equation (2), the focal volume can be equivalent to a cylinder, and h is the focal depth of the lens. S_{focus} is the cross-sectional area of the focus volume and $\eta_{probe} = \Omega_{obj}/4\pi$ is the receiving efficiency of the lens. Since the Raman light diverges in all directions, only the Raman light within the solid angle corresponding to the NA of the lens can be received by the probe.

$$P_{Probe} = \int_0^h \sigma m S_{focus} \eta_{probe} I_0 \exp(-\kappa_1 l) \exp(-\kappa_2 l) dl \quad (2)$$

In the case of using the structure with MLHCF, gold and silver have high reflectance for both 785 nm excitation light and its Raman light, so the MLHCF with gold or silver inside can effectively confine the light. As shown in Figure 4b, the Raman probe is coupled with the MLHCF through the optical fiber. The Raman light in the HCF with the same transmission direction as the excitation light is called transmitted Raman light, and the Raman signal propagating in the opposite direction is called reflected Raman light. Our research group selected the Raman spectrometer from Bidatek, and the structure of the Raman probe is shown in Figure 4a. The excitation light is irradiated onto the sample through the band-pass filter, dichroic mirror, and lens. If the excited Raman light is within the NA angle of the lens, it will be received by the lens, and will enter the receiving fiber through the dichroic mirror, mirror, and long-pass filter for Raman signal analysis. Band-pass filters and dichroic mirrors are used to prevent excitation light from entering the excitation fiber and damaging the laser. The long-pass filter is used to prevent the stronger excitation light from entering the receiving fiber.

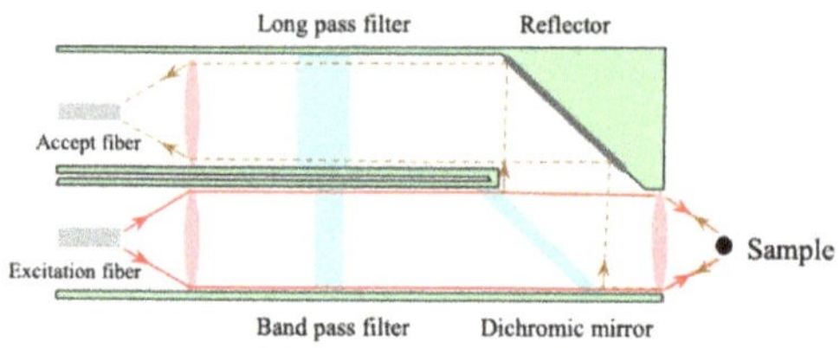

(a)

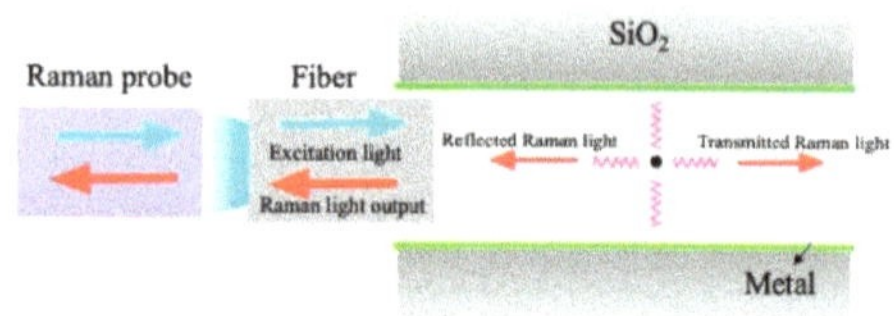

(b)

Figure 4. (**a**) Internal structure diagram of Bidatek Raman probe. (**b**) Detection principle diagram of MLHCF. Reproduced with permission [10].

When the excitation light is transmitted from the optical fiber to a position with a distance l from the optical fiber, the excitation light intensity I will be attenuated to $I * \exp(-\kappa_1 l)$. The reflected Raman light excited here will be attenuated to $I * \exp(-\kappa_1 l) * \exp(-\kappa_2 l)$. Considering that only the Raman light irradiated within the core diameter of the optical fiber and within its maximum receiving angle range can be received by the optical fiber, the receiving efficiency is $\eta = (d/D)^2 * (2\pi(1-\cos\alpha)/4\pi) * (\Omega_{obj}/\Omega_{fiber})$, where d is the fiber core diameter, D is the inner diameter of the HCF, α is the maximum acceptance angle of the fiber, Ω_{fiber} and Ω_{obj} are the solid angles corresponding to the NA of the lens and the fiber, respectively. The intensity of the reflected Raman signal and transmitted Raman signal received by the Raman probe in the structure of Figure 4b can be expressed by Equations (3) and (4). Compared with the direct detection, the Raman signal can be significantly enhanced by using the structure with MLHCF.

$$P_R = \int_0^L \sigma m\pi \left(\frac{D}{2}\right)^2 \eta I_0 \exp(-\kappa_1 l) \exp(-\kappa_2 l) dl \quad (3)$$

$$P_T = \int_0^L \sigma m\pi \left(\frac{D}{2}\right)^2 \eta I_0 \exp(-\kappa_1 l) \exp[-\kappa_2 (2L - l)] dl \quad (4)$$

Besides using MLHCF, there other structures which are used to enhance the interaction length of light objects and the efficiency of collecting Raman light, including a Sagnac loop, capillary with an inserted reflector, and capillary with an inserted FP cavity.

4.2.1. Sagnac Loop

As shown in Figure 5, the Sagnac loop allows the excitation light to enter and exit from both ends of the HCF simultaneously. In this way, both the transmitted Raman light and the reflected Raman light can be collected. The Raman light can be divided into four parts: the transmission and reflection Raman signals generated by the excitation of the two ends, respectively. The reflected Raman signal and the transmitted Raman signal output by the excitation light I_1 generated by the optical fiber 1 are:

$$P_R(L,) = \sigma m\pi \left(\frac{D}{2}\right)^2 \eta I_1 \int_0^L \exp(-\kappa_1 l) \exp(-\kappa_2 l) dl, \quad (5)$$

$$P_T(L, I_1) = \sigma m\pi \left(\frac{D}{2}\right)^2 \eta I_1 \int_0^L \exp(-\kappa_1 l) \exp[-\kappa_2 (L - l)] dl. \quad (6)$$

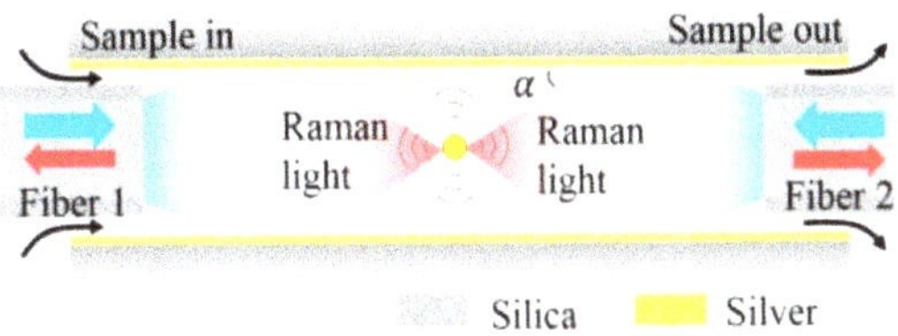

Figure 5. Diagram of light propagation in the Sagnac loop. from [32].

In the same way, $P_R(L, I_2)$ and $P_T(L, I_2)$ can be obtained, so the total Raman optical signal is:

$$P_{total}(L, I_1, I_1) = P_R(L, I_1) + P_T(L, I_1) + P_R(L, I_2) + P_T(L, I_2). \tag{7}$$

Comparing the intensity of Raman detection before and after adding Fiber 2 to the end, the Raman enhancement ratio of the structure of the Sagnac loop is:

$$\gamma = \frac{P_{total}}{P_R(L, I_1)}. \tag{8}$$

Jin et al. reported that the Raman enhancement ratio of the structure is about four times [44], which confirms that the all-fiber detection system with a Sagnac loop is a very effective way to enhance the Raman signal.

4.2.2. Capillary with an Inserted Reflector

In the capillary with an inserted reflector structure, a metal reflective film is placed on the other port opposite the laser in the HCF which can reflect the excitation light and Raman light back to the probe, thereby increasing the distance of the light object and the Raman light receiving efficiency. In this structure, the Raman signal is also divided into four parts as Figure 6 shows, the reflected Raman light intensity P_1 and the transmitted Raman light intensity P_2 generated by the excitation light transmitted to the HCF. After passing through the cavity again, the reflected Raman light intensity P_3 and the transmitted Raman light intensity P_4 are generated. The formulas are as follows:

$$P_1 = \sigma m \pi \left(\frac{D}{2}\right)^2 \eta I_0 \int_0^L \exp(-\kappa_1 l) \exp(-\kappa_2 l) dl, \tag{9}$$

$$P_2 = \sigma m \pi \left(\frac{D}{2}\right)^2 \eta I_0 R \left(\frac{d}{D}\right)^2 \int_0^L \exp[-\kappa_1 (2L - l)] \exp(-\kappa_2 l) dl, \tag{10}$$

$$P_3 = \sigma m \pi \left(\frac{D}{2}\right)^2 \eta I_0 R \left(\frac{d}{D}\right)^2 \int_0^L \exp(-\kappa_1 l) \exp[-\kappa_2 (2L - l)] dl, \tag{11}$$

$$P_4 = \sigma m \pi \left(\frac{D}{2}\right)^2 \eta I_0 R \left(\frac{d}{D}\right)^4 \int_0^L \exp[-\kappa_1 (2L - l)] \exp[-\kappa_2 (2L - l)] dl. \tag{12}$$

Therefore, the total light intensity is:

$$P_{total} = P_1 + P_2 + P_3 + P_4. \tag{13}$$

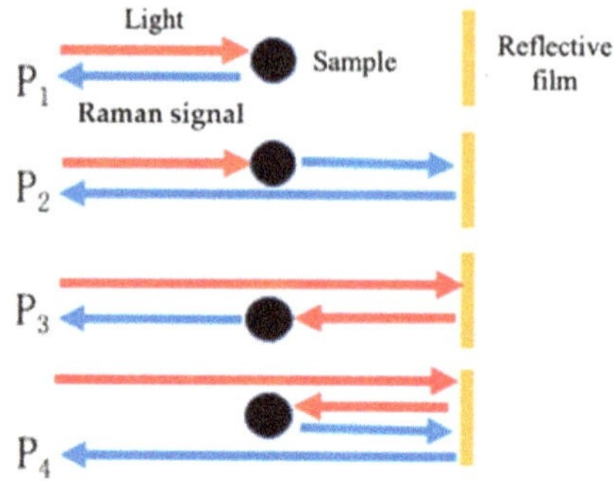

Figure 6. Raman signal transmission diagram in the capillary with an inserted reflector.

The enhancement by the inserted reflector is P_{total}/P_1 in comparison with a structure without the inserted reflector. Chu et al. reported the enhanced Raman signal to be around 1.73 times using the capillary with an inserted gold surface reflector [45].

4.2.3. Capillary with an Inserted FP Cavity

In the capillary with an inserted FP cavity structure, a long-pass filter and a metal reflector are introduced on both sides of the HCF cavity to confine the excitation light to reflect back and forth in the cavity until the light intensity attenuates to zero, as shown in Figure 7. It can be seen that the expression of light intensity is an infinite series, and we calculate the sum of it and obtain the light intensity at position x in the cavity as:

$$I_e(x,l) = \frac{I_1(x) + I_2(x,l)}{1 - R_1R_2e^{-2\alpha l}} \tag{14}$$

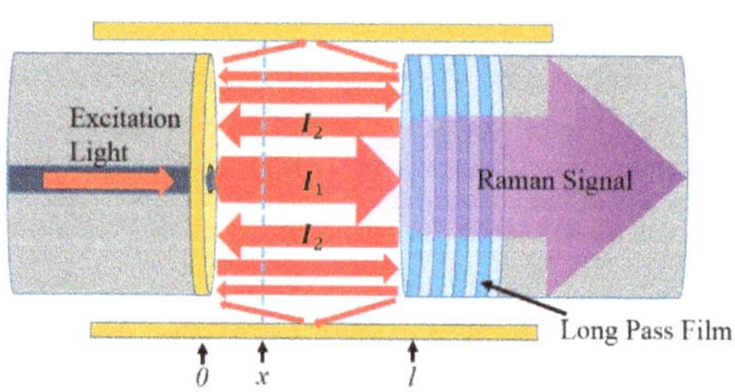

Figure 7. Schematic diagram of excitation light reflecting back and forth in the cavity, while the total light intensity is their scalar summation. Reprinted with permission from [46] © The Optical Society.

From the calculation formula of the Raman signal, the Raman signal can be obtained as:

$$P = \sigma m\pi\left(\frac{D}{2}\right)^2 \eta \int_0^l \left[e^{-\alpha(l-x)} + e^{-\alpha(l+x)}\right] I_e(x,l)dx. \tag{15}$$

Yu et al. reported that in the comparison with the structure without an inserted FP cavity, the enhancement ratio can reach about five times [46]. Here, we list the enhancement ratios of the three structures above in Table 1, and the specific experiment details will be described in Section 6.1.2.

Table 1. Comparison of the enhancement ratios of three different optical path configurations.

	Sagnac Loop	Capillary with an Inserted Reflector	Capillary with an Inserted FP Cavity
Enhancement Ratio	4 [44]	1.73 [45]	5 [46]

5. Photonic Crystal Fiber for Raman Sensing

5.1. Hollow Core Photonic Band Gap Fibers (HCPBGFs)

For HCPBGF, several capillaries are removed in the center, and the surrounding capillaries are arranged in a honeycomb structure to form a cladding [47]. The cladding is composed of SiO_2 and periodic arrangement of air holes. The transmission characteristics depend on the shape, size, and arrangement of air holes in the cladding. Different structures result in different photonic band gaps. When the light wave is incident on the core-cladding interface, the light wave is strongly scattered by periodic air holes in the cladding, and the light wave with specific wavelength satisfying the Bragg condition will return to the core by the photonic band gap effect [48]. Therefore, HCPBGF has strong wavelength selectivity for light transmission [49]. Only the light within the band gap can be transmitted, and the light beyond the band gap will leak into the cladding and redistribute in the cladding. Therefore, when designing the Raman experiment, it is necessary to design the structure of HCPBGF according to the Raman frequency shift range of the detected sample, so that the excitation light and Raman light are in the band gap of the fiber, so as to realize the desirable transmission of the signal. Only in this way can we strengthen the control of the excitation light and Raman light, and increase the interaction between light and matter.

The transmission loss of HCPBGF is very low for sensing applications, e.g., 1 dB/km [50], and the diameter of a hollow core hole is usually 5–30 μm. NA of HCPBGF is 0.2 [9], allowing transmission of a high-intensity laser in the hollow core for a long distance and collection of a strong Raman signal, with the enhancement factor being up to three orders of magnitude. Therefore, HCPBGF is very suitable for applications requiring high sensitivity and a low detection limit.

The band gap range of an HCPBGF is related to factors such as air filling ratio, effective refractive index, and air hole spacing. Based on the expression for the longitudinal wave vector of the light wave, $k_a = \frac{2\pi a}{\lambda}$ [51], where a is the aperture of the air hole, it can be seen that the shorter the wavelength, the smaller the aperture of the air hole that is needed. In other words, the photonic band gap period becomes smaller. At present, the wavelength in the photonic band gap of most HCPCFs is limited to the infrared communication band (0.85–1.55 μm). Due to the small wavelength of the visible light band, the microstructure size of the optical fiber needs to be greatly reduced compared with the transmission of infrared wave band light waves [52], which brings a certain degree of difficulty to the preparation of the optical fiber.

When the HCPBGF is filled non-selectively with liquid in all the air holes, the band gap of the photonic crystal fiber will move to a shorter wavelength due to the change in the refractive index of the cladding and the core, and the spectral width of the band gap will be narrow due to the decrease in the dielectric contrast, which limits the selection of the excitation wavelength and the available range of the Raman spectrum. In order to solve the problem of band gap narrowing caused by non-selective filling, it is necessary to selectively fill the photonic crystal fiber, that is, only the sample enters the hollow core hole, and the air holes in the cladding are not filled. One method is to use polymer to block the air holes of the cladding, but polymer will produce a strong fluorescence background, so this method is not suitable for FERS. Another method is to heat the cladding honeycomb structure to make it collapse, while the hollow core with bigger pore size remains hollow, so that only the sample is passed into the core, as shown in Figure 8. In this case, the HCPBGF can realize light transmission by total internal reflection in the whole visible and near-infrared spectrum range [53]. This novel fiber-enhanced method is used for the detection of the commonly used broad-spectrum antibiotic moxifloxacin, in which only 4 nL of sample is required, and the detection limit is as low as 1.7 μM, making it possible to detect the concentration of antibiotics in serum or other body fluids without damage and with high sensitivity and rapid detection.

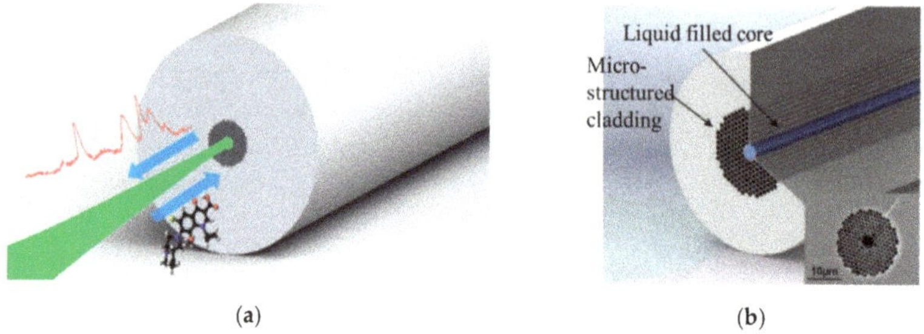

Figure 8. Raman detection principle diagram of hollow core photonic crystal fiber. (**a**) Raman light is generated by the interaction between the substance and the laser in the HCPCF. The green represents the laser transmitted in the central hole, and the red represents the Raman spectrum. (**b**) A cross-sectional view of a hollow core photonic crystal fiber selectively filled with a central hole. Reprinted with permission from [53]. Copyright (2017) American Chemical Society.

There are two kinds of loss in HCPBGF: limiting loss and scattering loss [54]. The limiting loss refers to the optical loss caused by the leakage of some conducted light from the cladding during transmission. The limiting loss can be reduced by increasing the number of air holes in the cladding [55]. Scattering loss refers to the loss due to surface roughness induced by surface capillary wave inevitably due to thermodynamic reasons during the pulling of HCPBGF. The core wall can be designed to reduce the roughness of the inner wall to reduce the scattering loss [56]. The lowest reported loss of HCPBGF is about 1 dB/km [50]. Hanf et al. measured the Raman spectrum of H_2 in the HCPBGF, with the equipment as shown in Figure 9a, and studied the influence of different hollow core apertures (Figure 9b) on noise and SNR, and the result is shown in Figure 9c. The result shows that in a certain range, the larger the pinhole size is, the greater the noise is and the smaller the SNR is [57].

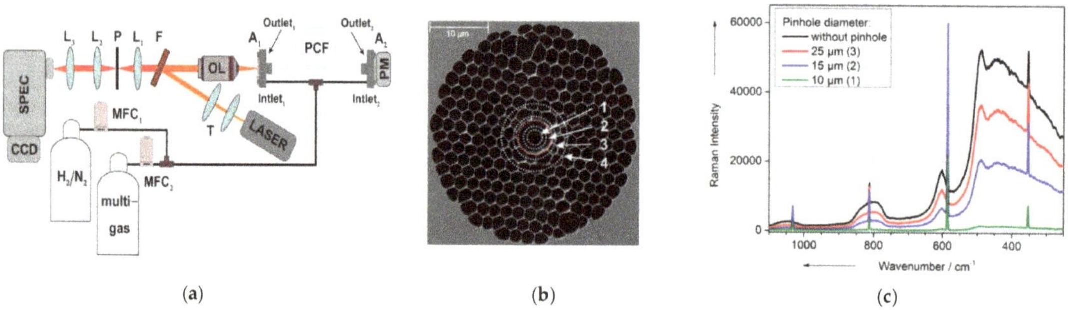

Figure 9. (**a**) Schematic sketch of the experimental setup for HCPBGF. (**b**) Electron microscopic picture of the fiber end face. The circular areas marked by '1', '2', '3', and '4' are the collection areas with pinholes sizes of 10, 15, 25, and 50 μm in the focal plane. (**c**) Raman spectrum of H_2 and the background silica Raman signal of different fibers. Reprinted with permission from [57]. Copyright (2015) American Chemical Society.

Khetani et al. [58] realized the detection of heparin concentration in bovine serum by using HCPBGF, and the Raman signal intensity was more than 90 times stronger than that of direct detection, and strong mode field overlap was achieved by non-selective filling.

However, the hollow hole of the HCPBGF is very small, which limits the speed of the sample in and out, and the sample cannot be switched quickly. Therefore, this kind of fiber is not suitable for high-throughput, online detection applications.

5.2. Hollow Core Anti-Resonant Fiber (HCARF)

The concept of anti-resonance comes from the anti-resonance planar waveguide proposed by Duguay et al. in 1986 [59]. The optical guiding principle of HCARF can be explained by the principle of an anti-resonant reflecting optical waveguide. The light that can be constrained in the fiber is mainly determined by the thickness of the quartz wall in the cladding. When the light is coupled in the core and transmitted to the interface between the core and cladding, the light near the resonant frequency will leak, and other frequencies of light can achieve low loss transmission in the core. That is, the light sandwiched between the specific wavelengths meeting the resonance conditions can be transmitted in the core region, so the region of anti-resonance wavelengths is very wide, which means that the wavelength range transmitted in the core is wide. As the HCARF has a large core and hole spacing, in addition to the advantage of a wide light guide band, it also has low transmission loss, a high damage threshold, single-mode light guide, and other advantages [60].

Due to the photon transmission in the air, the non-linear effect is greatly reduced in HCARF. In addition, HCARF has the characteristics of low transmission loss in the mid-infrared band and a flexible and controllable transmission spectrum [61,62], and is especially suitable for mid-infrared fiber gas lasers. At present, there have been experiments using HCARF for trace detection of gases. In 2018, a high peak power mid-infrared Raman laser output was reported based on a nodeless HCARF with a high-performance near-infrared and mid-infrared broad spectrum light guide [63,64], as shown in Figure 10. Knebl et al. successfully used HCARF for Raman gas sensing and demonstrated its application in environmental science [41,65]. In addition, the research of measuring Raman scattering spectra by HCARF for trace detection is also being developed further [66].

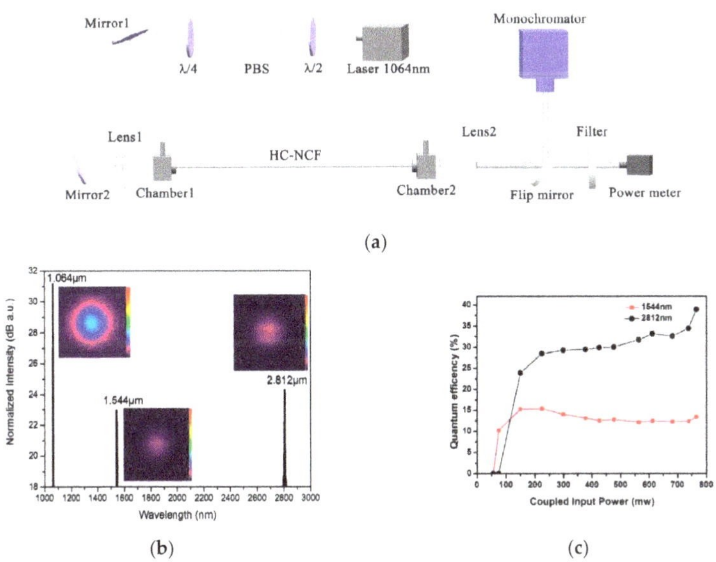

Figure 10. (**a**) Mid-infrared gas Raman laser device. (**b**) When the pressure of methane gas is 1.5 MPa and the pumping power is 381 mW, the Raman spectra obtained are inset for the near field mode at the wavelength of 1064, 1544, and 2812 nm. (**c**) Quantum conversion efficiencies versus the coupled input power at first and second Stokes. Reprinted with permission from [64] © The Optical Society.

6. Capillary-Based Hollow Core Fiber for Raman Sensing

6.1. Metal-Lined Hollow Core Fiber (MLHCF)

For application scenarios that require high time resolution and fast sample switching, HCF with a large core diameter is more suitable. For example, MLHCF has a hollow core aperture of several hundred microns, so there is no need to pressurize when injecting the sample, and its cross-section is shown in Figure 11a. The choice of the plated metal is based on the required transmission conditions. For infrared light waveband transmission, it is preferable to choose a gold-plated film. Silver-plated film and aluminized film are more suitable for visible light waveband transmission and ultraviolet light waveband transmission, respectively [41]. The transmission of light is based on the high reflectance of the metal film. The transmission principle is shown in Figure 11b. Due to multiple reflections of light in the waveguide, the surface roughness or defects of the metal film will cause high loss. The magnitude of the loss is related to the wavelength and the angle of incidence. The shorter the wavelength and the smaller the incident angle, the smaller the loss.

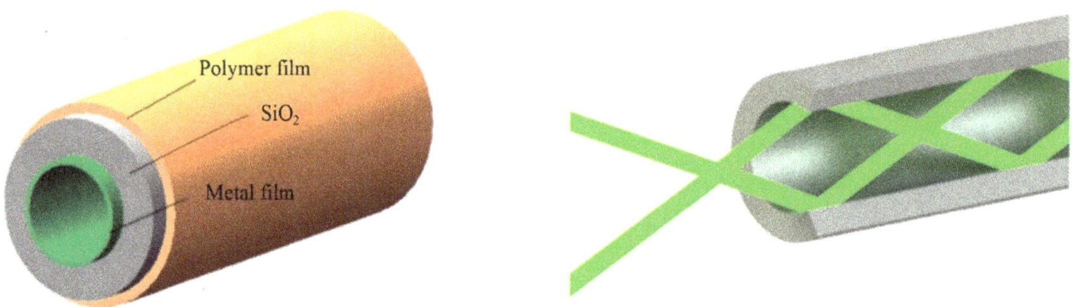

(a) (b)

Figure 11. The structure and light guiding principle diagram of MLHCF. (**a**) Structure diagram. (**b**) Transmission mechanism of light in MLHCF. Reproduced with permission [10].

Although the MLHCF has a larger transmission loss than the HCPBGF, it can enhance the signal collection efficiency and significantly increase the Raman signal strength, and is used in many detection scenarios. James et al. [67] used the internal silver-plated hollow core fiber (SLHCF) to dynamically monitor trace gases, and realized the real-time detection of various gases during the formation of HD by H_2 and D_2 under the action of a catalyst, with a detection limit of 100 ppm. The device is equipped with metal caps at both ends of the SLHCF to prevent laser light from entering the glass layer of the fiber and reduce the fluorescent background. Pearman et al. [68] used multiple optical fibers to receive the Raman light in the SLHCF, and detected the concentration of N_2, CO_2, and CH_4, and the device is shown in Figure 12a. Compared with the non-silver-plated HCF and the direct detection with a Raman probe, it can be seen from Figure 12b that this method enhanced the Raman signal of non-absorptive gases by 20 times.

In addition, the detection of Raman light in MLHCF needs to consider the effects of fluorescence and background light. The laser optical components and HCF will produce fluorescence, which affects the shot noise of the background spectrum and reduces the signal-to-noise ratio, thereby affecting the detection limit. The most straightforward method can reduce the fluorescence by reducing the number of optical devices, while preventing the spectrometer from receiving the fluorescence from the glass layer. In addition, Okita et al. [69] covered the end face of the HCF, and Mullen et al. [70] coated the end face of the HCF with a layer of silver, which not only prevented the laser from entering the HCF from the end face, but also prevented the glass fluorescence from being received by the spectrometer. In addition, the imperfect optical coupling and transmission loss of the optical fiber will cause background signals to be generated, causing great problems for trace detection. Rupp et al. [71] studied how to improve the Raman detection limit by reducing the fluorescent background of the MLHCF. The longer the wavelength of the excitation light, the smaller the fluorescence signal of the glass, and the noise can be reduced by adjusting the window structure of the optical device and the Raman cavity in the optical path.

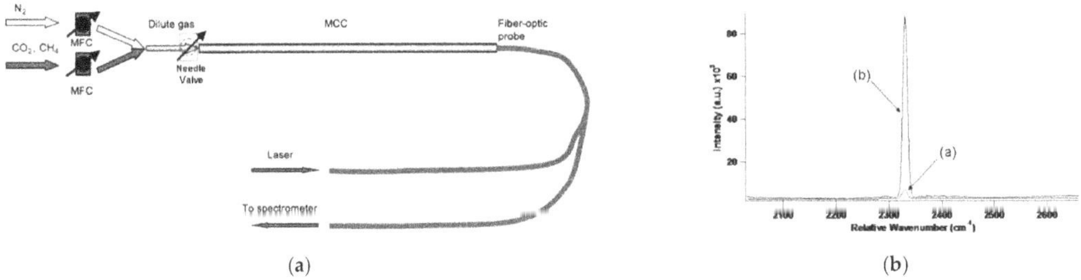

Figure 12. (**a**) Experimental setup for the collection of MCC Raman scatter. (**b**) Comparison of experimental results of H_2 in air and MCC. Reprinted with permission from [68] © The Optical Society.

6.1.1. Fabrication Method

The MLHCF used in our studies is the SLHCF produced by Doko Engineering in Japan [72]. The difficulty in manufacturing this kind of optical fiber lies in how to coat the inner wall of the hollow core glass optical fiber with uniform and smooth metallic silver and how to prevent the oxidation of the silver. The company's production principle is to use liquid deposition technology to form metal and dielectric coatings in silicon tubes. Firstly, polyimide film is coated on the outside of the thin-walled silicon tube to enhance the strength of the silicon tube. Secondly, the traditional electroless plating technique is used to deposit a layer of silver in the silicon tube, which includes a mixed silver solution and a retarding fluid. Finally, the dielectric film of silver iodide is formed by the iodine flowing in the tube, that is, the metal and dielectric coatings are formed in the silicon tube. The inner wall of the SLHCF manufactured by this method is smooth, with an average roughness of only 0.04 μm, and low transmission loss and bending loss. Using this method, waveguides with hollow core apertures of 250, 320, 530, and 700 μm and lengths of up to 6 m have been manufactured successfully. This method has simple production process, low cost, and is suitable for mass production.

At the same time, our research group has a set of equipment for MLHCF based on an evaporation method. We chose the VNANO/VZZ-300S high vacuum resistance evaporation coating equipment of Beijing Wiener Vacuum Technology Co., Ltd. to make the coating in house. This method can vapor-deposit the end surface of the part or the outer wall of the glass tube by rotary evaporation. Using this method, after evaporating metal on the outer wall of the glass tube, polyimide or ultraviolet glue is coated on the outside of the metal, and finally hydrofluoric acid is used to corrode the inner glass and the unevenly corroded parts at both ends are cut. With this method, we successfully manufactured gold-plated and aluminized capillaries, but the coating was uneven and the transmission loss was large.

To connect the capillary and the incident fiber (Fiber 1), there are two main methods: free space coupling and inserting coupling [73,74]. In the free space coupling configuration, as shown in Figure 13a, the incident laser is coupled by the convergence of the objective lens. When using this method, the focus position has to be readjusted every time, and the integration of the system is not optimal. In the inserting coupling configuration, as shown in Figure 13b, the incident laser is directly inserted into the capillary, which forms an all-fiber system with high integration. It should be noted that the difference between the inner diameter of the incident fiber and the capillary should be small to minimize the light leakage. Incident laser enters the capillary through Fiber 1 and interacts with the sample in the capillary to generate the Raman signal. The reflected Raman light is received and detected by Fiber 1 [44].

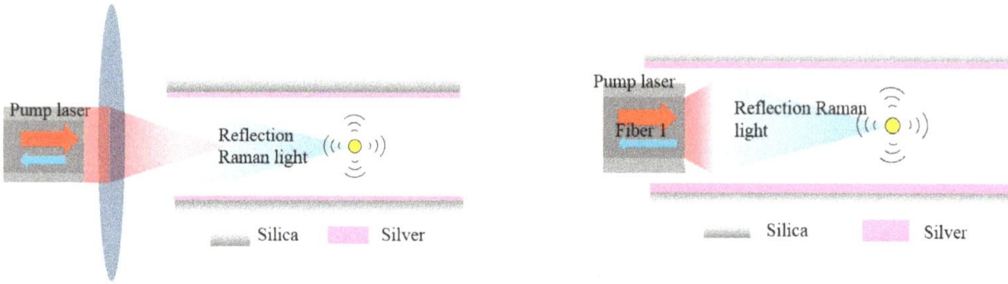

(a) (b)

Figure 13. (**a**) The schematic diagram of free space coupling. (**b**) The schematic diagram of inserting coupling [10].

6.1.2. Enhancement by Optical Path Configurations

In addition to the enhancement based on different types of optical fibers, further enhancement can be achieved by optical path configurations. As the Raman signal is very weak, MLHCF detection-based enhancement is not sufficient. Therefore, our research group proposed to several optical path configurations, including the addition of a Sagnac loop, capillary with an inserted reflector, and capillary with an inserted FP cavity to enhance the detection of Raman signals in the MLHCF.

- Sagnac Loop

In addition to the enhancement by MLHCF to increase the interaction distance of the light and object for improved Raman light collection efficiency, the Sagnac loop configuration can also collect the reflected and transmitted Raman signals at the same time to further improve the Raman light collection efficiency. The structure of the Sagnac loop is shown in Figure 14a. Two large-core optical fibers are inserted into opposite ends of the SLHCF to form a Raman cavity; the other ends of the optical fibers are tightly glued together to form a Sagnac loop, which is coupled with the Raman probe. The structure diagram of the SLHCF is shown in Figure 14b. The thickness of the silver film is 200 nm, the hollow core aperture is 320 μm, and the outer diameter is 450 μm. The large-core fiber structure is shown in Figure 14c.

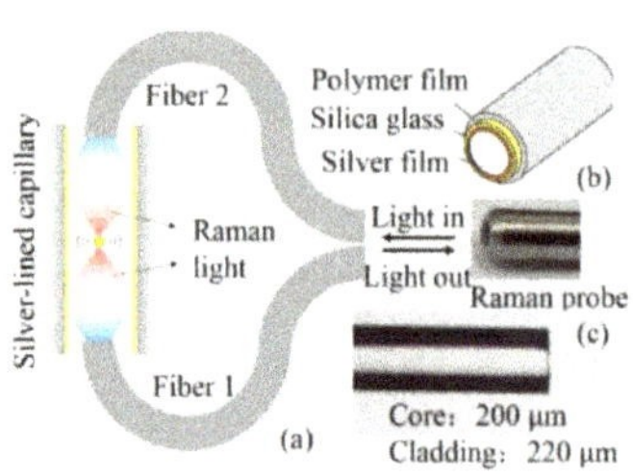

Figure 14. Sagnac ring experimental device diagram. (**a**) Sagnac loop diagram. (**b**) Capillary diagram. (**c**) Large-core diameter fiber after cutting.

To demonstrate the enhancement by the Sagnac loop structure, a comparison of detecting the Raman signal of isopropanol (ISO) from the Raman probe with and without the Sagnac loop was performed. Firstly, the reflection Raman spectrum from the Raman probe without the Sagnac loop is plotted as the blue line in Figure 15a. Only one fiber was inserted into the capillary and only the reflection Raman spectrum was measured. Next, equal power was pumped into both fibers in the Sagnac loop structure and the overall Raman spectrum consisting of both transmission and reflection Raman signals is plotted as the yellow line in Figure 15a. The yellow line shows a clear amplification of 3.72 times at 819 cm^{-1} (C-C-O band) while the total excitation optical power doubled in the Sagnac loop configuration [75]. The theoretical analysis suggested an increase of 3.99 times by the Sagnac loop, which matched well with the experimental results. In addition, the device was also demonstrated to detect samples with different concentrations, which proves that the all-fiber Raman cavity has the prospect of realizing online, portable, and fast Raman detection [69,70].

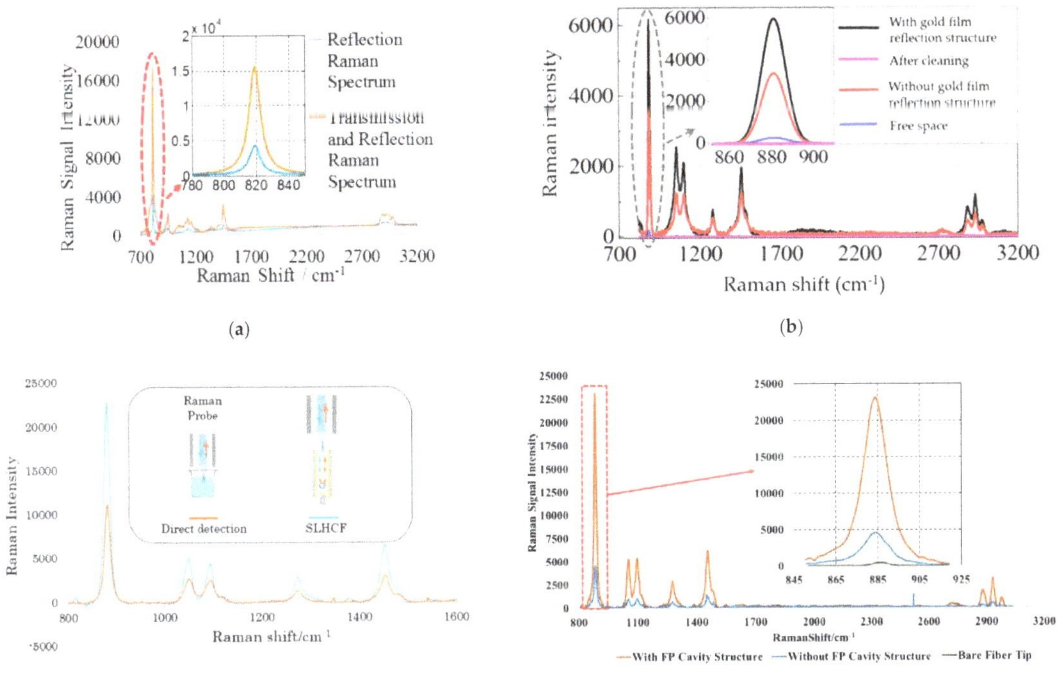

Figure 15. Raman signal enhancement experiment results of different structures. (**a**) Raman spectrum of ISO in Sagnac loop. © (2018) IEEE. Reprinted, with the permission, from [32]. (**b**) Raman spectrum of ethanol in capillary with an inserted reflector. Reproduced with permission [10]. (**c**) Comparison of Raman spectra for ethanol detected by SLHCF and direct detection [27]. (**d**) Raman spectrum of ethanol in an inserted FP cavity. Reprinted with permission from [46] © The Optical Society.

- Capillary with an inserted reflector

The second optical path configuration is the capillary with an inserted reflector as shown in Figure 16a. The structure used in the experiment is an SLHCF and a gold-plated glass fiber at the end. Two fibers were connected with the capillary by inserting coupling. The experiment used a capillary length of 3 cm and a capillary diameter of 320 μm. The diameter of the gold-plated inserted reflector on the end surface was 275 μm, and the attenuation coefficient was 23 m^{-1}. The experimental parameters and results were used in Formula (13) to calculate the theoretical enhancement ratio of the capillary with an inserted reflector compared with that of the structure without the reflector. The theoretical enhancement factor of 785 nm excitation light was 31.5 times when the cavity length was 3.1 cm. In order to verify the volume enhancement mechanism of the gold film reflection, the experimental parameters remained unchanged. In the case of 785 nm excitation light power of 36.2 mW and integration time of 2 s, we measured the Raman spectra of ethanol under four different conditions: the SLHCF with gold film reflection structure, the structure after cleaning, the SLHCF without gold film reflection structure, and the free space. The experimental results of the Raman spectrum are shown in Figure 15b.

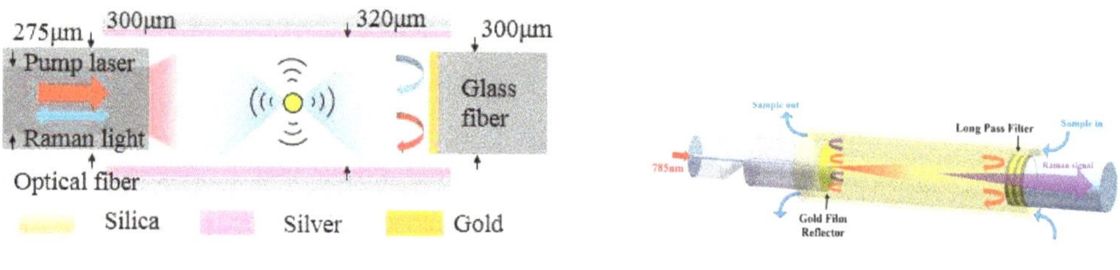

(a) (b)

Figure 16. (**a**) Capillary with an inserted reflector structure diagram [10]. (**b**) Capillary with an inserted FP cavity detection experimental device structure diagram. Reprinted with permission from [46] © The Optical Society.

First of all, the Raman peak intensity at the Raman frequency shift of 881 cm^{-1} (C-C-O band) based on the SLHCF without the gold film structure was 28 times stronger than that of direct detection in free space. Secondly, the intensity of the Raman peak at 881 cm^{-1} of the gold reflection structure was 1.76 times stronger than that of the SLHCF structure without gold film, so the total volume enhancement factor was 49.3 times. Next, the SLHCF was rinsed with ethanol and deionized water, and subsequently the Raman detection was performed. It can be seen that the sample was completely flushed out of the SLHCF without any residue. In summary, the gold reflection Raman detection device with an inserted reflector has several desirable properties. It can significantly enhance the collected Raman signal compared to the direct detection method. The device can be reused. Furthermore, both the sample entry and exit rates were also fast, which is very suitable for rapid detection application scenarios.

- Capillary with an inserted FP cavity

The capillary with an inserted FP cavity detection is established on the basis of the SLHCF and the end surface reflection structure. The structure is presented in Figure 16b. The FP cavity used a large-core diameter fiber with a gold-plated reflective surface and a 792 nm long-pass coating at both ends. At the output end of the fiber, 46 layers of film were plated on the end face of the large-core fiber with a thickness of 5300 nm, and multilayer materials with different refractive indices separate light with different wavelengths, so as to achieve the effect of separating excitation light and Raman light. The gold-plated reflective surface can reflect the Raman light and the excitation light back into the cavity at the same time, and the long-pass coating only allowed the Raman light whose wavelength is longer relative to the excitation light to pass through, so that the Raman light directly went to the detection unit. The excitation light was reflected back into the cavity, thereby increasing the interaction distance between the light and the substance, and at the same time increasing the collection efficiency of the detection signal.

In order to prove the enhancement effect of SLHCF on signal detection, the comparison of signal strength between SLHCF and direct detection was verified by experiments, and the results are shown in Figure 15c [27].

The experiment detected the Raman signal of ethanol liquid under the conditions of 20 mW excitation light and 10 s integration time. Three different conditions were measured: the SLHCF with FP cavity structure, the SLHCF without FP cavity structure, and bare fiber tip. The experimental results are shown in Figure 15d. The orange lines represent the Raman signal intensity detected in the FP structure, while the blue lines represent no FP cavity structure. The Raman characteristic peak with a bare fiber tip is represented by the green curve. Compared with the structure without FP cavity, the ethanol Raman signal of the structure with FP cavity was increased by five times. Compared with bare fiber tip, Raman light intensity of the structure with FP cavity was increased by 86 times, which had a good gain effect [46].

6.2. Liquid Core Optical Fiber (LOF)

LOF is a new type of optical transmission element that uses liquid material as the core material and polymer or quartz material as the sheath layer [76]. When the refractive index of the core liquid is higher than the refractive index of the skin tube, the sample can be confined in the waveguide to increase the interaction distance between the light and the substance [77], and increase the optical sensitivity and the efficiency of spectrum collection. LOF also has the advantages of large core diameter, large NA, wide spectrum transmission range, and high light transmission efficiency.

When the cladding material is quartz or glass (the refractive index is about 1.46), the refractive index of the liquid as the core must be greater than 1.46, which severely limits the application of LOF [78]. The new polymer LOF can be designed with a low refractive index cladding. At present, polytetrafluoroethylene (PTFE) material is commonly used as the cladding in LOF. PTFE has excellent chemical stability and corrosion resistance. The refractive index of the PTFE material (Teflon AF2400) produced by DuPont in the United States can be as low as 1.29, and the refractive index of domestic PTFE material is 1.35 [79].

Altkoen et al. designed a tubular LOF entirely made of Teflon AF2400 to measure the Raman spectra of low refractive index liquids [80], and demonstrated LOF filled with water, methanol, ethanol, and acetonitrile. The optical loss was also affected by the defects of the capillary wall and different sections of capillary. When the optical fiber was filled with water, ethanol, and alcohol, the optical loss could be lower than 3 dB/m, 2 dB/m, and 1.9 dB/m, respectively. It was proven that Teflon AF fiber has great advantages in different spectral applications and, even if the loss is high, it can significantly increase the intensity of the Raman spectrum. In 2016, Chiara et al. used 3 m long LOF to measure the linear relationship between ethanol concentration and Raman intensity, and demonstrated that LOF can be used to detect liquid concentration [81], as shown in Figure 17.

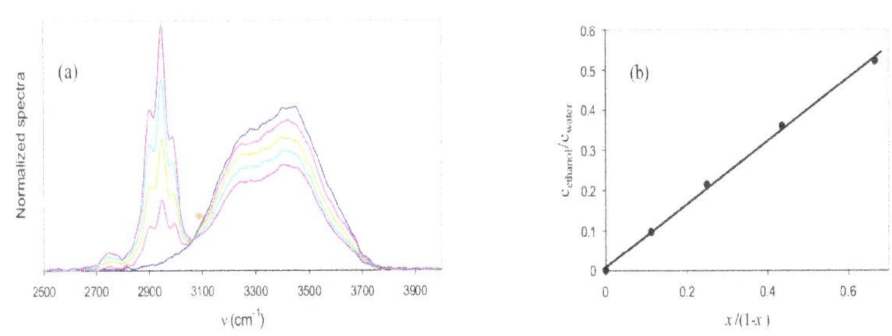

(a) (b)

Figure 17. (**a**) Using LOF to measure the Raman spectra of ethanol at 0, 0.1, 0.2, 0.3, and 0.4 concentrations. The peaks at 2905, 2949, and 2994 cm^{-1} correspond to ethanol. (**b**) Ethanol/water ratio as a function of the true volume fraction of ethanol x. Errors on the score ratios are less than 1%. Reproduced with permission [81].

Pelletier et al. used LOF made from Teflon AF to measure protein samples and proved that the LOF Raman cavity has a sensitivity enhancement factor of 5009 in aqueous solution under 532 nm excitation light [82]. At present, LOF has been successfully applied to fields including UV curing, UV degradation, UV lithography, forensics, fluorescence detection, UV medical treatment, spectral diagnosis and treatment, etc. [83,84]. Table 2 summarizes the parametric characteristics of the four types of fiber.

Table 2. Comparison of parameters of four optical fibers.

	HCPBGF	HCARF	LOF	MLHCF
Core diameter	Small (5–30 μm)	Medium (20–100 μm) [63]	Large (300–1000 μm)	Large (300–1000 μm)
Bandwidth	Narrow band [85]	Narrow band	All band [86]	All band
Loss	Low (1 dB/km)	Low (50 dB/km)	High (3 dB/m) [80]	Medium (1.2 dB/m) [72]
NA	0.12	0.03	0.54 [87]	0.22 [10]
Detection limit	4.7 ppm [57]	2 ppm [66]	6 ppm [88]	100 ppm [67]
Enhancement factor	10^4 [39]	10^4 [60]	10^3–10^4 [89]	10^3 [90]

7. HCF-based Raman Sensor Applications

7.1. Applications in Gas Detection

In recent years, the application research of gas Raman scattering in various fields has achieved rapid development, and it has broad applications in the fields of sensing, environmental monitoring, gas chemistry research, and medical treatment [91]. As HCF technology becomes more mature, many types of optical fibers become suitable for gas detection, and the performances of optical fibers are getting better. At present, the HCF-based gas Raman light source has achieved multiwavelength Raman output, and the wavelength range also covers the ultraviolet to the mid-infrared. The ultraviolet band has reached 184 nm, and the mid-infrared band has expanded to 4.4 μm. Additionally, distributed gas sensing using stimulated Raman scattering in hydrogen gas-filled HCPCFs has been reported. The distributed Raman gain measurement approach is highly sensitive with a large dynamic range, and both time and spatially resolved. Yang et al. investigated label-free optical fiber distributed Raman hydrogen sensors operating based on stimulated Raman spectroscopy, potentially allowing distributed chemical analysis in gas or liquid phase with high sensitivity and selectivity [92]. Table 3 summarizes the research status of Raman spectroscopy in gas detection in the research community.

Table 3. Examples of gas Raman spectroscopy with HCF.

Year	Research Team	Reference	Pump Wavelength	Fiber Type	Gas	Core Diameter	Enhance-ment
2008	National Energy Technology Laboratory	[93]	514.5 nm	HCPBGF	N_2, O_2	4.9 μm	Several hundreds
2009		[94]	1302–1637 nm	MLHCF	CO, C_3H_8, etc.	300 μm	N.A.
2014	Leibniz Institute of Photonic Technology	[57]	670 nm	HCPCF	H_2, CH_4	10, 20, 30 μm	N.A.
2014	University of Bath	[95]	1064 nm	HCPCF	H_2	53 μm	N.A.
2015	Fusion Science and Technology	[96]	532 nm	MLHCF	H_2	1 mm	10
2016	Ocean University of China	[97]	532 nm	HCF	O_2, N_2	500 μm	60
2019	The Hong Kong Polytechnic University	[93]	1532 nm	HCPCF	H_2	10 μm	N.A.
2021	Chongqing University	[67]	532 nm	HCARF	H_2, CO, etc.	26 μm	7

7.2. Applications in Liquid Detection

At present, trace detection of liquids is of great significance in the fields of industry, agriculture, and medical treatment, such as monitoring of industrial wastewater discharge, pesticide residues in crops, food additives, body fluid components, pharmaceutical content, etc. It plays a very critical role in maintaining public health and safety and ensuring people's quality of life. However, most of the current liquid detection instruments are large and expensive, and are not suitable for fast and convenient online detection. The volume enhancement effect of HCF can realize efficient and rapid detection of Raman spectroscopy of substances. LOF, MLHCF, and HCPCF have all made progress in liquid Raman spectroscopy detection, as summarized in Table 4.

Table 4. Examples of liquid Raman spectroscopy with HCF.

Year	Research Team	Reference	Pump Wavelength	Fiber Type	Liquid	Core Diameter	Enhancement
1997	Northwestern University	[80]	632.8 nm	LOF	Water, methanol, ethanol, acetonitrile	250 μm	N.A.
2001	Northwestern University	[82]	532 nm	LOF	Water protein	50 μm	500
2011	University of Ottawa	[58]	785 nm	HCPCF	Heparin	10.6 μm	90
2017	Leibniz Institute of Photonic Technology	[53]	532, 676, 752 nm	HCPCF	Ethanol	20 μm	10
2020	Nanjing University	[10]	785 nm	MLHCF	Ethanol	125 μm	4.83

7.3. Application in Medicine

Biological liquids provide a wealth of information about human health. The HCF-based Raman detection technology for detection of biological liquids can achieve applications in immunity and biochemistry. In 2017, Qi et al. reported a method using LOF to measure the chemical concentration in clinical serum and urine samples based on Raman spectra and absorption spectra in the near-infrared region [98]. The experimental setup is shown in Figure 18. The LOF used therein has a length of 30 cm and an inner and outer diameter of 600 μm and 800 μm, so the total volume is less than 0.1 mL. The laser power at the entrance of the LOF is about 160 mW. This method can automatically analyze the samples and predict the concentration of most samples within the accuracy range of the clinical reference analyzer, which has a huge potential in biochemical medicine. In 2019, Xiao et al. designed a device for detecting bisphenol A in blood and environmental samples using the superior physicochemical properties of the HCARF and the two-dimensional material black phosphorus. The detection limit increased by more than two orders of magnitude compared to traditional devices [99]. In addition, Azkune et al. used a 66 μm LOF to achieve the quantitative detection of low-concentration glucose, with a detection limit as low as 0.0186 mol/L, proving that this method is suitable for detecting clinically relevant glucose concentrations and, in the treatment of SGLT2 inhibitors, it has shown great potential for urine glucose monitoring [100].

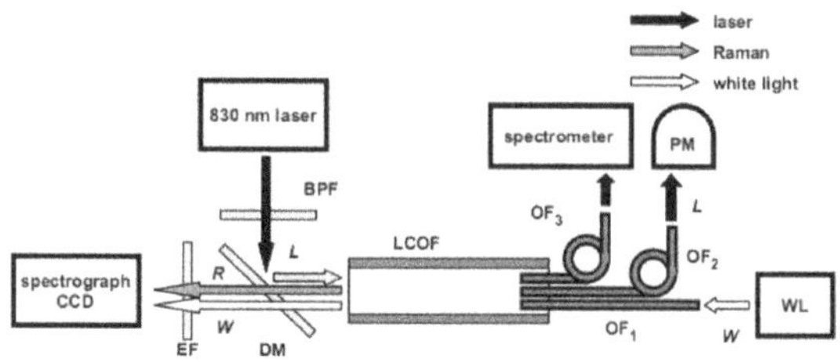

Figure 18. The experimental device for detecting Raman spectra and absorption spectra using LOF. Reprinted with permission from [98] © The Optical Society.

8. Prospect and Conclusions

This article summarizes the research of all-fiber online Raman sensors with hollow core microstructured optical fibers. In comparison with other Raman signal enhancement techniques, FERS has demonstrated attractive capabilities and great potential to achieve rapid online Raman detection. Four different types of HCFs are reviewed in detail, showing their configurations, theoretical analyses, fabrication, experimental results, and application demonstrations for fast online Raman spectroscopy with significant signal enhancement. The HCF detection method is used to make up for the weak Raman signal and the signal enhancement is up to 10,000 times higher than that of direct detection. In addition, the system can be used many times, is easy to clean and does not have residual crosstalk, and the detection speed is as fast as a few seconds, so it is suitable for online real-time material detection, such as for safety inspection, food detection, water quality detection, etc.

In the future, the all-fiber online Raman sensor with hollow core microstructured optical fiber system has a lot of room for improvement in signal enhancement and usage scenarios. In order to enhance the resolution of the Raman signal and reduce the background noise during detection, the hollow fiber detection technology can be combined with other detection enhancement methods such as SERS, related Raman methods such as stimulated Raman, and new fiber types. In terms of application scenarios, this technology can be combined with microfluidic technology and installed in the detection part after sample pretreatment; it can also be combined with magnetic beads and external magnetic force to achieve magnetic enrichment and purification. Combined with artificial intelligence and machine learning to analyze spectral data, a complete set of sample detection procedures can be realized.

Furthermore, recent innovations and continuous development of the laser technologies with new wavelengths and time resolutions open up new opportunities for Raman sensors. For example, UV Raman can effectively reduce the fluorescence interference in the Raman signal. Nanosecond and picosecond lasers improve the time resolution of Raman spectroscopy and further expand the application range of Raman spectroscopy. In addition to the qualitative and quantitative analysis of the substances mentioned in this article, Raman spectroscopy technology has certain advantages in the fields of medical biology, molecular microdynamics, archaeology, etc. Moving forward, with the further innovation and development of experimental and theoretical methods, Raman spectroscopy technology will become a powerful tool for biological and chemical technology and applications.

Author Contributions: Conceptualization: D.J.J.H., G.W. and X.Y. writing—original draft preparation, H.D. and X.Y.; writing—review and editing: H.D., D.J.J.H., G.W., X.L. and Y.Z. All authors have read and agreed to the published version of the manuscript.

Funding: This work was sponsored by National Natural Science Foundation of China (nos. 61875083, 61535005), and Social Development Project of Jiangsu Province (BE2019761).

Institutional Review Board Statement: Not applicable.

Informed Consent Statement: Not applicable.

Data Availability Statement: Not applicable.

Conflicts of Interest: The authors declare no conflict of interest.

References

1. Lee, K.S.; Landry, Z.; Pereira, F.C.; Wagner, M.; Berry, D.; Huang, W.E.; Taylor, G.T.; Kneipp, J.; Popp, J.; Zhang, M.; et al. Raman microspectroscopy for microbiology. *Nat. Rev. Methods Primers* **2021**, *1*, 80. [CrossRef]
2. Ho, C.-S.; Jean, N.; Hogan, C.A.; Blackmon, L.; Jeffrey, S.S.; Holodniy, M.; Banaei, N.; Saleh, A.A.E.; Ermon, S.; Dionne, J. Rapid identification of pathogenic bacteria using Raman spectroscopy and deep learning. *Nat. Commun.* **2019**, *10*, 4927. [CrossRef]
3. Kassa-Baghdouche, L.; Cassan, E. Mid-infrared refractive index sensing using optimized slotted photonic crystal waveguides. *Photon. Nanostruct. Fundam. Appl.* **2018**, *28*, 32–36. [CrossRef]
4. Kassa-Baghdouche, L.; Cassan, E. Sensitivity analysis of ring-shaped slotted photonic crystal waveguides for mid-infrared refractive index sensing. *Opt. Quantum Electron.* **2019**, *51*, 328. [CrossRef]
5. Wu, L.; Ouyang, Z.H.; Cao, S.C. Research development and application of Raman scattering technology. *Chin. J. Light Scatt.* **2005**, *17*, 180–186.
6. Schwab, S.D.; Mccreery, R.L. Remote, long-pathlength cell for high-sensitivity Raman spectroscopy. *Appl. Spectrosc.* **1987**, *41*, 126–130. [CrossRef]
7. Raman, C.V.; Krishnan, K.S. The optical analogue of the compton effect. *Nature* **1928**, *121*, 711. [CrossRef]
8. Kudelski, A. Analytical applications of Raman spectroscopy. *Talanta* **2008**, *76*, 1–8. [CrossRef]
9. Rostron, P.; Gerber, D. Raman spectroscopy, a review. *Int. J. Eng. Tech. Res.* **2016**, *6*, 50–64.
10. Chu, Q. Enhanced Raman Detection Based on the Metal-lined Hollow-Core Fiber. Master's Thesis, Nanjing University, Nanjing, China, 2019.
11. Harvey, B.; Coherent, A. Anti-stokes Raman spectroscopy. *Anal. Chem.* **1978**, *50*, 905A–912A. [CrossRef]
12. Hercher, M.; Mueller, W.; Klainer, S.; Adamowicz, R.F.; Schwartz, S.E. An efficient intracavity laser Raman spectrometer. *Appl. Spectrosc.* **1978**, *32*, 298–302. [CrossRef]
13. Fleischmann, M.P.; Hendra, P.J.; Mcquillan, A.J. Raman spectra of pyridine adsorbed at a silver electrode. *Chem. Phys. Lett.* **1974**, *26*, 163–166. [CrossRef]
14. Johannessen, C.; White, P.C.; Abdali, S. Resonance Raman optical activity and surface enhanced resonance Raman optical activity analysis of cytochrome C. *J. Phys. Chem. A* **2007**, *111*, 7771–7776. [CrossRef]
15. Spiro, T.G.; Strekas, T.C. Resonance Raman spectra of heme proteins. Effects of oxidation and spin state. *J. Am. Chem. Soc.* **1974**, *96*, 338–345. [CrossRef] [PubMed]
16. Maker, P.D.; Terhune, R.W. Study of optical effects due to an induced polarization third order in the electric field strength. *Phys. Rev.* **1965**, *137*, 801–818. [CrossRef]
17. Cheng, J.X.; Xie, X.S. Coherent anti-stokes Raman scattering microscopy: Instrumentation, theory, and applications. *J. Phys. Chem. B* **2004**, *108*, 827–840. [CrossRef]
18. Roy, S.; Gord, J.R.; Patnaik, A.K. Recent advances in coherent anti-stokes Raman scattering spectroscopy: Fundamental developments and applications in reacting flows. *Prog. Energy Combust. Sci.* **2010**, *36*, 280–306. [CrossRef]
19. King, D.A.; Pittaro, R.J. Simple diode pumping of a power-buildup cavity. *Opt. Lett.* **1998**, *23*, 774. [CrossRef] [PubMed]
20. Taylor, D.J.; Glugla, M.; Penzhorn, R.D. Enhanced Raman sensitivity using an actively stabilized external resonator. *Rev. Sci. Instrum.* **2001**, *72*, 1970–1976. [CrossRef]
21. Campion, A.; Kambhampati, P. Surface-enhanced Raman scattering. *Chem. Soc. Rev.* **1998**, *27*, 241–250. [CrossRef]
22. Stockle, R.M.; Suh, Y.D.; Deckert, V.; Zenobi, R. Nanoscale chemical analysis by tip-enhanced Raman spectroscopy. *Chem. Phys. Lett.* **2000**, *318*, 131–136. [CrossRef]
23. Pettinger, B.; Picardi, G.; Schuster, R.; Ertl, G. Surface-enhanced and STM tip-enhanced Raman spectroscopy of CN—Ions at gold surfaces. *J. Electroanal. Chem.* **2003**, *554–555*, 293–299. [CrossRef]
24. Neacsu, C.C.; Dreyer, J.; Behr, N.; Raschke, M.B. Scanning-probe Raman spectroscopy with single-molecule sensitivity. *Phys. Rev. B* **2006**, *73*, 193406. [CrossRef]
25. Moskovits, M. Surface-enhanced Raman spectroscopy: A brief retrospective. *J. Raman Spectrosc.* **2010**, *36*, 485–496. [CrossRef]
26. Pettinger, B.; Ren, B.; Picardi, G.; Schuster, R.; Ertl, G. Nanoscale probing of adsorbed species by tip-enhanced Raman spectroscopy. *Phys. Rev. Lett.* **2004**, *92*, 096101. [CrossRef]
27. Cai, H.; Yu, X.T.; Chu, Q.; Jin, Z.Q.; Lin, B.; Wang, G.H. Hollow-core fiber-based Raman probe extension kit for in situ and sensitive ultramicro-analysis. *Chin. Opt. Lett.* **2019**, *17*, 5–9. [CrossRef]
28. Zhang, N.Q.; Qin, T.L.; Wang, Z.F.; Liu, W.B.; Cao, J.Q.; Chen, Z.L. Low-loss coupling between tapered fibersand anti-resonant hollow-core photonic crystal fibers. *Laser Optoelectron. Prog.* **2017**, *54*, 100608. [CrossRef]
29. Shephard, J.D.; Macpherson, W.N.; Maier, R.R.J.; Jones, J.D.C.; Knight, J.C. Single-mode mid-IR guidance in a hollow-core photonic crystal fiber. *Opt. Express* **2005**, *13*, 7139–7144. [CrossRef]
30. Zhang, W.; Lou, S.; Wang, X.; Yan, S.; Xing, Z. A broadband single mode single polarization metal wires-embedded hollow core anti-resonant fiber for polarization filter. *Opt. Fiber Technol.* **2019**, *53*, 102011. [CrossRef]

31. Salter, R.; Chu, J.; Hippler, M. Cavity-enhanced Raman spectroscopy with optical feedback cw diode lasers for gas phase analysis and spectroscopy. *Analyst* **2012**, *137*, 4669–4676. [CrossRef]
32. Jin, Z.; Chu, Q.; Xu, W.; Cai, H.; Ji, W.; Wang, G.; Lin, B.; Zhang, X. All-fiber Raman biosensor by combining reflection and transmission mode. *IEEE Photon. Technol. Lett.* **2018**, *30*, 387–390. [CrossRef]
33. Zhang, Z.M.; Chen, S.; Liang, Y.Z.; Liu, Z.X.; Zhang, Q.M.; Ding, L.X.; Ye, F.; Zhou, H. An intelligent background-correction algorithm for highly fluorescent samples in Raman spectroscopy. *J. Raman Spectrosc.* **2010**, *41*, 659–669. [CrossRef]
34. Gambling, W.A.; Payne, D.N.; Matsumura, H. Gigahertz bandwidths in multimode, liquid-core, optical fibre waveguide. *Opt. Commun.* **1972**, *6*, 317–322. [CrossRef]
35. Stone, J. Optical transmission in liquid-core quartz fibers. *Appl. Phys. Lett.* **1972**, *20*, 239–241. [CrossRef]
36. Ogilvie, G.J.; Esdaile, R.J.; Kidd, G.P. Transmission loss of tetrachloroethylene-filled liquid-core-fibre light guide. *Electron. Lett.* **1972**, *8*, 533–534. [CrossRef]
37. Ippen, E.P. Low-power quasi-cw Raman oscillator. *Appl. Phys. Lett.* **1970**, *16*, 303–305. [CrossRef]
38. Walrafen, G.E.; Stone, J. Intensification of spontaneous Raman spectra by use of liquid core optical fibers. *Appl. Spectrosc.* **1972**, *26*, 585–589. [CrossRef]
39. Fuwa, K.; Lei, W.; Fujiwara, K. Colorimetry with a total-reflection long capillary cell. *Anal. Chem.* **1984**, *56*, 1640–1644. [CrossRef]
40. Fujiwara, K.; Simeonsson, J.B.; Smith, B.W.; Winefordner, J.D. Waveguide capillary flow cell for fluorometry. *Anal. Chem.* **1988**, *60*, 1065–1068. [CrossRef]
41. Knebl, A.; Yan, D.; Popp, J.; Frosch, T. Fiber enhanced Raman gas spectroscopy. *TrAC Trends Anal. Chem.* **2017**, *103*, 230–238. [CrossRef]
42. Frosch, T.; Di, Y.; Popp, J. Ultrasensitive fiber enhanced UV resonance Raman sensing of drugs. *Anal. Chem.* **2013**, *85*, 6264–6271. [CrossRef] [PubMed]
43. Chu, Q.; Jin, Z.; Yu, X.; Li, C.; Zhang, W.; Ji, W.; Lin, B.; Shum, P.P.; Zhang, X.; Wang, G. Volumetric enhancement of Raman scattering for fast detection based on a silver-lined hollow-core fiber. *Opt. Express* **2019**, *27*, 10370–10382. [CrossRef] [PubMed]
44. Jin, Z.Q.; Chu, Q.; Fan, D.Y.; Xu, W.H.; Cai, H.; Wang, G.H.; Xu, F.; Lin, B.; Zhang, X. All fiber on-line Raman detection system with silver-coated hollow fiber. In Proceedings of the 2017 16th International Conference on Optical Communications & Networks (ICOCN), Wuzhen, China, 7–10 August 2017; IEEE: Manhattan, NY, USA, 2017.
45. Chu, Q.; Wang, G.H.; Jin, Z.Q.; Tan, J.; Cai, H.; Lin, B.; Zhang, X. All-fiber, portable, online Raman biosensor with enhancement of signal excitation and collection efficiency. In Proceedings of the 2018 Conference On Lasers and Electro-Optics (CLEO), San Jose, CA, USA, 13–18 May 2018; IEEE: Manhattan, NY, USA, 2018.
46. Yu, X.T.; Li, C.X.; Hu, D.J.J.; Milenko, K.; Wang, G.H.; Shum, P.; Xu, F.; Lu, Y.Q.; Zhang, X.P. All-fiber online Raman sensor with enhancement via a fabry-perot cavity. *Opt. Lett.* **2020**, *45*, 5760–5763. [CrossRef]
47. Russell, P. Photonic crystal fibers. *Science* **2003**, *299*, 358–362. [CrossRef]
48. Benabid, F. Hollow-core photonic bandgap fibre: New light guidance for new science and technology. *Philos. Trans. Math. Phys. Eng. Sci.* **2006**, *364*, 3439–3462. [CrossRef]
49. Cheng, T.L.; Li, S.G.; Zhou, G.Y.; Hou, L.T. Relation between power fraction in the core of hollow-core photonic crystal fibers and their bandgap property. *Chin. J. Lasers* **2007**, *34*, 249.
50. Roberts, P.J.; Couny, F.; Sabert, H.; Mangan, B.J.; Russell, P.S.J. Ultimate low loss of hollow-core photonic crystal fibres. *Opt. Express* **2005**, *13*, 236–244. [CrossRef]
51. Guo, X.R.; Yang, D.X. Propagating light with the full cladding of hollow-core photonic crystal fiber. *J. Appl. Opt.* **2011**, *32*, 744–748.
52. Yuan, J.H.; Hou, L.T.; Wei, D.B.; Wang, H.Y.; Zhou, G.Y. Experimental investigation and the application of hollow-core photonic crystal fiber in visible range. *J. Optoelectron. Laser* **2008**, *09*, 1150–1153.
53. Yan, D.; Popp, J.; Pletz, M.W.; Frosch, T. Highly sensitive broadband Raman sensing of antibiotics in step-index hollow-core photonic crystal fibers. *ACS Photonics* **2017**, *4*, 138–145. [CrossRef]
54. He, D.D.; Liu, M.; Jian, D.; Li, D.; Liao, Z.Y. Study on loss in hollow-core photonic bandgap fibers. *Laser Technol.* **2013**, *37*, 243–246.
55. Saitoh, K.; Koshiba, M. Confinement losses in air-guiding photonic bandgap fibers. *IEEE Photonics Technol. Lett.* **2003**, *15*, 236–238. [CrossRef]
56. Kim, H.K.; Digonnet, M.; Kino, G.S.; Shin, J.; Fan, S. Simulations of the effect of the core ring on surface and air-core modes in photonic bandgap fibers. *Opt. Express* **2004**, *12*, 3436–3442. [CrossRef]
57. Hanf, S.; Bogozi, T.; Keiner, R.; Frosch, T.; Popp, J. Fast and highly sensitive fiber-enhanced raman spectroscopic monitoring of molecular H2 and CH4 for point-of-care diagnosis of malabsorption disorders in exhaled human breath. *Anal. Chem.* **2015**, *87*, 982–988. [CrossRef]
58. Khetani, A.; Tiwari, V.S.; Harb, A.; Anis, H. Monitoring of heparin concentration in serum by Raman spectroscopy within hollow core photonic crystal fiber. *Sch. Inf. Technol. Eng. (SITE)* **2011**, *19*, 15244–15254. [CrossRef]
59. Duguay, M.A.; Kokubun, Y.; Koch, T.L.; Pfeiffer, L. Antiresonant reflecting optical waveguides in SiO_2-Si multilayer structures. *Appl. Phys. Lett.* **1986**, *49*, 13–15. [CrossRef]
60. Cao, L. Mid-Infrared Gas Raman Laser Source Based on Hollow-Core Anti-Resonant Fiber. Master's Thesis, Beijing University of Technology, Chaoyang, China, 2018.
61. Fei, Y.; Wadsworth, W.J.; Knight, J.C. Low loss (34 dB/km) silica hollow core fiber for the 3 μm spectral region. In Proceedings of the Specialty Optical Fibers, Colorado Springs, CO, USA, 17–21 June 2012.

62. Yu, F.; Knight, J.C. Spectral attenuation limits of silica hollow core negative curvature fiber. *Opt. Express* **2013**, *21*, 21466–21471. [CrossRef]
63. Gao, S.F.; Wang, Y.Y.; Wang, P. Research progress on hollow-core anti-resonant fiber and gas raman laser technology. *Chin. J. Lasers* **2019**, *46*, 183–200.
64. Ling, C.; Gao, S.F.; Peng, Z.G.; Wang, X.C.; Pu, W. High peak power 2.8 μm Raman laser in a methane-filled negative-curvature fiber. *Opt. Express* **2018**, *26*, 5609–5615.
65. Knebl, A.; Domes, R.; Wolf, S.; Domes, C.; Popp, J.; Frosch, T. Fiber-enhanced Raman gas spectroscopy for the study of microbial methanogenesis. *Anal. Chem.* **2020**, *92*, 12564–12571. [CrossRef]
66. Wang, J.X.; Chen, W.G.; Wan, F.; Wang, P.; Zhang, R. Fiber-enhanced Raman spectroscopic monitoring of fault characteristic gases dissolved in transformer oil by hollow-core photonic crystal fiber. In Proceedings of the 2018 IEEE International Conference on High Voltage Engineering and Application (ICHVE), Athens, Greece, 10–13 September 2018.
67. James, T.M.; Rupp, S.; Telle, H.H. Trace gas and dynamic process monitoring by Raman spectroscopy in metal-coated hollow glass fibres. *Anal. Methods* **2015**, *7*, 2568–2576. [CrossRef]
68. Pearman, W.F.; Carter, J.C.; Angel, S.M.; Chan, W.J. Quantitative measurements of CO_2 and CH_4 using a multipass Raman capillary cell. *Appl. Opt.* **2008**, *47*, 4627–4632. [CrossRef] [PubMed]
69. Okita, Y.; Gannot, I.; Katagiri, T.; Matsuura, Y. A Raman cell based on hollow optical fibers for breath analysis. In Proceedings of the Optical Fibers and Sensors for Medical Diagnostics and Treatment Applications X, San Francisco, CA, USA, 23–28 January 2010; Volume 7559, pp. 42–46.
70. Mullen, J.C.; Buric, M.P.; Chorpening, B.T.; Woodruff, S.D. Azimuthal polarization for Raman enhancement in capillary waveguides. *Opt. Eng.* **2013**, *52*, 117103. [CrossRef]
71. Simone, R.; Andreas, O.; Hendrik, S.M.; Timothy, J.; Helmut, T. Improving the detection limit in a capillary raman system for in situ gas analysis by means of fluorescence reduction. *Sensors* **2015**, *15*, 23110–23125.
72. Matsuura, Y.; Abel, T.; Harrington, J.A. Optical properties of small-bore hollow glass waveguides. *Appl. Opt.* **1995**, *34*, 6842–6847. [CrossRef] [PubMed]
73. Fan, D.; Jin, Z.; Wang, G.; Xu, F.; Lu, Y.; Hu, D.J.J.; Wei, L.; Shum, P.; Zhang, X. Extremely high-efficiency coupling method for hollow-core photonic crystal fiber. *IEEE Photonics J.* **2017**, *9*, 1–8. [CrossRef]
74. Fan, D.Y. Research on On-Line Raman Detecting System Based on Hollow-Core Fiber. Master's Thesis, Nanjing University, Nanjing, China, 2017.
75. Liu, Z.Q.; Wang, Y.J.; Lu, X.B.; Kan, M.X. Determination of isopropanol content by laser Raman spectroscopic internal standard method. In Proceedings of the National Conference on Chemistry and Spectroscopic Analysis, Chengdu, China, 29 July–2 August 2011.
76. Xu, Y.; Chen, X.; Zhu, Y. High sensitive temperature sensor using a liquid-core optical fiber with small refractive index difference between core and cladding materials. *Sensors* **2008**, *8*, 1872–1878. [CrossRef]
77. Huo, L.; Lin, C.; Suen, Y.K.; Kong, S.K. Raman signal enhancement in a liquid-core optical fiber based on hollow-core photonic crystal fiber. In Proceedings of the 2007 Asia Optical Fiber Communication and Optoelectronics Conference, Shanghai, China, 17–19 October 2007; pp. 233–235.
78. Wang, W.; Feng, M.Z. Characteristics of low-refractive-index liquid-core optical fibers. *J. Transcluction Technol.* **1996**, *01*, 15–18.
79. Fan, Y.; Wu, R.M.; Ai, S.R.; Liu, M.H.; Yang, H.F.; Zheng, J.H. Identification study of edible oil species with laser induced fluorescence technology based on liquid core optical fiber. *Spectrosc. Spectr. Anal.* **2016**, *36*, 3202–3206.
80. Altkorn, R.; Koev, I.; Duyne, R.; Litorja, M. Low-loss liquid-core optical fiber for low-refractive-index liquids: Fabrication, characterization, and application in Raman spectroscopy. *Appl. Opt.* **1997**, *36*, 8992–8998. [CrossRef]
81. Meneghini, C.; Caron, S.; Proulx, A.; Emond, F.; Paradis, P.; Pare, C.; Fougeres, A. Ethanol concentration measurement by Raman spectroscopy in liquid-core microstructured optical fiber. In *Third European Workshop on Optical Fibre Sensors*; Cutolo, A., Culshaw, B., LopezHiguera, J.M., Eds.; SPIE: Bellingham, WA, USA, 2007; Volume 6619.
82. Pelletier, M.J.; Altkorn, R. Raman sensitivity enhancement for aqueous protein samples using a liquid-core optical-fiber cell. *Anal. Chem.* **2001**, *73*, 1393–1397. [CrossRef] [PubMed]
83. Ramza, H.; Arsad, N.; Abdurrahman, F.; Supian, L.S.; Ab-Rahman, M.S. Optical fiber pressure sensor using extrinsic Fabry-Perot interferometry (EFPI); A theoretical study. *J. Optoelectron. Adv. Mater.* **2015**, *17*, 545–551.
84. Bethoux, O.; Godoy, E.; Roche, I.; Naccari, B.; Taleb, M.A.; Koteiche, M.; Nassif, Y. A new state-observer of the inner PEM fuel cell pressures for enhanced system monitoring. *Eur. Phys. J. Appl. Phys.* **2014**, *66*, 30901. [CrossRef]
85. Amezcuacorrea, R.; Leonsaval, S.; Birks, T.; Knight, J.C. Control of surface modes in low loss hollow-core photonic bandgap fibers. *Opt. Express* **2008**, *16*, 1142–1149. [CrossRef]
86. Shen, L.B.; Zhang, X.N. A study on novel liquid-core fiber and its properties. *J. Jinling Inst. Technol.* **2018**, *34*, 25–28.
87. Ishiharada M, N.K.; Tanuma, I.; Matsumuro, Y.; Honda, T. Properties of flexible light guide made of liquid core and polymer clad. *Plast. Opt. Fibres Appl.* **1993**, 49–53.
88. Marquardt, B.J.; Turney, K.P.; Burgess, L.W. Raman waveguide detector for low analyte concentrations in liquid samples. In Proceedings of the SPIE—The International Society for Optical Engineering, Boston, MA, USA, 19–22 September 1999; Volume 3860, pp. 239–249.

89. Gao, S.Q.; Li, Z.W.; Zhang, W.; Li, J.N. Liquid core optical fiber preresonance and resonance raman spectroscopy and its application. *Chin. J. Anal. Chem.* **1997**, *25*, 4.
90. Jin, Z.Q. On-Line Raman Liquid Detecting Methods Based on Hollow-Core Fiber. Master's Thesis, Nanjing University, Nanjing, China, 2018.
91. Pearman, W.F.; Carter, J.C.; Angel, S.M.; Chan, W.J. Multipass capillary cell for enhanced raman measurements of gases. *Appl. Spectrosc.* **2015**, *62*, 285–289. [CrossRef]
92. Yang, F.; Zhao, Y.; Qi, Y.; Tan, Y.Z.; Ho, H.L.; Jin, W. Towards label-free distributed fiber hydrogen sensor with stimulated Raman spectroscopy. *Opt. Express* **2019**, *27*, 12869–12882. [CrossRef]
93. Buric, M.P.; Chen, K.P.; Falk, J.; Woodruff, S.D. Enhanced spontaneous Raman scattering and gas composition analysis using a photonic crystal fiber. *Appl. Opt.* **2008**, *47*, 4255–4261. [CrossRef]
94. Buric, M.P.; Chen, K.; Falk, J.; Velez, R.; Woodruff, S. Raman sensing of fuel gases using a reflective coating capillary optical fiber. *Proc. SPIE* **2009**, *7316*, 433–443.
95. Wang, Z.; Yu, F.; Wadsworth, W.J.; Knight, J.C. Efficient 1.9 μm emission in H2-filled hollow core fiber by pure stimulated vibrational Raman scattering. *Laser Phys. Lett.* **2014**, *11*, 105807. [CrossRef]
96. Rupp, S.; James, T.M.; Telle, H.H.; Schloesser, M.; Bornschein, B. Enhanced sensitivity of raman spectroscopy for tritium gas analysis using a metal-lined hollow glass fiber. *Fusion Sci. Technol.* **2015**, *67*, 547–550. [CrossRef]
97. Guo, J.J.; Yang, D.W.; Liu, C.H. Raman signal enhancement for gas detection using a hollow core optical fiber. *Spectrosc. Spectr. Anal.* **2016**, *36*, 96–98.
98. Qi, D.; Berger, A.J. Chemical concentration measurement in blood serum and urine samples using liquid-core optical fiber Raman spectroscopy. *Appl. Opt.* **2007**, *46*, 1726–1734. [CrossRef]
99. Qiao, P.; Wang, X.H.; Gao, S.; Yin, X.; Wang, P. Integration of black phosphorus and hollow-core anti-resonant fiber enables two-order magnitude enhancement of sensitivity for bisphenol A detection. *Biosens. Bioelectron.* **2019**, *149*, 111821. [CrossRef]
100. Azkune, M.; Frosch, T.; Arrospide, E.; Aldabaldetreku, G.; Bikandi, I.; Zubia, J.; Popp, J.; Frosch, T. Liquid-core microstructured polymer optical fiber as fiber enhanced raman spectroscopy probe for glucose sensing. *J. Lightwave Technol.* **2019**, *37*, 2981–2988. [CrossRef]

photonics

MDPI

Review

Recent Advancement of Anti-Resonant Hollow-Core Fibers for Sensing Applications

Wenjun Ni [1], Chunyong Yang [1,*], Yiyang Luo [2], Ran Xia [3], Ping Lu [3], Dora Juan Juan Hu [4], Sylvain Danto [5], Perry Ping Shum [6] and Lei Wei [7,*]

1. Hubei Key Laboratory of Intelligent Wireless Communications, Hubei Engineering Research Center of Intelligent Internet of Things Technology, College of Electronics and Information Engineering, South-Central University for Nationalities, Wuhan 430074, China; wjni@mail.scuec.edu.cn
2. Key Laboratory of Optoelectronic Technology & Systems (Ministry of Education), Chongqing University, Chongqing 400044, China; yyluo@cqu.edu.cn
3. School of Optical and Electronic Information, Huazhong University of Science and Technology, Wuhan 430074, China; ranxia@hust.edu.cn (R.X.); pluriver@mail.hust.edu.cn (P.L.)
4. Institute for Infocomm Research, 1 Fusionopolis Way, #21-01 Connexis, Singapore 138632, Singapore; jjhu@i2r.a-star.edu.sg
5. CNRS, University of Bordeaux, Bordeaux INP, ICMCB, UMR 5026, F-33600 Pessac, France; sylvain.danto@u-bordeaux.fr
6. Department of Electrical and Electronic Engineering, College of Engineering, Southern University of Science and Technology, Shenzhen 518055, China; shenp@sustech.edu.cn
7. School of Electrical and Electronic Engineering, Nanyang Technological University, 50 Nanyang Avenue, Singapore 639798, Singapore

* Correspondence: cyyang@mail.scuec.edu.cn (C.Y.); wei.lei@ntu.edu.sg (L.W.)

Citation: Ni, W.; Yang, C.; Luo, Y.; Xia, R.; Lu, P.; Hu, D.J.J.; Danto, S.; Shum, P.P.; Wei, L. Recent Advancement of Anti-Resonant Hollow-Core Fibers for Sensing Applications. *Photonics* **2021**, *8*, 128. https://doi.org/10.3390/photonics8040128

Academic Editor: Carlos Marques

Received: 21 March 2021
Accepted: 15 April 2021
Published: 19 April 2021

Abstract: Specialty fibers have enabled a wide range of sensing applications. Particularly, with the recent advancement of anti-resonant effects, specialty fibers with hollow structures offer a unique sensing platform to achieve highly accurate and ultra-compact fiber optic sensors with large measurement ranges. This review presents an overview of recent progress in anti-resonant hollow-core fibers for sensing applications. Both regular and irregular-shaped fibers and their performance in various sensing scenarios are summarized. Finally, the challenges and possible solutions are briefly presented with some perspectives toward the future development of anti-resonant hollow-core fibers for advanced sensing.

Keywords: specialty fiber; anti-resonant effect; hollow structure; sensing applications

1. Introduction

Optical fiber, as a transmission medium, has enabled the world to move into a rapid and high-capacity communication age. Particularly, ultra-long distance transmission has been made possible by ultralow propagation loss (~0.17 dB/km) of silica fiber [1]. Even so, faster speed, larger capacity, and longer distance have become inevitable trends of communication requirements, particularly concerning 5G and 6G. Thus, a promising fiber is urgently needed to rival or exceed the conventional single-mode fiber (SMF) in transmission performance. Fortunately, anti-resonant hollow-core fibers (AR-HCFs) have the potential to achieve the aforementioned communication goals, due to their lowest attenuation, optical nonlinearity, and chromatic dispersion over a broad bandwidth [2]. Thanks to breakthroughs at the specialty fiber manufacturing level over the past few years, AR-HCFs have gradually become a research hotspot. However, the transmission loss of AR-HCFs over long distances (>50 km) has not been addressed and this limits applications in novel optical fiber communication systems. Therefore, most of the reported research work concerning specialty fibers with hollow structures is concentrated on transmission loss reduction.

Recently, AR-HCFs with variously irregular cladding structures have been designed and fabricated to reduce transmission loss, and the hollow-core nested anti-resonant nodeless fiber (HC-NANF) is the most common [3]. Figure 1 shows the gradually decreasing transmission loss of AR-HCFs in the past five years. In 2015, Walter Belardi from the University of Bath demonstrated the attenuation of hollow-core anti-resonant fiber (HC-ARF) of 175 dB/km at 480 nm [4]. The year after, a hollow-core revolver fiber with a double-capillary reflective cladding was reported by A. F. Kosolapov et al., which reduced the transmission loss to 75 dB/km at 1850 nm [5]. In 2018, Wang et al. proposed a hollow-core conjoined-tube negative-curvature fiber with a low transmission loss of 2 dB/km at 1512 nm and a <16 dB/km bandwidth spanning across the O, E, S, C, L telecom bands (1302–1637 nm) [6]. Subsequently, based on HC-NANF, Thomas D. Bradley et al. achieved lower transmission losses of 1.3 dB/km [7] and 0.65 dB/km [8] corresponding to the transmission distance of 0.5 km and 1.2 km, respectively. Quickly, they reported that using HC-NANF reached the currently lowest attenuation of 0.28 dB/km over the C and L bands [9]. The transmission loss of HC-NANF is approaching the conventional SMF, which is located in the same order of magnitude. Additionally, the nonlinearities of AR-HCFs are 3~4 orders of magnitude lower than SMF, resulting in ultralow dispersion. On the basis of ultralow transmission loss and chromatic dispersion, Lumenisity limited experimentally demonstrated that HC-NANF realized 10 Gbit dense wavelength division multiplexing (DWDM) transmission over 10 km links [10]. Nevertheless, AR-HCF-based novel fiber devices still have a long way to overcome the difficulties of ultra-long transmission distance. Moreover, AR-HCFs also face limitations of volume production and preparation technology. Though AR-HCFs have unsolved issues in fiber communications, they have shone in optical fiber sensing applications owing to their unique anti-resonant effect and inline hollow-core platform.

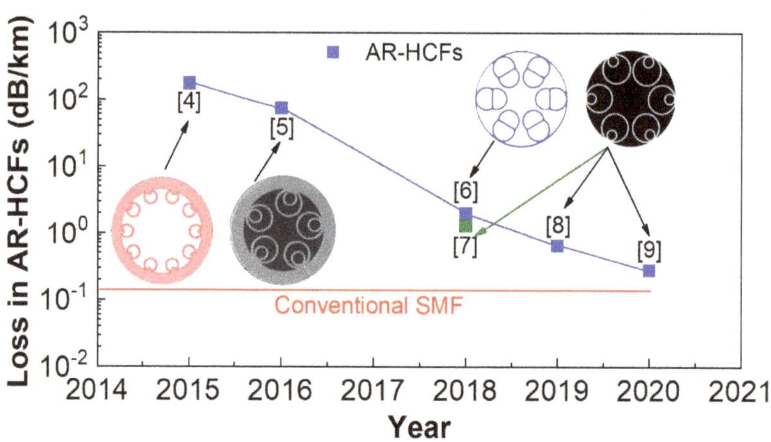

Figure 1. Transmission loss of AR-HCFs in different years.

This article illustrates a comprehensive review of the recent advancement of AR-HCFs for sensing applications. In the past five years, AR-HCFs have been widely employed in the sensing of various physical parameters, including temperature, strain, curvature, displacement, liquid level, and mechanical force, etc. All these measurands fully exploit the intrinsic advantages of anti-resonant effects that can generate specific resonant wavelengths over a broadband window. The sensitivity is then acquired by monitoring the shift or intensity fluctuation of the specific resonant wavelength. Additionally, AR-HCF is also a promising candidate for optofluidic applications because the hollow structure can offer a platform for fluid–light interaction. Consequently, AR-HCFs have attracted more attention for crucial applications in biomedical and biochemical fields, such as the detection of antibiotics, bisphenol, nitrous oxide, methane, and acetylene. Particularly, multiple gas detection will have significant potential applications in the fields of petrochemical and environmental protection, since AR-HCFs can act as a cell for gas–light interaction over extended lengths. In view of the aforementioned superiorities, AR-HCFs could pave the way for highly accurate, ultracompact, and practical sensing applications. In this text, we present the state-of-the-art of AR-HCF-based sensing applications. Firstly, we introduce the basic principle and some typical structures of AR-HCFs. Section 3 then focuses on recent advancements in sensing applications. Finally, we provide an outlook of development prospects and our related insights and viewpoints.

2. Principle and Various Structure of AR-HCFs

2.1. Principle of Light Propagation in AR-HCFs

In general, AR-HCFs features irregular silica cladding and an air core, which induces a special light propagation path. Here, AR-HCFs obey the light-guiding mechanisms of the anti-resonant reflecting optical waveguide (ARROW) [11], rather than the principle of total internal reflection of SMF. ARROW exploits coherent reflections at the air–silica interface to effectively guide the forward propagation light into the central hollow-core. To intuitively reflect the light propagation in the AR-HCFs, the hollow-core capillary with regular silica cladding is selected as the representative to describe the anti-resonant theory. As shown in Figure 2a, hollow-core capillary with a length L is sandwiched by the two segments of SMFs. The hollow-core capillary with regular ring cladding can be regarded as the FP cavity along the radial direction. The light beams, I_1 and I_2, represent the first reflection of their incident light at the air–silica and silica-air interface, respectively. Here, the incident light of I_2 is the refract light generated by the initial incident light. Then, the two beams I_1 and I_2 will form interference in the air core on the condition that the initial incident angle θ meets the grazing incidence (θ~90°). When the incident light of I_2 reaches the phase matching condition, it will leak out of the cladding. The leaked light is called resonant light, similarly, the reflected part is AR light. In short, AR-HCFs exploit coherent reflections from cladding to confine the guided light and propagate in the central air core. Thus, these light transmission mechanisms reveal the basic principle of the AR effect in HCFs. In addition, it is worth noting that the generation of the AR effect is limited by the length of the HCF. Obviously, if L is short enough, the Fabry–Perot (FP) cavity formed by the two fusion splicing interfaces dominates the whole transmission spectrum, as exhibited in Figure 2b. With an increasing L, however, the FP effect will gradually be weak or disappear caused by the increasing space loss. At this moment, nearly all of the light follows the ARROW transmission mechanisms. It can be seen that a critical length exists between FP and AR effect. The critical length L_c corresponds to the axial transmission length of beam I_2, it can be expressed as follows [12]:

$$L_c = \sqrt{n_0^2 + n_2^2 - n_1^2}\left(\frac{r}{\sqrt{n_1^2 - n_2^2}} + \frac{2d}{\sqrt{n_1^2 - n_0^2}}\right) \tag{1}$$

where n_0, n_1, and n_2 represent the refractive index of air, fiber core of SMF, and cladding, respectively. r and d denote the radius of the air core and the thickness of the ring cladding, respectively. If the capillary length is longer than L_c, the AR effect will be excited in the whole process. Otherwise, the sandwich structure only induces the FP effect. Figure 2b shows that several resonant wavelengths with a periodic distribution that is located in the AR effect-based transmission spectrum. According to the phase matching condition, the resonant wavelength λ_r can be given as follows [13]:

$$\lambda_r = \frac{2\left(d\sqrt{n_2^2 - n_0^2}\right)}{m} \tag{2}$$

where m is the resonance order. All of the aforementioned statements about AR-HCFs are based on the regular shape of the silica cladding, while most of the application scenes rely on an irregular structure to enhance the sensitivity. In view of diversified sensing applications, the next part will accordingly introduce the various structure of AR-HCFs.

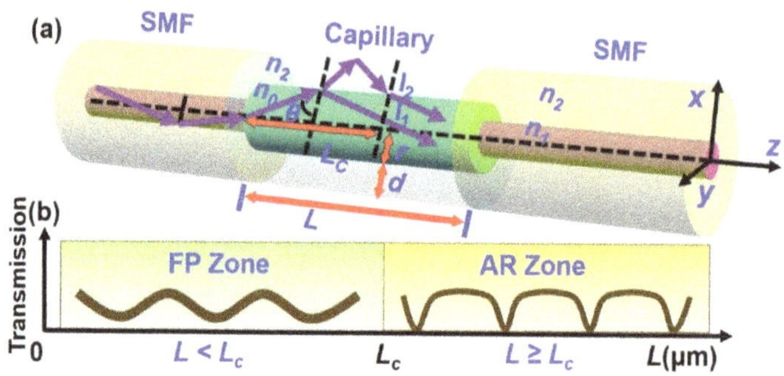

Figure 2. Schematic illustration of (**a**) optical pathways in the SMF–capillary–SMF structure and (**b**) transmission spectra as the increase of capillary length. [Reprinted/Adapted] with permission from [12] © The Optical Society.

2.2. Various Structures of AR-HCFs

In the past decade, AR-HCFs have experienced rapid development, mainly concentrated on structure improvement. The simplest structure is the hollow-core capillary exhibited in Figure 2a; thus, it is regarded as fundamental to exploiting novel structures. The optimization progress can be generally divided into three stages. Firstly, the negative curvature core [14] with a dense arrangement is introduced to surround the air core [15–18], as presented in Figure 3a. Then, several capillary tubes are designed to suspend on the silica cladding to form a negative curvature structure [19–30], as shown in Figure 3b. Finally, the suspended capillary tubes are optimized to be nested structures, as displayed in Figure 3c,d. In this way, the number of coherent reflections at the air–silica interface is increased and, thus, confinement loss is reduced, which is the latest HC-NANF [31,32].

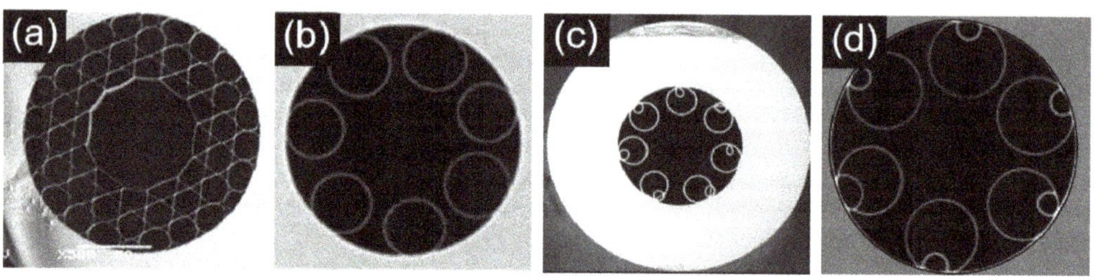

Figure 3. Scanning electron micrographs (SEMs) of AR-HCFs. (**a**) [Reprinted/Adapted] with permission from [16] © The Optical Society. (**b**) © 2017 IEEE. Reproduced with permission from Hayes et al., J. Lightwave Technol. 35, 437–442 (2017) [22]. (**c**) © 2017 IEEE. Reproduced with permission from Jose E. Antonio-Lopez et al., IEEE Photonics Conference (2017) [31]. (**d**) © 2020 IEEE. Reproduced with permission from Y. Hong, J. Lightwave Technol. 38, 2849–2857 (2020) [32].

All these AR-HCFs utilize the negative curvature effect in the surrounding region of the central core to reduce attenuation and promote longer propagation distance, so they are also referred to as negative curvature fibers (NCF). In other words, in most cases, AR-HCFs focus the performance enhancement on the optimization of negative curvature structure. In general, the performance of novel structures depends on significant indicator parameter improvements after optimization. Accordingly, Table 1 lists several key parameters of various AR-HCFs as contrasts.

Table 1. Parameters in various AR-HCFs.

Type	Core Diameter	Silica Thickness	Transmission Loss	Reference
Densely arrangement	45.8 μm	0.51 μm	300 dB/km	[15]
	50 μm	0.28 μm	180 dB/km	[16]
	60 μm	1.4 μm	17 dB/km	[18]
Suspended capillary tubes	30 μm	0.44 μm	180 dB/km	[19]
	30 μm	0.83 μm	30 dB/km	[21]
	41 μm	0.545 μm	7.7 dB/km	[26]
Nested Suspended tubes	25 μm	2.3 μm	75 dB/km	[5]
	~33 μm	~0.78 μm	2 dB/m	[31]
	~35 μm	~0.5 μm	6.6 dB/km	[32]

Table 1 illustrates that the silica thickness is one or two orders of magnitude lower than the core diameter. Here, the geometric size of the air core and silica cladding are in accordance with the AR theory in Section 2.1. Meanwhile, it can be seen from Table 1 that the transmission loss of various AR-HCFs is gradually decreasing with optimization stages approaching the nested topological structure. Indeed, the latest HC-NANFs based on nested structures achieve low attenuation in wide bandwidth ranges, single-mode transmission in hollow-core regions, and relatively long-distance propagation (~10 km) [10]. Currently, most sensing applications are limited by the aforementioned advantages, and, on the contrary, they require high loss structure modification, multimode coexistence, and micro or small size. It is obvious from Figure 3 that the most popular AR-HCFs still concentrate on the type of suspended capillary tubes. On the one hand, they have appropriate transmission loss values that can be satisfied in most sensing application scenarios [33–39]. On the other hand, their distinct advantage is that they can offer a versatile multichannel and lab-on-a-fiber platform [40–45]. As a consequence, the following sections will emphasize the novel and advanced sensing applications based on the most common AR-HCFs at present.

3. AR-HCF-Based Sensing Applications

Concerning the diversified sensing applications of AR-HCFs, not only can they be defined as discrete specialty fiber sensing devices, but can also be treated as a versatile platform for lab-on-a-fiber. The former is mainly used for common physical or chemical measurand detection, which is also the most reported field. The latter pays more attention to biomedical or online fluidic applications, which are relatively frontier research fields for now. All of the mentioned sensing application mechanisms can be broadly classified into two types, namely, special devices and versatile platforms. Thus, we will discuss the AR-HCF-based sensing applications in the light of these classifications.

3.1. Special Device-Based Sensing Applications

Similar to the Bragg wavelength for fiber Bragg grating (FBG) [46–48] and loss peak for long period grating (LPG) [49,50], the resonant wavelength is a remarkable characterization for AR-HCFs. Accordingly, AR-HCFs can be also regarded as one of the functional fiber sensing devices. As analyzed in Section 2.1, the unique AR effect can provide a series of resonant wavelengths with the periodic distribution. The measurand sensitivity can be acquired by interrogating the wavelength shift or intensity. In recent years, most research works have reported sensing applications for generic parameters, including strain, displacement, curvature, static pressure, temperature, liquid level, and mechanical force, et al. Moreover, some modified structures based on AR-HCFs can achieve the simultaneous measurement of dual or multiple parameters. Figure 4 provides a comprehensive review of various sensing applications based on single- or double-layer AR effects.

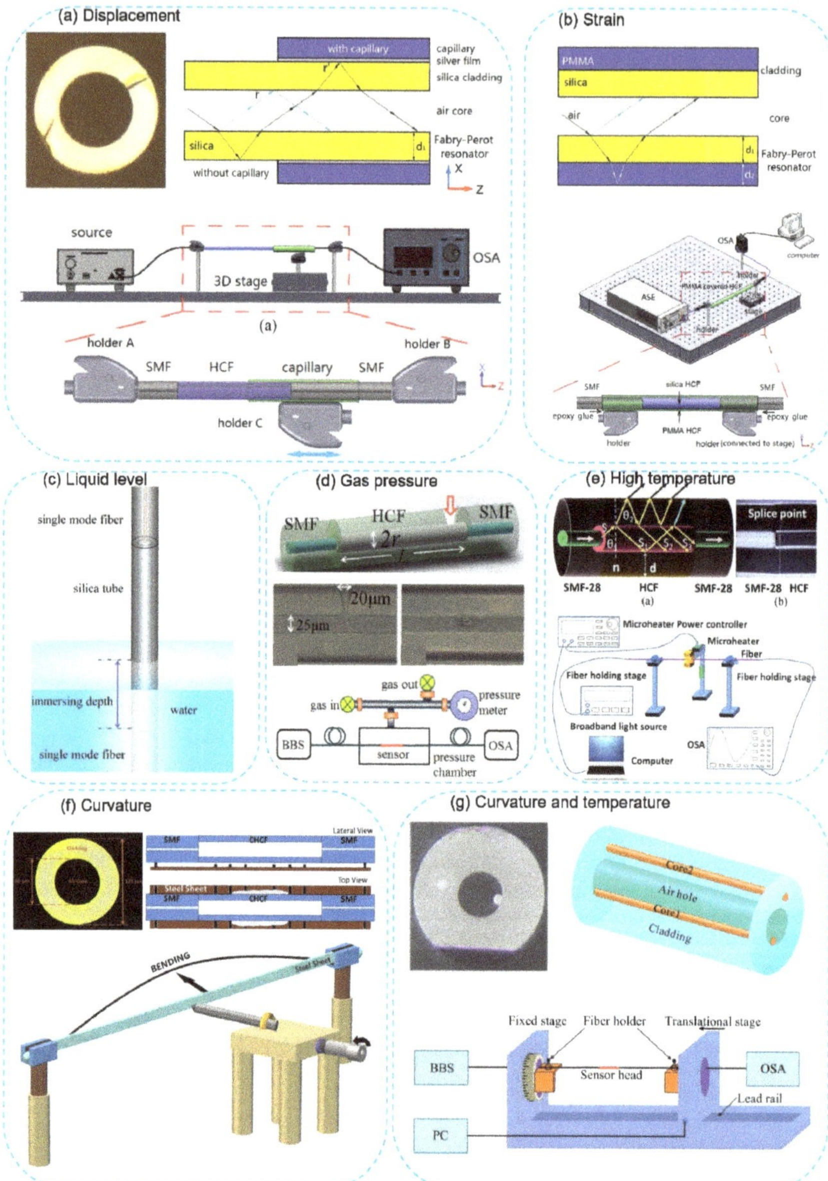

Figure 4. Special device-based sensing applications. (**a**) © 2016 IEEE. Reproduced with permission from R. Gao, J. Sel. Top Quant. 23, 5,600,106 (2016) [51]. (**b**) © 2017 IEEE. Reproduced with permission from R. Gao, IEEE Photonic. Tech. L. 29, 857-860 (2017) [52]. (**c**) © 2015 IEEE. Reproduced with permission from S. Liu, J. Lightwave Technol. 33, 5239-5243 (2016) [53]. (**d**) [Reprinted/Adapted] with permission from [54] © The Optical Society. (**e**) © 2018 IEEE. Reproduced with permission from D. Liu, J. Lightwave Technol. 36, 1583-1590 (2018) [55]. (**f**) Reproduced with permission [56]. Copyright 2020 MDPI. (**g**) [Reprinted/Adapted] with permission from [57] © The Optical Society.

Figure 4a shows that the inner capillary is coated with silver, and then wraps the HCF [51]. This most common HCF agrees with the theory of Section 2.1, so it will form an FP cavity in the cladding. The leaky mode of the HCF actually features the resonant wavelengths in the transmission spectrum. The initial leaky mode will be reflected back into cladding with the capillary location movement because the covered part is transformed to the silica–silver interface. Thus, the intensity of the resonant wavelength is sensitive to the displacement, and it finally obtains a high sensitivity of 0.578 dB/μm. Likewise, as shown in Figure 4b, HCF is inserted into the polymethyl methacrylate (PMMA) hollow-core fiber [52]. This is a double-layer AR structure since the refractive index of PMMA is between air and silica. The distinct difference between the single- and double-layer AR structure is the free spectrum range (FSR), and the latter is larger than the former owing to the longer optical path. The tensile strain will reduce the thickness of the PMMA resulted in the optical path change, then the resonant wavelength will shift. Using this double-layer AR structure, it achieves the strain sensitivity of 27.9 pm/με. As shown in Figure 4c, the HCF is used for liquid level monitoring [53], and the sensing mechanism is similar to that in Figure 4a. The leaky mode intensity is influenced by the change of the surrounding refractive index. More specifically, the contrast of resonant wavelength will increase or decrease with the liquid level rising or falling. On the basis of this simple HCF structure, a liquid-level sensitivity of 0.4 dB/mm has been realized. Furthermore, the temperature cross-sensitivity is only 0.004 mm/°C, which can be regarded as insensitive. It is apparent that all the sensing mechanisms from Figure 4a to Figure 4c are based on the external refractive index change. On the contrary, Figure 4d devotes to the inner refractive index variation to induce the wavelength shift [54]. As displayed in Equation (2), the resonant wavelength not only depends on the silica thickness and refractive index, but also the air core refractive index. It exploits femtosecond laser drilling on the ring cladding to form a microchannel, which keeps the air-core pressure equivalent to the external environment. The manufactured part is placed into the airtight container, then the increasing pressure will induce the air core refractive index change. The gas pressure sensitivity reaches 3.592 nm/MPa by the wavelength demodulation in the range of 0 to 2 MPa. It can be seen from Figure 4a to Figure 4d, that they only cause a single parameter change in the whole sensing process, and all of these structures transfer to the refractive index variation. While temperature is treated as a sensing variable, which can simultaneously induce parameter changes in multiple structures in the AR-HCFs. As shown in Figure 4e, HCF is applied to high-temperature measurement attributed to the AR effect [55]. It is difficult to use common multimode interference for high-temperature sensing because the employed high order cladding modes have a nonlinearity variation with the temperature increasing. For the HCF, the AR effect is only dependent on the geometric size and material refractive index. Thereby, AR-HCFs have a linear relationship with temperature in a large dynamic range. Figure 4e shows a high-temperature sensitivity of 33.4 pm/°C with the working range from room temperature to 1000 °C. The sensitivities of all the aforementioned structures are determined by their own material properties. Figure 4f enhances the curvature sensitivity assisted by coating the sensitive material [56]. After the polydimethylsiloxane (PDMS) coated on the HCF, the mode confinement is reduced and bending losses are significantly higher. By the external structure modifications, it acquires the curvature sensitivity of -5.26 dB/m^{-1}, which is several times higher than without PDMS coated. In addition, the structure modifications can be also conducted by the special design of HCF. Figure 4g illustrates a novel fiber with a single-hole twin eccentric core, which simultaneously has AR effect and inline Mach-Zehnder interference (IMMI) functions [57]. In this case, the dual or multiple parameters can be measured simultaneously based on different mechanisms and without crosstalk. A curvature sensitivity of -1.54 dB/m^{-1} and the temperature sensitivity of 70.71 pm/°C is realized based on the AR effect and IMMI, respectively. The listed seven examples reveal that the most common HCF mainly rely on the AR effect for sensing at present.

In summary, AR effect-based HCFs have been widely used for all kinds of sensing applications, covering solids, gas, and liquid. Among the AR-HCFs, some have been used for practical applications. Nonetheless, the detection of most parameters is part of mechanisms research, and, thus, remains at a laboratory stage.

3.2. Versatile Platform Based Sensing Applications

The HCFs in 3.1 are relatively simple structures without negative curvature topological configurations. This section introduces AR-HCFs with negative curvature structures. Here, the dominant AR effect, like the Figure 4 examples, will still exist in the irregular hollow-core region. The AR-HCFs in this section mainly exploit the multichannel structure for fluidic sensing. In contrast to common HCFs, these AR-HCFs need to fill various gases, liquids, or solid materials in the air channel to enhance the light–matter interaction. This versatile platform can offer a promising route for compact, integral, and biocompatible all-fiber multifunctional optofluidic devices for in-situ applications.

As shown in Figure 5a, light is guided inside an air core filled with the analyte gas, and the AR-HCF serves as a low-volume and robust absorption cell. Laser-based gas sensors utilize borosilicate fiber because the absorption wavelength of the monitoring nitrous oxide is located at 5.26 μm. The feasibility of exploiting the eight-hole AR-HCF with 1.15 m length as a gas cell is verified, and the minimum detection limitation (MDL) of 20 ppbv with 70 s signal interrogation time is achieved [58]. Similarly, Figure 5b replaces the eight-hole negative curvature with a nested nodeless structure for nitrous oxide sensing [59]. Compared with Figure 5a, the whole length of AR-HCF is increased to 3.2 m, so the time and area of light–gas interaction is accordingly enhanced. Finally, the experimental results show that the response time shortens to 1s and the MDL decreases to ~5.4 ppbv. Figures 5a and 5b are used for single gas detection, which is a waste of resources. In practical applications, most occasions require simultaneous measurement of multi-component gas, for example, gas micro-leakage in a transformer. As shown in Figure 5c, laser-based dual gas sensing of methane and carbon dioxide is achieved by a length of 1m silica AR-HCF. The absorption wavelengths of these two gasses are 3.334 μm and 1.574 μm, respectively, corresponding to the near- and mid-infrared spectral region [60]. That is, the monitoring crosstalk between methane and carbon dioxides can be well avoided, which paves the way for multi-component gas detection. The experimental results show that the MDL of methane and carbon reaches 24 ppbv and 144 ppmv, respectively. From Figure 5a to Figure 5c, all the utilized AR-HCFs retain the single-mode transmission in the air core region. In order to acquire lower MDL, Figure 5d demonstrates a mode-phase-difference (MPD) photothermal spectroscopy for acetylene detection [61]. In a 4-m-long AR-HCF, it supports LP_{01}-like and LP_{11}-like transmission in a broadband. The photothermal effect induces variations of the dual-mode refractive index. The MPD is sensitive to gas absorption and insensitive against the external environment perturbation. The proof-of-concept realizes the MDL down to pptv. AR-HCFs are treated as microcells to provide a gas–light interaction platform in the aforementioned four examples. More importantly, AR-HCFs have a great potential for multi-component gas simultaneous detection owing to the wide transmission broadband in the air core region, nearly covering all molecule absorption lines. In addition to gas detection, AR-HCFs are also attracting attention for their potential in liquid sensing. The distinct difference of absorption spectroscopy measurements between gas and liquid is the different filling medium. As shown in Figure 5e, the hollow-core region is filled with antibiotic, which will cause the absorption of the guided light. Using a 1-m-long AR-HCF achieves the detection of sulfamethoxazole (SMX) and sodium salicylate (SS) down to 0.1 μM (26 ppb) and 0.4 μM (64 ppb), respectively [62]. Figure 5f conducts a proof-of-concept experiment of Raman spectroscopy using ethanol, which provides an original idea for noninvasive biochemical analysis [63]. In contrast to Figure 5e–g principally proposes a low-loss micro-machining method for optofluidic applications [64]. They exploit the focused ion beam to mill the cladding of AR-HCFs, and no additional loss is generated in the whole process. Once the hole of the cladding is opened, it will shine new light on the inline optofluidic applications. This method is similar to femtosecond laser drilling on the ring cladding in Figure 4d. Both contribute to forming the inline channel. Figure 5h utilizes a 40-cm-long AR-HCF to reach the MDL of 1.69 pM for bisphenol A detection [65], which is the experimental verification for Figure 5f. The applications depicted in Figure 5a–h are all employed in fluidic detection, including gas and liquid. While in Figure 5i, HC-NANF is wound into a coil to be integrated into a resonator fiber optic gyroscope [66]. This structure not only breaks through the limitation of nonlinear influence, but also considerably improves spatial mode purity. It is a promising candidate for civil aircraft navigation usage. In recent years, some AR-HCF-based optofluidic applications have been used for practical applications based on similar methods developed for microstructured fiber-based devices [67–72].

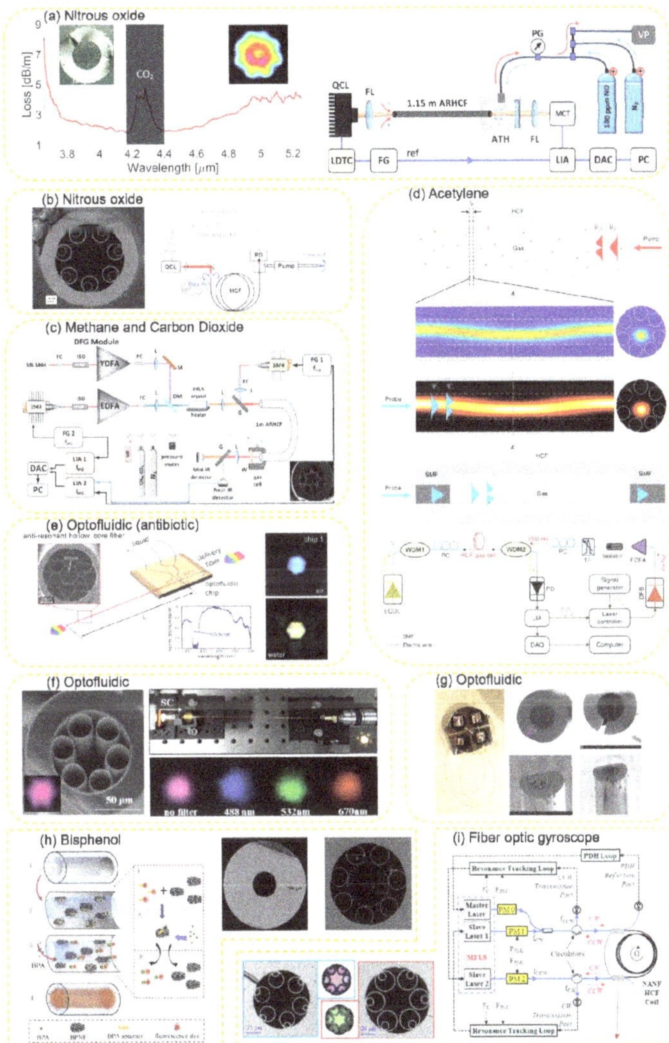

Figure 5. Versatile platform based sensing applications. (**a**) [Reprinted/Adapted] with permission from [58] © The Optical Society. (**b**) [Reprinted/Adapted] with permission from [59] © The Optical Society. (**c**) Reproduced with permission [60]. Copyright 2020 MDPI. (**d**) Reproduced with permission [61]. Copyright 2020, Springer Nature. (**e**) Reproduced with permission [62]. Copyright 2018 MDPI. (**f**) [Reprinted/Adapted] with permission from [63] © The Optical Society. (**g**) [Reprinted/Adapted] with permission from [64] © The Optical Society. (**h**) Reproduced with permission [65]. Copyright 2020, Elsevier. (**i**) [Reprinted/Adapted] with permission from [66] © The Optical Society.

4. Prospects and Conclusions

Enabled by the specialty fiber manufacturing industry, AR-HCFs have shown great potential in optical fiber communication and sensing. AR-HCFs have very low transmission loss, optical nonlinearity, and chromatic dispersion over a broad bandwidth. They also have intrinsic advantages of high sensitivity, compact structures, and robust operation. All these remarkable advantages promote diversified sensing applications of AR-HCF. As a functionalized device, it has been extensively used for common parameter sensing, including solid, gas, and liquid. Meanwhile, as a versatile platform, it sheds new light on optofluidic fields, mainly including gas micro-leakage, multi-component gas, pharmaceutical substances, and tissue fluid detection. Particularly, the versatile platform has made a significant contribution to providing a new route for photoacoustic spectroscopy (PAS) or photothermal spectroscopy (PTS). With the rapid development of micro-machining techniques, there have been substantial laboratory-stage research achievements, inspiring hope for practical applications. Furthermore, it is believed that AR-HCF will become the ultimate optical fiber once there has been a breakthrough in long-distance single-mode transmission loss.

Author Contributions: W.N. and L.W. contributed to the idea. W.N. contributed to the writing of the manuscript. W.N., C.Y., Y.L., R.X., P.L., D.J.J.H., S.D., P.P.S. and L.W. contributed to the reviewing and editing of the manuscript. C.Y. and L.W. supervised the project. All authors have read and agreed to the published version of the manuscript.

Funding: This work was supported by the Key Technology R&D Program of Hubei province under Grant 2020BBB097, Singapore Ministry of Education Academic Research Fund Tier 2 (MOE2019-T2-2-127 and T2EP50120-0005), A*STAR under AME IRG (A2083c0062), the Singapore Ministry of Education Academic Research Fund Tier 1 (RG90/19 and RG73/19) and the Singapore National Research Foundation Competitive Research Program (NRF-CRP18-2017-02). This work was also supported by South-Central University for Nationalities and Nanyang Technological University.

Data Availability Statement: Data sharing not applicable.

Conflicts of Interest: The authors declare no conflict of interest.

References

1. Sagae, Y.; Matsui, T.; Sakamoto, T.; Nakajima, K. Ultra-Low Crosstalk Multi-Core Fiber with Standard 125-μm Cladding Diameter for 10,000km-Class Long-Haul Transmission. *IEICE Trans. Commun.* **2020**, *E103.B*, 1199–1205. [CrossRef]
2. Habib, M.S.; Bang, O.; Bache, M. Low-loss single-mode hollow-core fiber with anisotropic anti-resonant elements. *Opt. Express* **2016**, *24*, 8429–8436. [CrossRef]
3. Francesco, P. Nested antiresonant nodeless hollow core fiber. *Opt. Express* **2014**, *22*, 23807–23828.
4. Belard, W. Design and Properties of Hollow Antiresonant Fibers for the Visible and Near Infrared Spectral Range. *J. Lightwave Technol.* **2015**, *33*, 4497–4503. [CrossRef]
5. Kosolapov, A.F.; Alagashev, G.K.; Kolyadin, A.N.; Pryamikov, A.D.; Biryukov, A.S.; Bufetov, I.A.; Dianov, E.M. Hollow-core revolver fibre with a double-capillary reflective cladding. *Quantum Electron.* **2016**, *46*, 267–270. [CrossRef]
6. Gao, S.; Wang, Y.; Ding, W.; Jiang, D.; Gu, S.; Zhang, X.; Wang, P. Hollow-core conjoined-tube negative-curvature fibre with ultralow loss. *Nat. Commun.* **2018**, *9*, 1–6. [CrossRef]
7. Bradley, T.D.; Hayes, J.R.; Chen, Y.; Jasion, G.T.; Sandoghchi, S.R.; Slavik, R.; Fokoua, E.N.; Bawn, S.; Sakr, H.; Davidson, I.A.; et al. Record low-loss 1.3 dB/km data transmitting antiresonant hollow core fibre. In Proceedings of the 2018 European Conference on Optical Communication (ECOC), Rome, Italy, 23–27 September 2018.
8. Bradley, T.D.; Jasion, G.T.; Hayes, J.R.; Chen, Y.; Hooper, L.; Sakr, H.; Gouveia, M.A.; Taranta, A.; Saljoghei, A.; Mulvad, H.C.; et al. Antiresonant hollow core fibre with 0.65 dB/km attenuation across the C and L telecommunication bands. In Proceedings of the 45th European Conference on Optical Communication (ECOC 2019), Dublin, Ireland, 22–26 September 2019.
9. Jasion, G.T.; Bradley, T.D.; Harrington, K.; Sakr, H.; Chen, Y.; Fokoua, E.N.; Davidson, I.A.; Taranta, A.; Hayes, J.R.; Richardson, D.J.; et al. Hollow core NANF with 0.28 dB/km attenuation in the C and L bands. In Proceedings of the Optical Fiber Communication Conference, San Diego, CA, USA, 10–12 March 2020; Optical Society of America: Washington, DC, USA, 2020.
10. Laser Focus World Reported on 01 Oct, 2020, and Sources from Lumenisty Company. Available online: https://www.laserfocusworld.com/fiber-optics/article/14184480/lumenisity-unveils-hollowcore-fiberoptic-cable-for-10-gbit-dwdm-transmission-over-10-km-links (accessed on 14 April 2021).
11. Litchinitser, N.M.; Abeeluck, A.K.; Headley, C.; Eggleton, B.J. Antiresonant reflecting photonic crystal optical waveguides. *Opt. Lett.* **2002**, *27*, 1592–1594. [CrossRef]

12. Xiaobei, Z.; Haiyang, P.; Huawen, B.; Ming, Y.; Jiawei, W.; Chuanlu, D.; Tianxing, W. Transition of Fabry–Perot and antiresonant mechanisms via a SMF-capillary-SMF structure. *Opt. Lett.* **2018**, *43*, 2268–2271.
13. Cai, N.; Xia, L.; Wu, Y. Multiplexing of anti-resonant reflecting optical waveguides for temperature sensing based on quartz capillary. *Opt. Express* **2018**, *26*, 33501–33509. [CrossRef]
14. Wei, C.; Weiblen, R.J.; Menyuk, C.R.; Hu, J. Negative curvature fibers. *Adv. Opt. Photonics* **2017**, *9*, 504–561. [CrossRef]
15. Couny, F.; Benabid, P.S. Light. Large-pitch kagome-structured hollow-core photonic crystal fiber. *Opt. Lett.* **2006**, *31*, 3574–3576. [CrossRef] [PubMed]
16. Wang, Y.Y.; Wheeler, N.V.; Couny, F.; Roberts, P.J.; Benabid, F. Low loss broadband transmission in hypocycloid-core Kagome hollow-core photonic crystal fiber. *Opt. Lett.* **2011**, *36*, 669–671. [CrossRef] [PubMed]
17. Wheeler, N.V.; Heidt, A.M.; Baddela, N.K.; Fokoua, E.N.; Hayes, J.R.; Sandoghchi, S.R.; Poletti, F.; Petrovich, M.N.; Richardson, D.J. Low-loss and low-bend-sensitivity mid-infrared guidance in a hollow-core–photonic-bandgap fiber. *Opt. Lett.* **2014**, *39*, 295–298. [CrossRef]
18. Debord, B.; Alharbi, M.; Bradley, T.; Fourcade-Dutin, C.; Wang, Y.Y.; Vincetti, L.; Gérôme, F.; Benabid, F. Hypocycloid-shaped hollow-core photonic crystal fiber Part I: Arc curvature effect on confinement loss. *Opt. Express* **2013**, *21*, 28597–28608. [CrossRef]
19. Uebel, P.; Günendi, M.C.; Frosz, M.H.; Ahmed, G.; Edavalath, N.N.; Ménard, J.M.; Russell P. St., J. Broadband robustly single-mode hollow-core PCF by resonant filtering of higher-order modes. *Opt. Lett.* **2016**, *41*, 1961–1964. [CrossRef]
20. Provino, L.; Haboucha, A.; Havranek, M.; Monteville, A.; Landais, D.; Goffic, O.; Insou, X.; Barbier, M.; Chartier, T.; Thual, M.; et al. Large-Core Hollow-Core Antiresonant Fiber with Low-Loss and Truly Single-Mode Guidance for N-IR Wavelengths. In Proceedings of the OSA Laser Congress, Boston, MA, USA, 4–8 November 2018; Optical Society of America: Washington, DC, USA, 2018.
21. Michieletto, M.; Jens, K.; Lyngsø, C.J.; Lægsgaard, J.; Bang, O.; Alkeskjold, T.T. Hollow-core fibers for high power pulse delivery. *Opt. Express* **2016**, *24*, 7103–7119. [CrossRef]
22. Hayes, J.R.; Sandoghchi, S.R.; Bradley, T.D.; Liu, Z.; Slavík, R.; Gouveia, M.A.; Wheeler, N.V.; Jasion, G.; Chen, Y.; Fokoua, E.N.; et al. Antiresonant Hollow Core Fiber with an Octave Spanning Bandwidth for Short Haul Data Communications. *J. Lightwave Technol.* **2017**, *35*, 437–442. [CrossRef]
23. Pryamikov, A.D.; Biriukov, A.S.; Kosolapov, A.F.; Plotnichenko, V.G.; Semjonov, S.L.; Dianov, E.M. Demonstration of a waveguide regime for a silica hollow—core microstructured optical fiber with a negative curvature of the core boundary in the spectral region > 3.5 μm. *Opt. Express* **2011**, *19*, 1441–1448. [CrossRef] [PubMed]
24. Kolyadin, A.N.; Kosolapov, A.F.; Pryamikov, A.D.; Biriukov, A.S.; Plotnichenko, V.G.; Dianov, E.M. Light transmission in negative curvature hollow core fiber in extremely high material loss region. *Opt. Express* **2013**, *21*, 9514–9519. [CrossRef]
25. Newkirk, A.V.; Antonio-Lopez, J.E.; Anderson, J.; Aguirre, R.A.; Eznaveh, Z.S.; Galmiche, G.L.; Correa, R.A.; Schülzgen, A. Modal analysis of antiresonant hollow core fibers using S2 imaging. *Opt. Lett.* **2016**, *41*, 3277–3280. [CrossRef]
26. Debord, B.; Amsanpally, A.; Chafer, M.; Baz, A.; Maurel, M.; Blondy, J.M.; Hugonnot, E.; Scol, F.; Vincetti, L.; Gérome, F.; et al. Ultralow transmission loss in inhibited-coupling guiding hollow fibers. *Optica* **2017**, *4*, 209–217. [CrossRef]
27. Belardi, W.; Knight, J.C. Hollow antiresonant fibers with low bending loss. *Opt. Express* **2014**, *22*, 10091–10096. [CrossRef] [PubMed]
28. Kosolapov, A.F.; Pryamikov, A.D.; Biriukov, A.S.; Shiryaev, V.S.; Astapovich, M.S.; Snopatin, G.E.; Plotnichenko, V.G.; Churbanov, M.F.; Dianov, E.M. Demonstration of CO_2-laser power delivery through chalcogenide-glass fiber with negative-curvature hollow core. *Opt. Express* **2011**, *19*, 25723–25728. [CrossRef] [PubMed]
29. Gattass, R.R.; Rhonehouse, D.; Gibson, D.; McClain, C.C.; Thapa, R.; Nguyen, V.Q.; Bayya, S.S.; Weiblen, R.J.; Menyuk, C.L.; Shaw, B.; et al. Infrared glass-based negative-curvature anti-resonant fibers fabricated through extrusion. *Opt. Express* **2016**, *24*, 25697–25703. [CrossRef] [PubMed]
30. Setti, V.; Vincetti, L.; Argyros, A. Flexible tube lattice fibers for terahertz applications. *Opt. Express* **2013**, *21*, 3388–3399. [CrossRef]
31. Antonio-Lopez, J.E.; Habib, M.S.; Newkirk, A.V.; Galmiche, G.L.; Eznaveh, Z.S.; Zacarisa, J.C.; Bang, O.; Bache, M.; Schülzgen, A.; Correa, R.A. Antiresonant hollow core fiber with seven nested capillaries. In Proceedings of the 2016 IEEE Photonics Conference (IPC), Waikoloa, HI, USA, 2–6 October 2016.
32. Hong, Y.; Sakr, H.; Taengnoi, N.; Bottrill, K.R.; Bradley, T.D.; Hayes, J.R.; Jasion, G.T.; Kim, H.; Thipparapu, N.K.; Wang, Y.; et al. Multi-Band Direct-Detection Transmission Over an Ultrawide Bandwidth Hollow-Core NANF. *J. Lightwave Technol.* **2020**, *38*, 2849–2857. [CrossRef]
33. Wang, S.; Shan, C.; Jiang, J.; Liu, K.; Zhang, X.; Han, Q.; Lei, J.; Xiao, H.; Liu, T. Temperature-insensitive curvature sensor based on anti-resonant reflection guidance and Mach–Zehnder interferometer hybrid mechanism. *Appl. Phys. Express* **2019**, *12*, 106503. [CrossRef]
34. Liu, S.; Ji, Y.; Cui, L.; Sun, W.; Yang, J.; Li, H. Humidity-insensitive temperature sensor based on a quartz capillary anti-resonant reflection optical waveguide. *Opt. Express* **2017**, *25*, 18929–18939. [CrossRef]
35. Gao, R.; Lu, D.; Cheng, J.; Qi, Z. Self-referenced antiresonant reflecting guidance mechanism for directional bending sensing with low temperature and strain crosstalk. *Opt. Express* **2017**, *25*, 18081–18091. [CrossRef]
36. Wang, S.; Wang, S.; Zhang, S.; Feng, M.; Wu, S.; Jin, R.; Zhang, L.; Lu, P. An inline fiber curvature sensor based on anti-resonant reflecting guidance in silica tube. *Opt. Laser Technol.* **2019**, *111*, 407–410. [CrossRef]

37. Gao, H.; Jiang, Y.; Zhang, L.; Cui, Y.; Jiang, Y.; Jia, J.; Jiang, L. Antiresonant mechanism based self-temperature-calibrated fiber optic Fabry–Perot gas pressure sensors. *Opt. Express* **2019**, *27*, 22181–22189. [CrossRef] [PubMed]
38. Suo, X.; Yu, H.; Li, J.; Wu, X. Transmissive resonant fiber-optic gyroscope employing Kagome hollow-core photonic crystal fiber resonator. *Opt. Lett.* **2020**, *45*, 2227–2230. [CrossRef] [PubMed]
39. Ni, W.; Lu, P.; Fu, X.; Sun, H.; Shum, P.; Liao, H.; Jiang, X.; Liu, D.; Yang, C.; Zhang, J.; et al. Simultaneous implementation of enhanced resolution and large dynamic range for fiber temperature sensing based on different optical transmission mechanisms. *Opt. Lett.* **2018**, *26*, 18341–18350. [CrossRef]
40. Ankan, I.M.; Mollah, M.A.; Sultana, J.; Islam, M. Negative Curvature Hollow-core Anti-resonant Fiber for Terahertz Sensing. *Appl. Opt.* **2020**, *59*, 8519–8525. [CrossRef]
41. Gao, R.; Lu, D.; Cheng, J.; Jiang, Y.; Jiang, L.; Ye, J.; Qi, Z. Magnetic Fluid-Infiltrated Anti-Resonant Reflecting Optical Waveguide for Magnetic Field Sensing Based on Leaky Modes. *J. Lightwave Technol.* **2016**, *34*, 3490–3495. [CrossRef]
42. Newkirk, A.V.; Lopez, J.E.A.; Correa, R.A.; Schülzgen, A.; Mazurowski, J. Anti-resonant hollow core fiber for precision timing applications. In *Astronomical Optics: Design, Manufacture, & Test of Space & Ground Systems*; Astronomical Optics: San Diego, CA, USA, 2017; Volume 10401, p. 104010F-1.
43. Jaworski, P.; Yu, F.; Boj, P.; Wu, D.; Koziol, P.E.; Dudzik, G.; Abramski, K.M.; Liao, M.; Krzempek, K. Antiresonant Hollow-Core Fiber for Multiple Gas Detection in the Mid-IR. In *CLEO: QELSFundamental Science*; Optical Society of America: Washington, DC, USA, 2020; p. JTu2G.33.
44. Liu, S.; Wang, Y.; Hou, M.; Guo, J.; Li, Z.; Lu, P. Anti-resonant reflecting guidance in alcohol-filled hollow core photonic crystal fiber for sensing applications. *Opt. Lett.* **2013**, *21*, 31690–31697. [CrossRef]
45. Gao, R.; Lu, D.; Zhang, M.; Qi, Z. Optofluidic Immunosensor Based on Resonant Wavelength Shift of a Hollow Core Fiber for Ultratrace Detection of Carcinogenic Benzo[a]pyrene. *ACS Photonics* **2018**, *5*, 1273–1280. [CrossRef]
46. Zheng, D.; Cai, Z.; Floris, I.; Madrigal, J.; Pan, W.; Zou, X.; Sales, S. Temperature-insensitive optical tilt sensor based on a single eccentric-core fiber Bragg grating. *Opt. Lett.* **2019**, *44*, 5570–5573. [CrossRef] [PubMed]
47. Guo, K.; He, J.; Shao, L.; Xu, G.; Wang, Y. Simultaneous Measurement of Strain and Temperature by a Sawtooth Stressor-Assisted Highly Birefringent Fiber Bragg Grating. *J. Lightwave Technol.* **2020**, *38*, 2060–2066. [CrossRef]
48. Zuowei, X.; Huang, B.; Shu, X. Dual interference effects in a line-by-line inscribed fiber Bragg grating. *Opt. Lett.* **2020**, *45*, 2950–2953.
49. Zhang, S.; Li, X.; Niu, H.; Yan, Q.; Sun, C.; Peng, F.; Ma, Y.; Zhang, K.; Geng, T.; Yang, W.; et al. Few-mode fiber-embedded long-period fiber grating for simultaneous measurement of refractive index and temperature. *Appl. Opt.* **2020**, *59*, 9248–9253. [CrossRef] [PubMed]
50. Ni, W.; Lu, P.; Luo, C.; Fu, X.; Liu, L.; Jiang, X.; Liao, H.; Liu, D.; Zhang, J. Bending direction detective fiber sensor for dual-parameter sensing based on an asymmetrical thin-core long-period fiber grating. *IEEE Photonics J.* **2016**, *8*, 6803811. [CrossRef]
51. Gao, R.; Lu, D.; Cheng, J.; Jiang, Y.; Jiang, L.; Qi, Z. Optical displacement sensor in a capillary covered hollow core fiber based on anti-resonant reflecting guidance. *IEEE J. Sel. Top. Quantum Electron.* **2016**, *23*, 5600106. [CrossRef]
52. Gao, R.; Lu, D.; Cheng, J.; Qi, Z. In-Fiber Double-Layered Resonator for High-Sensitive Strain Sensing. *IEEE Photonics Technol. Lett.* **2017**, *29*, 857–860. [CrossRef]
53. Liu, S.; Tian, J.; Liu, N.; Xia, J.; Lu, P. Temperature Insensitive Liquid Level Sensor Based on Antiresonant Reflecting Guidance in Silica Tube. *J. Lightwave Technol.* **2016**, *34*, 5239–5243. [CrossRef]
54. Hou, M.; Zhu, F.; Wang, Y.; Wang, Y.; Liao, C.; Liu, S.; Lu, P. Antiresonant reflecting guidance mechanism in hollow-core fiber for gas pressure sensing. *Opt. Lett.* **2016**, *24*, 27890–27898. [CrossRef] [PubMed]
55. Liu, D.; Wu, Q.; Mei, C.; Yuan, J.; Xin, X.; Mallik, A.K.; Wei, F.; Han, W.; Kumar, R.; Yu, C.; et al. Hollow Core Fiber Based Interferometer for High-Temperature (1000 °C) Measurement. *J. Lightwave Technol.* **2018**, *36*, 1583–1590. [CrossRef]
56. Herrera-Piad, L.A.; Hernández-Romano, I.; May-Arrioja, D.A.; Minkovich, V.P.; Cisneros, M.T. Sensitivity Enhancement of Curvature Fiber Sensor Based on Polymer-Coated Capillary Hollow-Core Fiber. *Sensors* **2020**, *20*, 3763. [CrossRef]
57. Ni, W.; Lu, P.; Zhang, J.; Yang, C.; Fu, X.; Sun, Y.; Liao, H.; Liu, D. Single hole twin eccentric core fiber sensor based on anti-resonant effect combined with inline Mach-Zehnder interferometer. *Opt. Lett.* **2017**, *25*, 12372–12380. [CrossRef] [PubMed]
58. Jaworski, P.; Krzempek, K.; Dudzik, G.; Sazio, P.J.; Belardi, W. Nitrous oxide detection at 5.26 µm with a compound glass antiresonant hollow-core optical fiber. *Opt. Lett.* **2020**, *45*, 1326–1329. [CrossRef] [PubMed]
59. Nikodem, M.; Gomółka, G.; Klimczak, M.; Pysz, D.; Buczyński, R. Demonstration of mid-infrared gas sensing using an anti-resonant hollow core fiber and a quantum cascade laser. *Opt. Lett.* **2019**, *27*, 36350–36357. [CrossRef] [PubMed]
60. Jaworski, P.; Koziol, P.E.; Krzempek, K.; Wu, D.; Yu, F.; Boj, P.; Dudzik, G.; Liao, M.; Abramski, K.M.; Knight, J.C. Antiresonant Hollow-Core Fiber-Based Dual Gas Sensor for Detection of Methane and Carbon Dioxide in the Near-and Mid-Infrared Regions. *Sensors* **2020**, *20*, 3813. [CrossRef] [PubMed]
61. Zhao, P.; Zhao, Y.; Bao, H.; Ho, H.; Jin, W.; Fan, S.; Gao, S.; Wang, Y.; Wang, P. Mode-phase-difference photothermal spectroscopy for gas detection with an anti-resonant hollow-core optical fiber. *Nat. Commun.* **2020**, *11*, 847. [CrossRef]
62. Mona, N.; Brenda, D.; Jonas, H.; Jens, K.; Karina, W.; Thomas, H.; Maekus, A.S. UV Absorption Spectroscopy in Water-Filled Antiresonant Hollow Core Fibers for Pharmaceutical Detection. *Sensors* **2018**, *18*, 478.
63. Liu, X.; Ding, W.; Wang, Y.; Gao, S.; Cao, L.; Feng, X.; Wang, P. Characterization of a liquid-filled nodeless anti-resonant fiber for biochemical sensing. *Opt. Lett.* **2017**, *42*, 863–866. [CrossRef]

64. Adamu, A.I.; Wang, Y.; Correa, R.A.; Bang, O.; Markos, C. Low-loss micro-machining of anti-resonant hollow-core fiber with focused ion beam for optofluidic application. *Opt. Mater. Express* **2021**, *11*, 338–344. [CrossRef]
65. Qiao, P.; Wang, X.; Gao, S.; Yin, X.; Wang, Y.; Wang, P. Integration of black phosphorus and hollow-core anti-resonant fiber enables two-order magnitude enhancement of sensitivity for bisphenol A detection. *Biosens. Bioelectron.* **2019**, *149*, 111821. [CrossRef]
66. Sanders, G.A.; Taranta, A.A.; Narayanan, C.; Fokoua, E.N.; Mousavi, S.A.; Strandjord, L.K.; Smiciklas, M.; Bradley, T.D.; Hayes, J.; Jasion, G.T.; et al. Hollow-core resonator fiber optic gyroscope using nodeless anti-resonant fiber. *Opt. Lett.* **2021**, *46*, 46–49. [CrossRef]
67. Zhang, N.; Humbert, G.; Gong, T.; Shum, P.; Li, K.; Auguste, J.R.; Wu, Z.; Hu, J.; Luan, F.; Dinh, Q.; et al. Side-channel photonic crystal fiber for surface enhanced Raman scattering sensing. *Sens. Actuators B. Chem.* **2016**, *223*, 195–201. [CrossRef]
68. Zhang, M.; Li, K.; Shum, P.; Yu, X.; Zeng, S.; Wu, Z.; Wang, Q.; Yong, K.; Wei, L. Hybrid graphene/gold plasmonic fiber-optic biosensor. *Adv. Mater. Technol.* **2017**, *2*, 1600185. [CrossRef]
69. Zhang, M.; Li, K.; Zhang, T.; Shum, P.; Wang, Z.; Wang, Z.; Zhang, N.; Zhang, J.; Wu, T.; Wei, L. Electron-rich two-dimensional molybdenum trioxides for highly integrated plasmonic biosensing. *ACS Photonics* **2018**, *5*, 347–352. [CrossRef]
70. Zhang, N.; Humbert, G.; Wu, Z.; Li, K.; Shum, P.; Zhang, M.; Cui, Y.; Auguste, J.L.; Dinh, X.; Wei, L. In-line optofluidic refractive index sensing in a side-channel photonic crystal fiber. *Opt. Lett.* **2016**, *24*, 27674–27682. [CrossRef] [PubMed]
71. Zhang, N.; Li, K.; Cui, Y.; Wu, Z.; Shum, P.; Auguste, J.L.; Dinh, X.; Humbert, G.; Wei, L. Ultra-sensitive chemical and biological analysis via specialty fibers with built-in microstructured optofluidic channels. *Lab Chip* **2018**, *18*, 655–661. [CrossRef] [PubMed]
72. Zhang, M.; Qi, M.; Wang, Z.; Wang, Z.; Chen, M.; Li, K.; Shum, P.; Wei, L. One-step synthesis of cyclodextrin-capped gold nanoparticles for ultra-sensitive and highly-integrated plasmonic biosensors. *Sens. Actuators B. Chem.* **2019**, *286*, 429–436. [CrossRef]

photonics

MDPI

Article

Sensitivity Enhanced Refractive Index Fiber Sensor Based on Long-Range Surface Plasmon Resonance in SiO_2-Au-TiO_2 Heterostructure

Wenyi Bu [1], Zhifang Wu [1,*], Perry Ping Shum [2], Xuguang Shao [3] and Jixiong Pu [1]

1 Fujian Key Laboratory of Light Propagation and Transformation, College of Information Science and Engineering, Huaqiao University, Xiamen 361021, China; 19014082023@stu.hqu.edu.cn (W.B.); jixiong@hqu.edu.cn (J.P.)
2 Department of Electrical and Electronic Engineering, Southern University of Science and Technology, Shenzhen 518055, China; shenp@sustech.edu.cn
3 School of Electrical and Electronic Engineering, Nanyang Technological University, 50 Nanyang Avenue, Singapore 639798, Singapore; XGshao@ntu.edu.sg
* Correspondence: zfwu@hqu.edu.cn

Abstract: Long-range surface plasmon resonance (LRSPR), generated from a coupled plasmon polariton in a thin metal slab sandwiched by two dielectrics, has attracted more and more attention due to its merits, such as longer propagation and deeper penetration than conventional single-interface surface plasmon resonance. Many useful applications related to light–medium interaction have been demonstrated based on the LRSPR effect, especially in the sensing area. Here, we propose and demonstrate an LRSPR-based refractive index sensor by using a SiO_2-Au-TiO_2 heterostructure, in which a D-shaped honeycomb-microstructure optical fiber (MOF) is designed as the silica substrate and then deposited with a gold film and thin-layer titanium dioxide (TiO_2). By using the full-vector finite-element method (FEM), this heterostructure is numerically investigated and demonstrated to excite LRSPR without a buffer layer, which is usually necessary in previous LRSPR devices. Through comprehensive discussion about the influence of structural parameters on the resonant wavelength, the excitation of the LRSPR in the proposed heterostructure is revealed to be highly related to the effective refractive index of MOF's fundamental core mode, which is mainly determined by the MOF's pitch, the thicknesses of the silica web and the planar-layer silica. Moreover, the thin-layer TiO_2 plays an important role in significantly enhancing the resonance and the sensitivity to analyte's refractive index as well, when it is coated on the top of the Au film rather than between the metal and waveguide. Finally, the proposed LRSPR sensor based on SiO_2-Au-TiO_2 heterostructure shows an ultra-high wavelength sensitivity of 20,100 nm/RIU and the corresponding minimum resolution is as low as 4.98×10^{-7} RIU. Thus, the proposed LRSPR device offers considerable potential for sensing applications in biomedical and biochemical areas.

Keywords: SiO_2-Au-TiO_2 heterostructure; long-range surface plasmon resonance; microstructured optical fiber; fiber sensor

Citation: Bu, W.; Wu, Z.; Shum, P.P.; Shao, X.; Pu, J. Sensitivity Enhanced Refractive Index Fiber Sensor Based on Long-Range Surface Plasmon Resonance in SiO_2-Au-TiO_2 Heterostructure. *Photonics* **2021**, *8*, 379. https://doi.org/10.3390/photonics8090379

Received: 29 July 2021
Accepted: 6 September 2021
Published: 9 September 2021

1. Introduction

With the rapid development of multi-disciplinary science and technology in recent years, optical measurement methods and tools play significant roles in many areas, especially in chemistry, biology, medicine, and so on. As a representative, surface plasmon resonance (SPR) based technology is receiving increasing attention thanks to its remarkably high sensitivity and accuracy. Surface plasmon—also called surface plasmon polariton (SPP)—refers to a quantum of charge density oscillations at the interface between a metal and a dielectric, and propagates along the interface. It can be excited through photon–electron coupling resonance, when the wave vector of incident light from the dielectric matches the wave vector of SPP. From the light transmission point of view, such coupling resonance is accompanied by the energy transference from incident light to SPP, generating a narrow-band resonant peak in the spectrum. The oscillations are very sensitive to any perturbation of the metal's boundary, especially when the thickness of metal layer is as thin as several tens of nanometers. It indicates that the SPR-based devices are intrinsically suitable for high-sensitivity detection [1–4].

For further enhancing their performance and functions, SPR-based devices with multi-layer heterostructures arouse researchers' considerable interest. For example, bimetallic layers are widely proposed to improve the sensitivity and stability of SPR biosensors [5–7]. A layer made of metal oxide, such as indium tin oxide (ITO) [8], tin oxide (SnO_2) [9], titanium dioxide (TiO_2) [10], hafnium oxide (HfO_2) [11], tantalum oxide (Ta_2O_5) [12], and so forth, is commonly used as an adhesive layer to strengthen the attachment of a metal film on a substrate, or deposited on top of the metal layer to protect metal from oxidization. Besides, some metal oxide films are demonstrated to have the capability to enhance sensitivity [8,12–15], or can even be used as the sensing layer by means of lossy mode resonance [16,17]. Moreover, 2D materials, such as graphene [18], molybdenum disulfide (MoS_2) [19], tungsten disulfide (WS_2) [20] and so forth, provide more choice and fascinating properties for SPR-based devices [20].

From the physics point of view, multi-layer structured systems extend the concept of SPR to a broader realm. Considering a typical three-layer heterostructure, that is, a dielectric–metal–dielectric system, SPP might be excited at both metallic–dielectric interfaces and interact with each other to give rise to two coupled modes, if the separation between two interfaces is close to the decay length of single-interface SPP and the effective refractive indices of two dielectrics are comparable. Based on the symmetry of the transverse electric field, these two coupled modes are distinguished as the symmetric mode and the asymmetric mode [21]. Compared with the conventional single-interface SPP, the symmetric mode features lower attenuation, longer propagating distance and deeper penetration—termed long-range surface plasmon polariton (LRSPP) mode—while the asymmetric one is denoted as short-range surface plasmon polariton (SRSPP) mode due to its contrary characteristics. Hence, the devices based on long-range surface plasmon resonance (LRSPR) promise to offer a better performance than conventional single-interface SPR based devices for applications in biomedical or chemical measurement [21–24]. As for the applications related to liquid analytes, a buffer layer (e.g., Teflon, Cytop, MgF_2) with a similar material refractive index (RI) to the analyte is usually necessary to balance the RI gap between the analyte and waveguide and then facilitate the generation of LRSPR [25,26]. Alternatively, by properly designing the waveguide structure, some microstructured optical fibers (MOFs) find another approach to achieving LRSPR excitation in a wide wavelength range without a buffer layer [27].

In this paper, an LRSPR refractive index sensor based on a SiO_2-Au-TiO_2 heterostructure is proposed and theoretically investigated. This heterostructure is composed of a D-shaped honeycomb-structure silica MOF coated with a thin-layer of Au and a TiO_2 film sequentially. By designing the specialty MOF with a high air-filling-ratio cladding and suitable core size, the effective refractive index (n_{eff}) of the fundamental core mode is comparable with the material RI of aqueous analyte. Therefore, the LRSPP mode in the proposed heterostructure is directly achieved without the assistance of an additional buffer layer. Furthermore, the structural parameters, including the cladding pitch, the thicknesses of silica web, the flatten silica layer, the Au layer and the TiO_2 film, are discussed in detail. Based on the wavelength interrogation method, the sensing performance of the different heterostructures (i.e., SiO_2-Au-TiO_2 and SiO_2-TiO_2-Au) is compared within the analyte's RI range from 1.33 to 1.39. The results illustrate that the sensor based on the SiO_2-Au-TiO_2 configuration has higher sensitivity and its maximum and average sensitivities reach 20,100 nm/RIU and 11,586 nm/RIU, respectively. The corresponding minimum resolution is as low as 4.98×10^{-7} RIU.

2. Structural and Theoretical Modeling

The cross-section of the proposed sensor is depicted in Figure 1a. The pure silica D-shaped MOF plays the role as a substrate with a waveguide, in which the microstructured cladding consists of five-layer hexagonal air holes with the honeycomb-structure arrangement and the central air hole is replaced by a silica rod. This hexagonal-air-holes structure provides a very high air-filling ratio. It not only significantly lowers the n_{eff} of the MOF's core mode but also reduces the optical field to spread in air-holes cladding, resulting in a strong interaction of the light between the waveguide and metal film. The flattened side of the D-shaped MOF can be treated as a very thin layer of silica and provides a planar surface, on which a layer of gold and a very thin-film of TiO_2 are coated sequentially. Gold is a commonly-used plasma material in many SPR devices due to its high resonance shift capability and chemical stability. The TiO_2 film is deposited on the gold layer to increase the electromagnetic field extension from the metal film to the analyte, enhancing the interaction of the evanescent wave with the analyte [28]. It is noteworthy that most of the planar-layer silica connects with the air holes directly, where the electromagnetic field tends to be partially reflected to the metal film because of the considerable RI difference at the silica–air interface. To mimic the scenario of practical sensing, the proposed device is surrounded by a liquid analyte which is usually viewed as another dielectric layer.

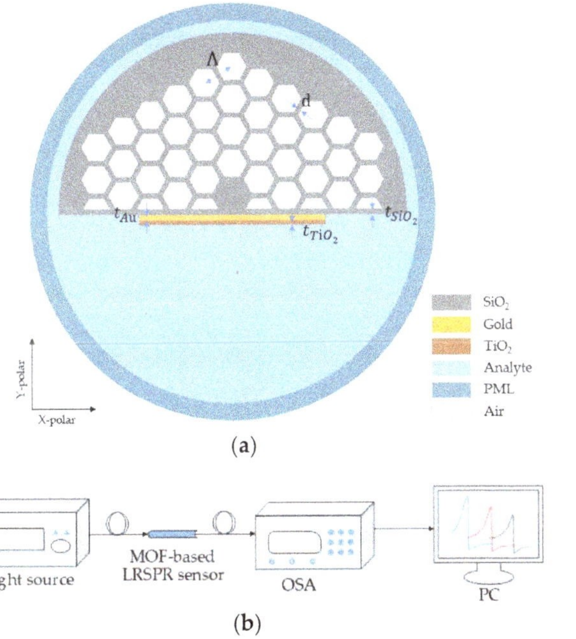

Figure 1. (**a**) Cross-sectional schematic structure of the D-shaped LRSPR refractive index sensor with SiO_2-Au-TiO_2 structure. (**b**) Diagram for the experimental setup for sensor characterization. Λ denotes the pitch of MOF's air-hole cladding; *d* denotes the thickness of silica web; t_{Au} denotes the thickness of gold film; t_{TiO_2} denotes the thickness of the tin oxide film; t_{SiO_2} denotes the thickness of the planar silica layer; PML: perfectly-matched layer; OSA: optical spectrum analyzer.

The numerical simulation and analysis were implemented by using a full-vector finite-element method (FEM), which was provided by COMSOL Multiphysics. As an example, the initial model parameters included the MOF's pitch size (Λ = 2600 nm), the thickness of the silica web (d = 500 nm), the thickness of the planar-layer silica (t_{SiO_2} = 150 nm), the thickness of the gold layer (t_{Au} = 70 nm), and the thickness of the TiO_2 film (t_{TiO_2} = 3 nm).

The diameters of the MOF's outer cladding, analyte and perfectly-matched layer (PML) are respectively set to 36 μm, 40 μm, and 43 μm.

The background material of the MOF is fused silica, and the refractive index of SiO_2 is calculated by the following Sellmeier equation as mentioned in [29]:

$$n_{\mathrm{SiO_2}} = \sqrt{1 + \frac{0.691663\lambda^2}{\lambda^2 - 0.004679} + \frac{0.407943\lambda^2}{\lambda^2 - 0.013512} + \frac{0.897479\lambda^2}{\lambda^2 - 97.934003}}, \tag{1}$$

where λ denotes the wavelength of the incident light and its unit is micrometer.

The complex dielectric permittivity of gold is described by the Drude–Lorentz model [30]:

$$\varepsilon_{\mathrm{Au}} = \varepsilon_\infty - \frac{\omega_D^2}{\omega(\omega + i\gamma_D)} - \frac{\Delta\varepsilon \cdot \Omega_L^2}{(\omega^2 - \Omega_L^2) + i\Gamma_L \omega}, \tag{2}$$

where ω is the angular frequency of transmitted light; ε_∞, ω_D, and γ_D stand for the permittivity at high frequency, the plasma frequency and the damping coefficient, respectively; Ω_L and Γ_L are the oscillator strength and the spectral width of the Lorentz oscillators, respectively; $\Delta\varepsilon$ is a weighting factor. Their values are listed as follows: $\varepsilon_\infty = 5.9673$, $\omega_D/2\pi = 2113.6$ THz, $\gamma_D/2\pi = 15.92$ THz, $\Omega_L/2\pi = 650.07$ THz, $\Gamma_L/2\pi = 104.86$ THz, and $\Delta\varepsilon = 1.09$ [31].

The refractive index of TiO_2 can be defined by the given equation [3]:

$$n_{\mathrm{TiO_2}} = \sqrt{5.913 + \frac{2.441 \times 10^7}{(\lambda^2 - 0.803 \times 10^7)}}. \tag{3}$$

The RI of air is fixed at 1.0, while that of the analyte keeps a constant value for wavelength sweep and then changes from 1.33 to 1.39 for discussing the sensing performance. The material RI of the PML is set to be the same as that of the analyte.

Regarding the feasibility of device fabrication, the specialty MOF can be fabricated by the stack and draw technique, which is similar to the fabrication of a side-channel MOF [32]. Then, the planar surface for coating metal film can be exposed by fiber post-processing methods, such as etching or side polish [33–37]. The deposition of Au and TiO_2 films is usually achieved by using thermal evaporation or sputtering techniques [2]. The practical scheme for wavelength interrogation of the proposed LRSPR sensor could be similar to that of the experimental setup for conventional SPR fiber sensors [35]. As shown in Figure 1b, the D-shaped MOF with the LRSPR sensor head is spliced with two single-mode fibers (SMFs) and is then connected with a broadband light source and an optical spectrum analyzer (OSA), respectively. The output spectrum is recorded by the OSA, and then transferred to a computer for data processing.

In order to achieve accurate enough results, the maximum element size of the meshes in the TiO_2 film and the gold layer are respectively set to be 1 nm and 10 nm, whereas the mesh size in the silica web and MOF's core are less than 50 nm. With frequency domain mode analysis, a series of complex eigenvalues (N_{eff}) are simulated. Each eigenvalue corresponds to a specific mode. The real part of a complex eigenvalue (Re[N_{eff}]) represents the effect refractive index (n_{eff}) of the corresponding mode, whereas its imaginary part (Im[N_{eff}]) determines the confinement loss $\alpha(\lambda)$ [30,36]:

$$\alpha(\lambda) = 8.686 \times \frac{2\pi}{\lambda} \mathrm{Im}[N_{\mathrm{eff}}] \times 10^4, \tag{4}$$

where λ is the wavelength in micrometers.

Based on the aforementioned method and the parameters of the example model, the dispersion and confinement loss curves of the y-polarized core mode and LRSPP mode are calculated and plotted in Figure 2a. The blue and red solid line curves stand for the n_{eff} of y-polarized core mode and LRSPP mode, respectively. Moreover, the loss spectra of the core mode and LRSPP mode are respectively illustrated as the red and blue dot curves. From this figure, it is noteworthy that the dispersion curve of the core mode declines more steeply than that of the LRSPP mode. This abnormal dispersion phenomenon is different from that in conventional SPR devices and might be essentially related to the mode field distribution of the LRSPP mode. The pictures shown in Figure 2c,f illustrate the electric field distribution of the MOF's core mode and LRSPP mode at the wavelength of 1900 nm, respectively. For further in-depth analysis, the field distribution along y-axis of the LRSPP mode is plotted as shown in Figure 2b. We find that most of electric field of the LRSPP mode distributes in the analyte region. Thus, the n_{eff} of LRSPP mode will be close to the material RI of analyte according to effective-medium theory [38,39]. It changes more slowly than that of the MOF's core mode, which is simultaneously determined by the material dispersion of silica and the MOF's waveguide dispersion [39]. Moreover, there is a break around the wavelength of 1925 nm in their dispersion curves, in which the avoided-crossing phenomenon occurs. As the pictures show in Figure 2c–e, the electric field of the core mode is transferred step-by-step from the core region to the analyte with the increment of wavelength, indicating a conversion from the core mode to the LRSPP mode. On the other hand, the LRSPP mode is transformed into the core mode gradually as the pictures show in Figure 2f–h. The attenuation spectrum consists of the confinement loss of the core mode before the avoided-crossing point and that of the converted "core mode" after the avoided-crossing point, as the blue dot curve shows in Figure 2a. It has a narrow-band peak with a wavelength of 1925 nm, which corresponds to the strongest mode coupling.

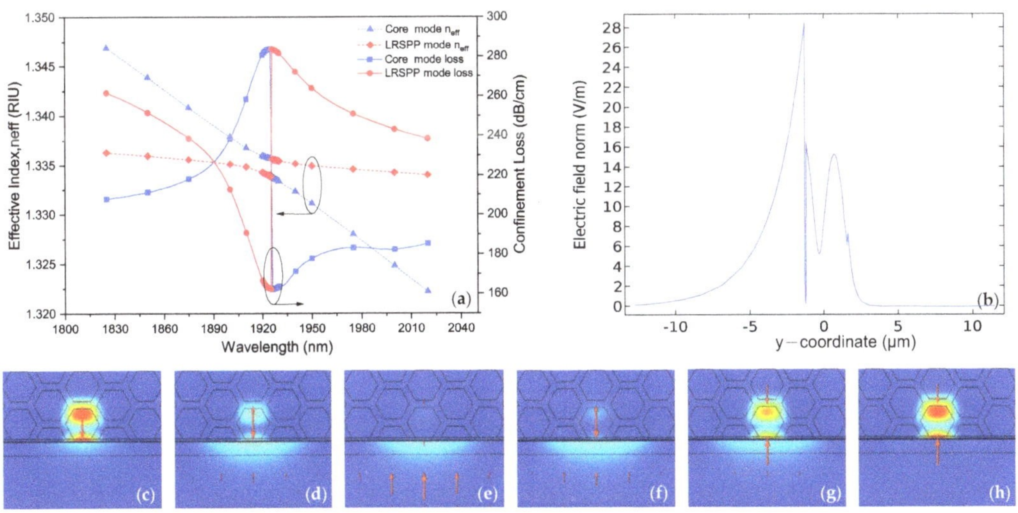

Figure 2. (**a**) Dispersion relationship (solid lines) and attenuation spectra (dot lines) of the core mode (blue), LRSPP mode (red). (**b**) The electric field profile along y axis of the LRSPP mode at 1900 nm. (**c**–**e**) and (**f**–**h**) are the electric field distributions of the core mode and LRSPP mode at the wavelength of 1900 nm, 1925 nm and 1950 nm, respectively. They illustrate the mode conversion around the avoided-crossing region. The red arrows in (**c**–**h**) denote the directions of electric field.

Based on the wavelength interrogation method, the performance of the sensor can be evaluated by the "so-called" wavelength sensitivity S_λ [40]:

$$S_\lambda = \frac{\Delta\lambda_{\text{peak}}}{\Delta n_{\text{a}}}, \ (\text{nm/RIU}), \tag{5}$$

where $\Delta\lambda_{\text{peak}}$ is the resonant wavelength shift and Δn_{a} is the variation of analyte RI.

Additionally, the resolution (R) and figure of merit (FOM) are widely used to further evaluate the performance of SPR-based fiber sensors. They are respectively defined as follows [41,42]:

$$R = \frac{\Delta\lambda_{\text{min}}}{S_\lambda}, \ (\text{RIU}) \tag{6}$$

$$\text{FOM} = \frac{S_\lambda}{\text{FWHM}}, \ (\text{RIU}^{-1}), \tag{7}$$

where $\Delta\lambda_{\text{min}}$ for the minimum detectable wavelength of an OSA. It is usually set to 0.01 nm [42]; FWHM is the abbreviation of full width half maximum.

3. Discussion

By tracking the change of the LRSPR peak, the influence of the structural parameters on the prosed sensor is analyzed in detail as follows. In order to maintain consistency, except for the discussed parameter, other parameters are kept the same as those in the example model in Section 2.

3.1. Influence of MOF's Pitch on LRSPR

Firstly, the pitch Λ of the D-shaped MOF is increased from 2000 nm to 2600 nm with the step of 200 nm, while the other parameters remain from the initial model. As shown in Figure 3a, with the increases of pitch, the resonant wavelength shifts towards the longer wavelengths. Actually, the core size is mainly determined by the pitch; the increase of pitch is equivalent to the increase of the core diameter. Hence, the n_{eff} of the core mode at each wavelength decreases accordingly, while the corresponding n_{eff} of the LRSPP mode is hardly affected and still around the value of the analyte's RI. This means that the phase-matching condition can only be fulfilled at the longer wavelength if the MOF's pitch is increasing. By extracting the resonant wavelengths under different pitches and then fitting linearly, as shown in Figure 3b, the dependence of the resonant wavelength on the MOF's pitch is highly linear.

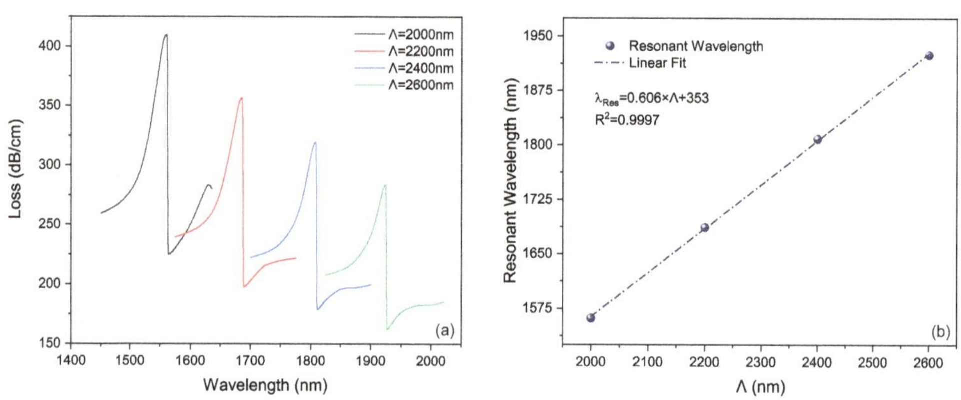

Figure 3. (**a**) The variation of loss spectrum with respect to different pitches of honeycomb structure cladding (d = 500 nm, t_{SiO_2} = 150 nm, t_{Au} = 70 nm, t_{TiO_2} = 3 nm, n_{a} = 1.33). (**b**) The resonant wavelength and the corresponding linear fitting under different pitches of honeycomb structure cladding.

3.2. Influence of Silica-Web Thickness on LRSPR

Secondly, how the LRSPR peak wavelength is affected by the silica-web thickness d is evaluated by varying the thickness from 200 nm to 600 nm with a step of 100 nm. With the increment of silica-web thickness, the resonant wavelength shifts to the longer wavelength as shown in Figure 4a. Similar to the pitch, the influence of silica-web thickness on the resonant wavelength is also related to the fiber's core size. As silica-web thickness increases, the core size will increase accordingly, resulting in the red shift of the LRSPR peak. The tendency of the resonant wavelength with respect to the silica-web thickness is almost linear, as shown in Figure 4b.

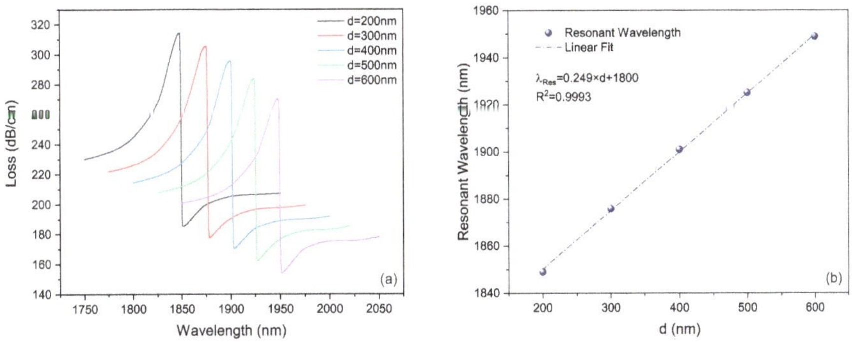

Figure 4. (**a**) The variation of loss spectrum with respect to the change of the silica-web thickness ($\Lambda = 2600$ nm, $t_{\mathrm{SiO_2}} = 150$ nm, $t_{\mathrm{Au}} = 70$ nm, $t_{\mathrm{TiO_2}} = 3$ nm, $n_a = 1.33$); (**b**) The dependence between the resonant wavelength and the thickness of the silica web.

3.3. Influence of Planar-Silica Thickness on LRSPR

In our proposed model, the planar silica layer acts not only as the flat plane for depositing gold film, but also contributes to the LRSPR excitation. Figure 5a shows that the resonant peak shifts to the shorter wavelength as the thickness of planar-layer silica decreases from 250 nm to 100 nm with a step of 50 nm. The wavelength of the resonant peak shows a linear relationship with respect to the thickness of the planar-layer silica, as shown in Figure 5b. The impact of the planar-silica thickness on the resonance can be attributed to the variation of the MOF's core size as well. In other words, when the thickness of the planar silica is decreased but the MOF's pitch and silica-web thickness are fixed, the core size will still decrease accordingly, causing the blue shift of the LRSPR peak. Moreover, the FWHM becomes larger as the thickness of the silica layer decreases, which means the FWHM can be tuned by changing the thickness of the planar-layer silica.

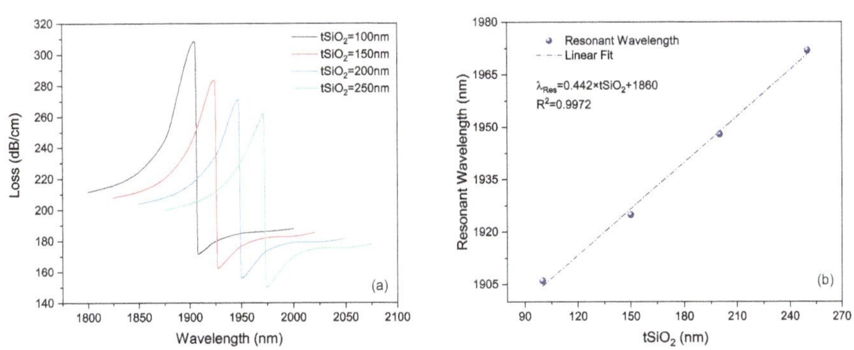

Figure 5. (**a**) The variation of loss spectrum with respect to the change of the thickness of planar-layer silica ($\Lambda = 2600$ nm, $d = 500$ nm, $t_{\mathrm{Au}} = 70$ nm, $t_{\mathrm{TiO_2}} = 3$ nm, $n_a = 1.33$); (**b**) The relationship between the resonant wavelength and the thickness of the planar-layer silica.

3.4. Comparison between SiO_2-Au-TiO_2 and SiO_2-TiO_2-Au Heterostructures

In order to figure out how a thin-layer dielectric affects the LRSPR resonance, we designed the device with different heterostructures: SiO_2-Au-TiO_2 and SiO_2-TiO_2-Au. They are respectively denoted as S1 and S2 in short. In the first design, that is, S1, the Au film is deposited on the surface of the planar-layer silica and then coated with a layer of TiO_2 to protect the metal film. In S2, on the contrary, the planar-layer silica of D-shaped fiber is first coated with the TiO_2 layer and then plated with the gold film. Here, the TiO_2 layer is considered an adhesive layer between the waveguide and gold film to enhance the attachment of the metal film on waveguide surface. For the convenience of comparison, the parameters of the waveguide are fixed as follows: $\Lambda = 2600$ nm, $d = 500$ nm, $t_{SiO_2} = 150$ nm.

As shown in Figure 6a,c, both these heterostructures can be used to excite LRSPR, but the resonant wavelengths in S1 and S2 show a different tendency. By increasing the thickness of the TiO_2 layer from 0 to 11 nm, the resonant wavelength of the LRSPR sensor made of the S1 structure shifts towards the shorter wavelengths as shown in Figure 6b, whereas the LRSPR peak wavelength of the S2 configuration moves to the longer wavelength, as shown in Figure 6d, when the TiO_2 layer is becoming thicker. Based on effective medium theory, the TiO_2 layer can be treated as a part of a waveguide in the S2 configuration. It will help to increase the n_{eff} of the core mode since the material RI of TiO_2 is higher than that of SiO_2 for the same wavelength. This indicates that the resonance occurs at longer wavelengths as the thickness of the TiO_2 layer increases. On the contrary, the contribution of the TiO_2 layer lies in increasing the n_{eff} of LRSPP mode in the S1 heterostructure, pushing the phase matching towards the shorter wavelength. Moreover, the amplitude of the resonant peak in the S1 structure is enhanced as well with the assistance of the TiO_2 layer.

Figure 6e,g illustrates the influence of the Au-film thickness on the LRSPR in two heterostructures. As shown, the resonant peaks in both S1 and S2 configurations shift very slightly when the thickness of Au film is increased from 50 nm to 80 nm. Their corresponding dependence of the peak wavelength is no more than 1 nm when the thickness of the Au film varies 10 nm. From the device fabrication point of view, the requirement of the coating process for the proposed sensor will not be as critical as it is for conventional SPR devices. Meanwhile, as the the thickness of the Au film increases, the resonance coupling between the upper and lower parts of the gold layer is weakened, leading to the decrement of the FWHM and amplitude in both heterostructures.

A further comparison of sensing performance between S1 and S2 heterostructures is addressed by following the variation of their resonant peak wavelengths under different analytes. The thicknesses of the Au film and the TiO_2 layer are set to 70 nm and 3 nm, respectively. By increasing the analyte's RI from 1.33 to 1.39, as shown in Figure 7a, the resonant peaks of both heterostructures undergo blue shifts. However, it is apparent that the wavelength shifts in S1 are larger than those in S2. This indicates that the sensor based on the S1 heterostructure has a higher sensitivity to ambient RI than does the S2 based sensor. Their average sensitivities within the RI range from 1.33 to 1.39, as the linearly fitting curves show in Figure 7b, are about 11,586 nm/RIU and 10,639 nm/RIU, respectively. The corresponding average resolutions are about 8.63×10^{-7} RIU and 9.40×10^{-7} RIU, respectively, based on Equation (6).

Additionally, their wavelength sensitivities are recalculated piecewise based on Equation (5), as the bar chart shown in Figure 8a. Both of them increase monotonically for higher RI interval. The maximum sensitivity reaches 20,100 nm/RIU for S1 based sensor, while that of S2 based sensor is about 15,500 nm/RIU. The corresponding minimum resolutions are about 4.98×10^{-7} RIU and 6.45×10^{-7} RIU, respectively. The LRSPR sensor based on SiO_2-Au-TiO_2 heterostructure is found with a superior performance as a RI sensor. Its FWHM and FOM are calculated and plotted in Figure 8b. In the Figure, the FWHM is defined as the average of the FWHMs of adjacent two resonant peaks as shown in red curve and the corresponding FOMs are drawn as well and shown as the blue curve. The highest FOM is of 374 RIU^{-1}, achieved in the RI range of 1.33–1.34.

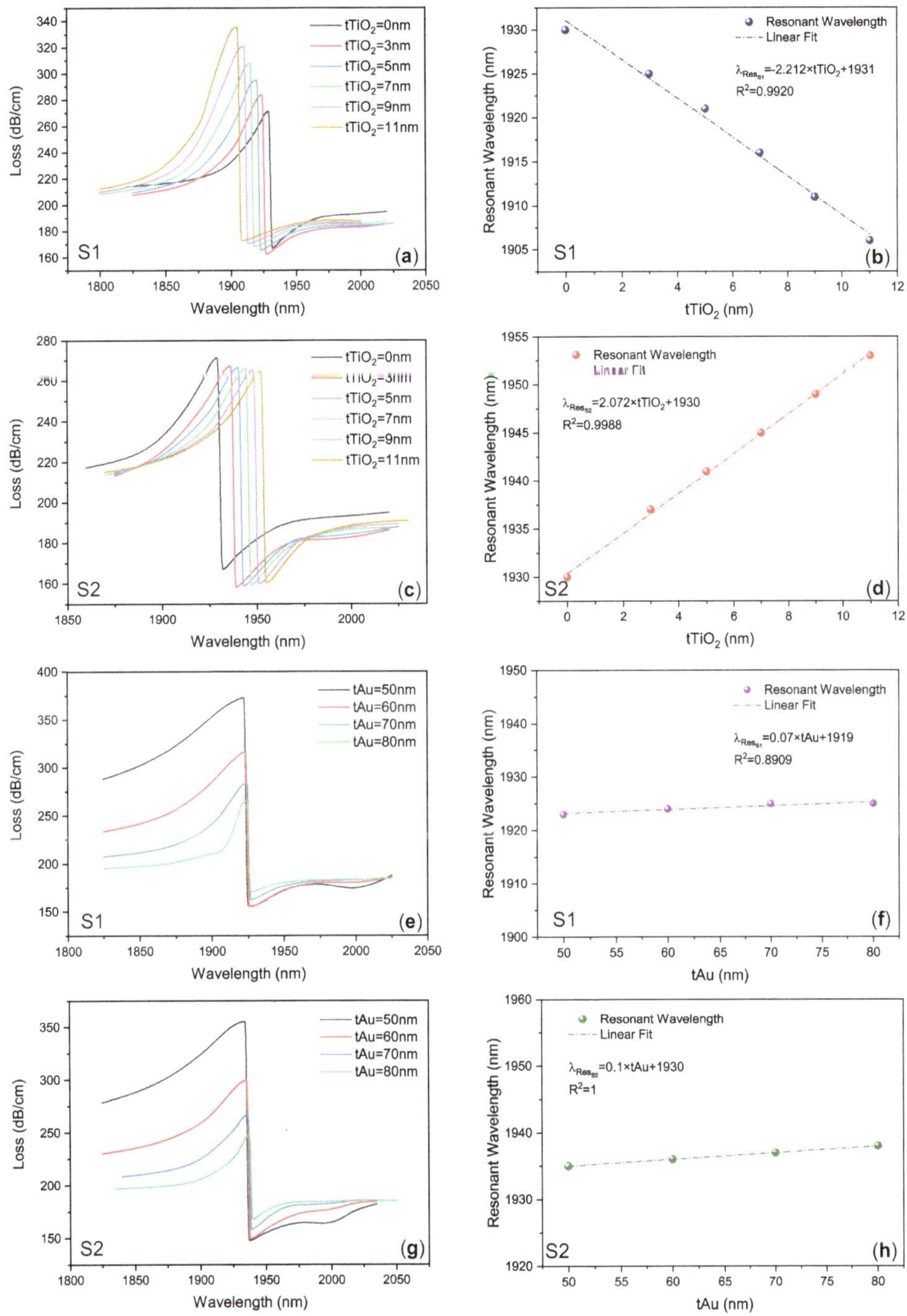

Figure 6. (**a–d**) The variation of loss spectrum and the corresponding peak wavelength with respect to the change of the thickness of TiO_2 film in S1 and S2, respectively ($\Lambda = 2600$ nm, $t_{SiO_2} = 150$ nm, $d = 500$ nm, $t_{Au} = 70$ nm, $n_a = 1.33$). (**e–h**) The variation of loss spectrum and the corresponding peak wavelength with respect to the change of the thickness of gold film in S1 and S2, respectively ($\Lambda = 2600$ nm, $t_{SiO_2} = 150$ nm, $d = 500$ nm, $t_{TiO_2} = 3$ nm, $n_a = 1.33$).

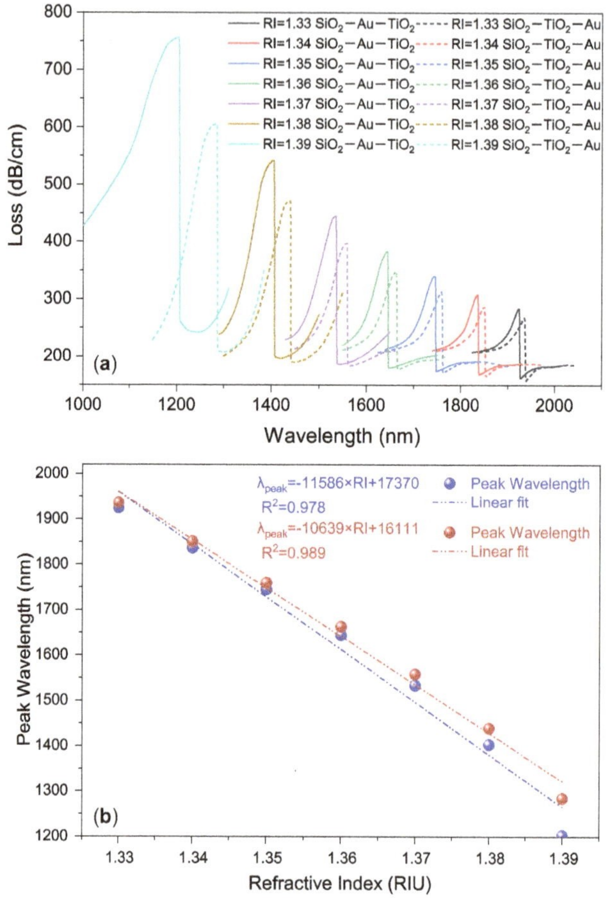

Figure 7. (**a**) The variation of loss spectrum of two heterostructures with respect to the change of the analyte's RI. (**b**) The wavelength sensitivities to analyte's RI for S1 and S2 heterostructures.

3.5. Sensing Performance Comparison with Previous LRSPR-Based Fiber Sensors

Finally, the performance of the proposed LRSPR sensor is compared with those recently-reported LRSPR-based fiber sensors. As shown in Table 1, the LRSPR sensor proposed in this work exhibits a very competitive performance. It has advantages such as higher sensitivity, higher FOM and lower resolution.

Table 1. Comparison of the performance of the proposed sensor with recently reported LRSPR sensors.

Description	RI Range	S_λ (nm/RIU)	Resolution (RIU)	FOM (RIU^{-1})	Ref.
GK570/silver in hollow fiber	1.4772–1.5116	12,500	0.8×10^{-5}	150	[43] (Experiment)
Liquid-core PTFE-based POF 3	1.325–1.35	16,750	5.97×10^{-6}	310.6	[4] (Simulation)
H-shaped MOF with DMDW	1.33–1.39	7540	1.3×10^{-5}	280	[44] (Simulation)
D-type LRSPR sensor with high Q	1.33–1.38	3500.6	2.86×10^{-7}	107.52	[45] (Experiment)
Dual-channel fiber with Au/TiO_2	1.39–1.43	13,200	7.58×10^{-6}	363	[28] (Simulation)
Gold-and MgF_2-Coated RI sensor	1.32–1.38	27,959	3.70×10^{-5}	–	[46] (Simulation)
This work	1.33–1.39	20,100	4.98×10^{-7}	374	(Simulation)

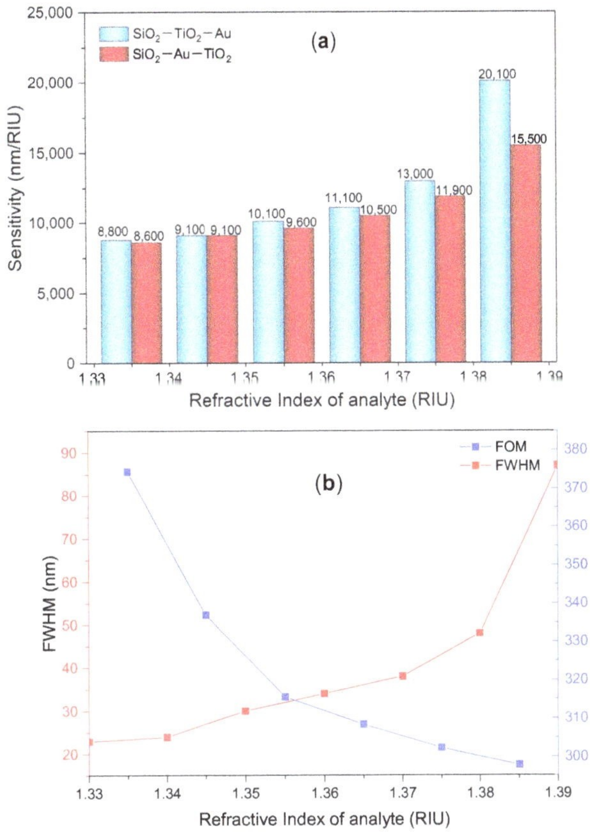

Figure 8. (**a**) The wavelength sensitivities of S1 and S2 heterostructures in each step of analyte's RI; (**b**) The calculated FWHWs and FOMs of S1 and S2 in each step of analyte's RI.

4. Conclusions

In summary, an LRSPR sensor based on a SiO_2-Au-TiO_2 heterostructure on a D-shaped MOF is proposed and numerically investigated. The specialty D-shaped silica MOF is designed with a honeycomb structure, which is arranged with hexagonal air holes and a solid silica core. By means of advanced fiber fabrication and post-processing techniques, part of the cladding is removed and then a plane silica surface is generated due to surface tension during the fiber drawing. Then, the planar surface of the D-shaped MOF is coated in a layer of Au film followed by a thin-layer TiO_2 film, forming the SiO_2-Au-TiO_2 heterostructure. In this heterostructure, a strong LRSPR is achieved when the n_{eff} of the fundamental core mode is comparable with the material RI of the analyte. Based on a series of detailed discussions about the structural parameters, the resonant wavelength is highly related to the MOF's core size but is affected trivially by the thickness of the Au film. Furthermore, through the comparison between SiO_2-Au-TiO_2 and SiO_2-TiO_2-Au heterostructures, the former shows higher wavelength sensitivity with the analyte's RI varying from 1.33 to 1.39. Its maximum and average sensitivities reach 20,100 nm/RIU and 11,586 nm/RIU, respectively. The corresponding minimum resolution is as low as 4.98×10^{-7} RIU. Finally, compared with the LRSPR based RI sensor reported recently, our design is characterized by high sensitivity and low resolution. Therefore, the proposed LRSPR sensor might find significant potential for high-sensitivity detection in biomedicine and biochemical fields.

Author Contributions: Conceptualization, Z.W. and W.B.; methodology, Z.W.; software, X.S. and P.P.S.; formal analysis, Z.W. and W.B.; investigation, Z.W. and W.B.; resources, X.S., P.P.S. and J.P.; writing—original draft preparation, W.B. and Z.W.; writing—review and editing, Z.W. and W.B.; supervision, Z.W., X.S., P.P.S. and J.P.; project administration, Z.W.; funding acquisition, Z.W. and J.P. All authors have read and agreed to the published version of the manuscript.

Funding: This work is partially supported by National Natural Science Foundation of China (11774102), the Scientific Research Funds and Promotion Program for Young and Middle-aged Teacher in Science and Technology Research of Huaqiao University (ZQN-YX504, 17BS412).

Institutional Review Board Statement: Not applicable.

Informed Consent Statement: Not applicable.

Conflicts of Interest: The authors declare no conflict of interest.

References

1. Rahman, M.S.; Anower, M.S.; Hasan, M.R.; Hossain, M.B.; Haque, M.I. Design and numerical analysis of highly sensitive Au-MoS_2-graphene based hybrid surface plasmon resonance biosensor. *Opt. Commun.* **2017**, *396*, 36–43. [CrossRef]
2. Noman, A.A.; Haque, E.; Hossain, M.A.; Hai, N.H.; Namihira, Y.; Ahmed, F. Sensitivity enhancement of modified D-shaped microchannel PCF-based surface plasmon resonance sensor. *Sensors* **2020**, *20*, 6049. [CrossRef] [PubMed]
3. Rifat, A.A.; Ahmed, R.; Mahdiraji, G.A.; Adikan, F.R.M. Highly sensitive D-shaped photonic crystal fiber-based plasmonic, biosensor in visible to Near-IR. *IEEE Sens. J.* **2017**, *17*, 2776–2783. [CrossRef]
4. Shui, X.; Gu, Q.; Jiang, X.; Si, G. Surface plasmon resonance sensor based on polymer liquid-core fiber for refractive index detection. *Photonics* **2020**, *7*, 123. [CrossRef]
5. Zynio, S.; Samoylov, A.; Surovtseva, E.; Mirsky, V.; Shirshov, Y. Bimetallic layers increase sensitivity of affinity sensors based on surface plasmon resonance. *Sensors* **2002**, *2*, 62–70. [CrossRef]
6. Kashyap, R.; Chakraborty, S.; Zeng, S.; Swarnakar, S.; Kaur, S.; Doley, R.; Mondal, B. Enhanced Biosensing activity of bimetallic surface plasmon resonance sensor. *Photonics* **2019**, *6*, 108. [CrossRef]
7. Ong, B.H.; Yuan, X.; Tjin, S.C.; Zhang, J.; Ng, H.M. Optimised film thickness for maximum evanescent field enhancement of a bimetallic film surface plasmon resonance biosensor. *Sens. Actuators B Chem.* **2006**, *114*, 1028–1034. [CrossRef]
8. Liu, C.; Wang, J.; Wang, F.; Su, W.; Yang, L.; Lv, J.; Fu, G.; Li, X.; Liu, Q.; Sun, T.; et al. Surface plasmon resonance (SPR) infrared sensor based on D-shape photonic crystal fibers with ITO coatings. *Opt. Commun.* **2020**, *464*, 125496. [CrossRef]
9. Gonzalez-Valencia, E.; Villar, I.D.; Torres, P. Novel Bloch wave excitation platform based on few-layer photonic crystal deposited on D-shaped optical fiber. *Sci. Rep.* **2021**, *11*, 11266. [CrossRef]
10. Shafkat, A.; Rashed, A.N.Z.; El-Hageen, H.M.; Alatwi, A.M. The effects of adding different adhesive layers with a micro-structure fiber sensor based on surface plasmon resonance: A Numerical Study. *Plasmonics* **2021**, *16*, 819–832. [CrossRef]
11. Wang, X.; Wang, Q.; Song, Z.; Qi, K. Simulation of a microstructure fiber pressure sensor based on lossy mode resonance. *AIP Adv.* **2019**, *9*, 095005.
12. Takagi, K.; Watanabe, K. Near infrared characterization of hetero-core optical fiber SPR sensors coated with Ta_2O_5 film and their applications. *Sensors* **2012**, *12*, 2208–2218. [CrossRef] [PubMed]
13. Sharma, A.K.; Dominic, A.; Kaur, B.; Popescu, V.A. Fluoride fiber sensor with huge performance enhancement via optimum radiative damping at Ag–Al_2O_3–Graphene heterojunction on silicon. *J. Light. Technol.* **2019**, *37*, 5641–5646. [CrossRef]
14. Azad, S.; Sadeghi, E.; Parvizi, R.; Mazaheri, A.; Yousefi, M. Sensitivity optimization of ZnO clad-modified optical fiber humidity sensor by means of tuning the optical fiber waist diameter. *Opt. Laser Technol.* **2017**, *90*, 96–101. [CrossRef]
15. Hosoki, A.; Nishiyama, M.; Igawa, H.; Seki, A.; Choi, Y.; Watanabe, K. A surface plasmon resonance hydrogen sensor using Au/Ta_2O_5/Pd multi-layers on hetero-core optical fiber structures. *Sens. Actuators B Chem.* **2013**, *185*, 53–58. [CrossRef]
16. Del Villar, I.; Zamarreno, C.R.; Hernaez, M.; Arregui, F.J.; Matias, I.R. Lossy mode resonance generation with indium-tin-oxide-coated optical fibers for densing applications. *J. Light. Technol.* **2010**, *28*, 111–117. [CrossRef]
17. Kaur, D.; Sharma, V.K.; Kapoor, A. High sensitivity lossy mode resonance sensors. *Sens. Actuators B Chem.* **2014**, *198*, 366–376. [CrossRef]
18. Paul, A.K.; Mollah, M.A.; Hassan, M.Z.; Gomez-Cardona, N.; Reyes-Vera, E. Graphene-coated highly sensitive photonic crystal fiber surface plasmon resonance sensor for aqueous solution: Design and numerical analysis. *Photonics* **2021**, *8*, 155. [CrossRef]
19. Zeng, S.; Hu, S.; Xia, J.; Anderson, T.; Dinh, X.; Meng, X.; Coquet, P.; Yong, K. Graphene–MoS_2 hybrid nanostructures enhanced surface plasmon resonance biosensors. *Sens. Actuators B Chem.* **2015**, *207*, 801–810. [CrossRef]
20. Sharma, A.K.; Kaur, B.; Popescu, V.A. On the role of different 2D materials/heterostructures in fiber-optic SPR humidity sensor in visible spectral region. *Opt. Mater.* **2020**, *102*, 109824. [CrossRef]
21. Berini, P. Long-range surface plasmon polaritons. *Adv. Opt. Photonics* **2009**, *1*, 484–588. [CrossRef]
22. Jing, J.; Wang, Q.; Zhao, W.; Wang, B. Long-range surface plasmon resonance and its sensing applications: A review. *Opt. Laser Eng.* **2019**, *112*, 103–118. [CrossRef]

23. Liu, F.; Li, Y.; Wan, R.; Huang, Y.; Feng, X.; Zhang, W. Hybrid coupling between long-range surface plasmon polariton mode and dielectric waveguide mode. *J. Light. Technol.* **2011**, *29*, 1265–1273. [CrossRef]
24. Berini, P. Plasmon-polariton modes guided by a metal film of finite width bounded by different dielectrics. *Opt. Express* **2000**, *7*, 329–335. [CrossRef] [PubMed]
25. Vala, M.; Robelek, R.; Bocková, M.; Wegener, J.; Homola, J. Real-time label-free monitoring of the cellular response to osmotic stress using conventional and long-range surface plasmons. *Biosens. Bioelectron.* **2013**, *40*, 417–421. [CrossRef]
26. Vala, M.; Etheridge, S.; Roach, J.A.; Homola, J. Long-range surface plasmons for sensitive detection of bacterial analytes. *Sens. Actuators B Chem.* **2009**, *139*, 59–63. [CrossRef]
27. Chen, X.; Bu, W.; Wu, Z.; Zhang, H.; Shum, P.P.; Shao, X.; Pu, J. Near-infrared long-range surface plasmon resonance in a D-shaped honeycomb microstructured optical fiber coated with Au film. *Opt. Express* **2021**, *29*, 16455–16468. [CrossRef]
28. Wang, H.; Rao, W.; Luo, J.; Fu, H. A dual-channel surface plasmon resonance sensor based on dual-polarized photonic crystal fiber for ultra-wide range and high sensitivity of refractive index detection. *IEEE Photonics J.* **2021**, *13*, 1–11.
29. Hossain, M.A.; Namihira, Y.; Islam, M.A.; Razzak, S.M.A.; Hirako, Y.; Miyagi, K.; Kaijage, S.F.; Higa, H. Tailoring supercontinuum generation using highly nonlinear photonic crystal fiber. *Opt. Laser Technol.* **2012**, *44*, 1889–1896. [CrossRef]
30. Sehmi, H.S.; Langbein, W.; Muljarov, E.A. Optimizing the Drude-Lorentz model for material permittivity: Method, program, and examples for gold, silver, and copper. *Phys. Rev. B* **2017**, *95*, 115444. [CrossRef]
31. Vial, A.; Grimault, A.; Macias, D.; Barchiesi, D.; De la Chapelle, M.L. Improved analytical fit of gold dispersion: Application to the modeling of extinction spectra with a finite-difference time-domain method. *Phys. Rev. B* **2005**, *71*, 85411–85416. [CrossRef]
32. Zhang, N.; Humbert, G.; Gong, T.; Shum, P.P.; Li, K.; Auguste, J.; Wu, Z.; Hu, D.J.J.; Luan, F.; Dinh, Q.X.; et al. Side-channel photonic crystal fiber for surface enhanced Raman scattering sensing. *Sens. Actuators B Chem.* **2016**, *223*, 195–201. [CrossRef]
33. Zhang, N.M.Y.; Li, K.; Shum, P.P.; Yu, X.; Zeng, S.; Wu, Z.; Wang, Q.J.; Yong, K.T.; Wei, L. Hybrid graphene/gold plasmonic fiber-optic biosensor. *Adv. Mater. Technol.* **2017**, *2*, 1600185. [CrossRef]
34. Wu, T.; Shao, Y.; Wang, Y.; Cao, S.; Cao, W.; Zhang, F.; Liao, C.; He, J.; Huang, Y.; Huang, M.; et al. Surface plasmon resonance biosensor based on gold-coated side-polished hexagonal structure photonic crystal fiber. *Opt. Express* **2017**, *25*, 20313–20322. [CrossRef]
35. Xie, Q.; Chen, Y.; Li, X.; Yin, Z.; Wang, L.; Geng, Y.; Hong, X. Characteristics of D-shaped photonic crystal fiber surface plasmon resonance sensors with different side-polished lengths. *Appl. Opt.* **2017**, *56*, 1550–1555. [CrossRef]
36. Zhang, N.M.Y.; Hu, D.J.J.; Shum, P.P.; Wu, Z.; Li, K.; Huang, T.; Wei, L. Design and analysis of surface plasmon resonance sensor based on high-birefringent microstructured optical fiber. *J. Opt.* **2016**, *18*, 65005. [CrossRef]
37. Klantsataya, E.; François, A.; Ebendorff-Heidepriem, H.; Hoffmann, P.; Monro, T. Surface plasmon scattering in exposed core optical fiber for enhanced resolution refractive index sensing. *Sensors* **2015**, *15*, 25090–25102. [CrossRef]
38. Morales-Luna, G.; Herrera-Domínguez, M.; Pisano, E.; Balderas-Elizalde, A.; Hernandez-Aranda, R.I.; Ornelas-Soto, N. Plasmonic biosensor based on an effective medium theory as a simple tool to predict and analyze refractive index changes. *Opt. Laser Technol.* **2020**, *1314*, 106332. [CrossRef]
39. Tuniz, A.; Wieduwilt, T.; Schmidt, M.A. Tuning the effective PT phase of plasmonic eigenmodes. *Phys. Rev. Lett.* **2019**, *123*, 213903. [CrossRef] [PubMed]
40. Haque, E.; Hossain, M.A.; Ahmed, F.; Namihira, Y. Surface plasmon resonance sensor based on modified D-shaped photonic crystal fiber for wider range of refractive index detection. *IEEE Sens. J.* **2018**, *18*, 8287–8293. [CrossRef]
41. Dong, J.; Zhang, Y.; Wang, Y.; Yang, F.; Hu, S.; Chen, Y.; Zhu, W.; Qiu, W.; Guan, H.; Lu, H.; et al. Side-polished few-mode fiber based surface plasmon resonance biosensor. *Opt. Express* **2019**, *27*, 11348–11360. [CrossRef] [PubMed]
42. Momota, M.R.; Hasan, M.R. Hollow-core silver coated photonic crystal fiber plasmonic sensor. *Opt. Mater.* **2018**, *76*, 287–294. [CrossRef]
43. Zhao, X.; Zhang, X.; Zhu, X.; Shi, Y. Long-range surface plasmon resonance sensor based on the GK570/Ag coated hollow fiber with an asymmetric layer structure. *Opt. Express* **2019**, *27*, 9550–9560. [CrossRef]
44. Gomez-Cardona, N.; Reyes-Vera, E.; Torres, P. High sensitivity refractive index sensor based on the excitation of long-range surface plasmon polaritons in H-shaped optical fiber. *Sensors* **2020**, *20*, 2111. [CrossRef]
45. Jing, J.; Li, S.; Wang, X.; Zhu, Q.; Meng, F.; Wang, Q. A D-type fiber based symmetrical long-range surface plasmon resonance sensor with high quality factor. *Measurement* **2019**, *140*, 395–406. [CrossRef]
46. Ahmed, K.; AlZain, M.; Abdullah, H.; Luo, Y.; Vigneswaran, D.; Faragallah, O.; Eid, M.; Rashed, A. Highly sensitive twin resonance coupling refractive index sensor based on gold- and MgF_2-coated nano metal films. *Biosensors* **2021**, *11*, 104. [CrossRef] [PubMed]

photonics

Communication

Nanograting-Enhanced Optical Fibers for Visible and Infrared Light Collection at Large Input Angles

Ning Wang [1,2,*], Matthias Zeisberger [3], Uwe Hübner [3] and Markus A. Schmidt [3,4,5,*]

1 Institute of Advanced Technology, Westlake Institute for Advanced Study, 18 Shilongshan Road, Hangzhou 310024, China
2 School of Engineering, Westlake University, 18 Shilongshan Road, Hangzhou 310024, China
3 Leibniz Institute of Photonic Technology, Albert-Einstein-Str. 9, 07745 Jena, Germany; matthias.zeisberger@leibniz-ipht.de (M.Z.); uwe.huebner@leibniz-ipht.de (U.H.)
4 Faculty of Physics, Abbe School of Photonics, Max-Wien-Platz 1, 07743 Jena, Germany
5 Otto Schott Institute of Materials Research, Fraunhoferstr. 6, 07743 Jena, Germany
* Correspondence: wangning@westlake.edu.cn (N.W.); markus.schmidt@leibniz-ipht.de (M.A.S.)

Abstract: The efficient incoupling of light into particular fibers at large angles is essential for a multitude of applications; however, this is difficult to achieve with commonly used fibers due to low numerical aperture. Here, we demonstrate that commonly used optical fibers functionalized with arrays of metallic nanodots show substantially improved large-angle light-collection performances at multiple wavelengths. In particular, we show that at visible wavelengths, higher diffraction orders contribute significantly to the light-coupling efficiency, independent of the incident polarization, with a dominant excitation of the fundamental mode. The experimental observation is confirmed by an analytical model, which directly suggests further improvement in incoupling efficiency through the use of powerful nanostructures such as metasurface or dielectric gratings. Therefore, our concept paves the way for high-performance fiber-based optical devices and is particularly relevant within the context of endoscopic-type applications in life science and light collection within quantum technology.

Keywords: fiber optics; plasmonic gratings; nanodot; light collection

Citation: Wang, N.; Zeisberger, M.; Hübner, U.; Schmidt, M.A. Nanograting-Enhanced Optical Fibers for Visible and Infrared Light Collection at Large Input Angles. *Photonics* **2021**, *8*, 295. https://doi.org/10.3390/photonics8080295

Received: 2 June 2021
Accepted: 21 July 2021
Published: 24 July 2021

1. Introduction

Over the past half century, optical fiber has revolutionized modern telecommunication technology and industry from various perspectives [1–3]. Up to today, optical fiber research still represents a major scientific field within the optics community with applications ranging from highly efficient bio-sensor devices [4–7] to broadband light sources [8–10]. However, suffering from a small numerical aperture (e.g., the NA of single-mode fiber 28 (SMF-28) is only 0.14 at λ = 1550 nm), the light collection efficiency of commonly used step-index fibers is insufficient for many applications and cannot satisfy the demands of emerging applications such as wide-field endoscopes [11] or quantum photon probes [12]. To address this challenge, we have recently introduced a universal approach for improving the light-incoupling ability of optical fibers through integrating plasmonic nanodot arrays [13] and dielectric concentric rings [14] on end faces of SMF-28. Through exploiting different diffraction orders, this leads to significant improvements in coupling efficiency, especially at very large angles at infrared wavelengths. In the present work, we extend the spectral range of this concept and experimentally demonstrate a significant improvement of light-coupling efficiencies across a wide incident angle interval ($30° < \theta < 85°$) at selected visible and infrared wavelengths. We specifically choose 550 nm as one of the excitation wavelengths to investigate the coupling concept at such a short wavelength, whereas the measurements at 1650 nm serve as a reference. Note that, theoretically, at any individual wavelength within the spectral range (i.e., from 550 nm to 1650 nm), the coupling efficiencies can be improved by our approach.

The structure targeted here is schematically illustrated in Figure 1a. Here, an SMF-28 fiber tip functionalized by an array of hexagonally arranged gold nanodots is excited by either a p- or a s- polarized beam (indicated by a green arrow) under an incident angle of θ (wavelength λ). The grating deflects a certain amount of transmitted power to a defined diffraction order (for example, the -1st order is illustrated in Figure 1a by a dark red arrow). The diffracted light couples to the fiber modes, leading to an enhancement of the light-coupling efficiencies at angles that otherwise show little coupling efficiency in case of a fiber with an unstructured interface. The incoupling efficiency η is defined here as $\eta(\theta) = P_{out}/P_{in}$, where P_{out} and P_{in} are the fiber output and input powers, respectively (see labels 'P_{out}' and 'P_{in}' in Figure 1a). Note that, in the following, we use the normalized incoupling efficiency defined as $\eta_{norm}(\theta) = \eta(\theta)/\eta(\theta = 0°)$ to clearly highlight the increases in value enabled by the nanostructures.

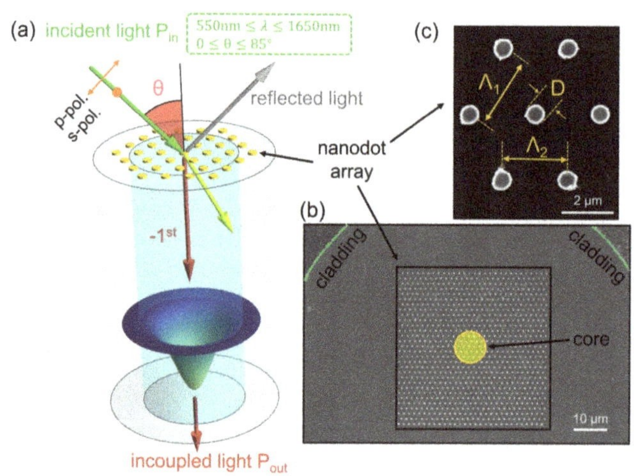

Figure 1. A schematic (**a**) illustrating the concept of the nanoarray-assisted fiber-based light collection at incoupling angles of θ. (**b**,**c**): The scanning electron microscopy (SEM) images of gold nanodot array (D: 480 nm, Λ_1: 1.9 µm, Λ_2: 1.7 µm) located at the core section of an SMF-28. In (**b**), the orange shadow denotes the fiber mode area.

2. Theoretical Toy Model Analysis

For understanding the light-coupling process from a semi-quantitative perspective, we utilize a strongly simplified one-dimensional scalar model [15] to describe the coupling efficiency/incident angle dependence $\eta = \eta(\theta)$ at a fixed wavelength. The model is based on the electric field integral between the fundamental fiber mode (HE_{11}-mode) and the exciting wave just below the nanostructure that is partially generated by the nanodot grating (the model is depicted in Figure 2a). In the case of SMF-28, the HE_{11} mode can be approximated by a Gaussian profile ($E \approx exp(\frac{x^2}{w^2})$) due to the weak guidance approximation, and the excitation field is assumed to be a plane wave (wavelength λ, wavenumber $k_0 = 2\pi/\lambda$). Note that, as mentioned above for theoretical analysis, we normalize the coupling efficiency to its value at perpendicular incidence $\eta_{norm}(\theta) = \eta(\theta)/\eta(\theta = 0°)$. As the first step, we defined the coupling strength $C_b(\theta)$ parameter to characterize the coupling into a fiber with no nanostructure:

$$C_b(\theta) \sim \cos\theta \left| \int_{-\infty}^{\infty} e^{\frac{-x^2}{w^2}} e^{ik_x x} dx \right|^2 \quad (1)$$

where k_x is the projection of wavenumber k_0 along the x-axis ($k_x = k_0 \sin\theta$). The normalized coupling efficiency of the bare fiber $\eta_b(\theta)$ is then defined as $\eta_b(\theta) = C_b(\theta)/C_b(0)$.

By placing a plasmonic nanodot array (interdot distance (pitch) Λ) on the fiber facet, the corresponding nth diffraction order modifies the projection of the wave vector on the x-axis, and thus replaces k_x by $(k_x + nG)$ (lattice vector $G = \frac{2\pi}{\Lambda}$). Therefore, the coupling strength $C_g(\theta)$ is changed to:

$$C_g(\theta) \sim \cos\theta \left| \int_{-\infty}^{\infty} e^{\frac{-x^2}{w^2}} \sum_n a_n e^{i(k_x+nG)x} dx \right|^2 \quad (2)$$

Here, the coefficients a_n represent the relative amplitudes of the diffracted waves. The summation covers all possible diffraction orders in the medium below the grating including the non-diffracted wave (n = 0). As a result of this modification, the nanograting-enhanced coupling efficiency follows the equation $\eta_g(\theta) = C_g(\theta)/C_g(0)$. Note that, for the shortest wavelength used in this work (λ = 550 nm), both -1st and -2nd orders (i.e., $n = -1$ or $n = -2$) contribute to efficiency enhancement, while for λ = 1650 nm only the -1st order (i.e., $n = -1$) is relevant. In addition to the two mentioned wavelengths, the toy model suggests that nanograting enhanced light-incoupling concept should principally work for any wavelength within the demonstrated spectral range.

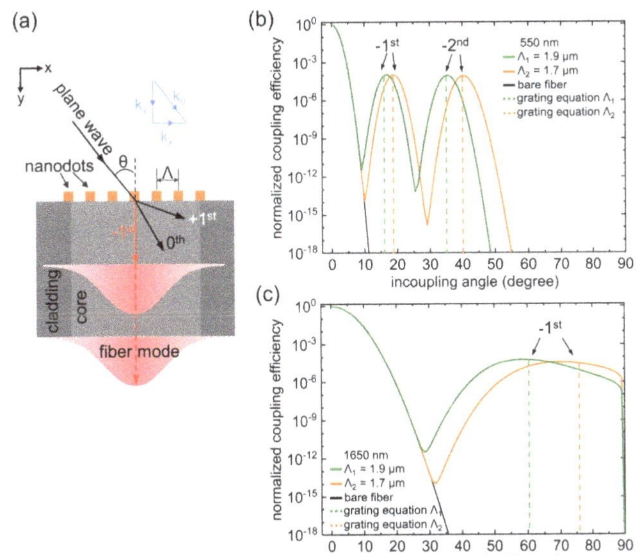

Figure 2. A theoretical model (**a**) of fiber light-coupling efficiencies for calculations at wavelengths of 550 nm (**b**) and 1650 nm (**c**), respectively. The dashed vertical lines in (**b**,**c**) correspond to additional enhancement position calculated by the grating equation.

The theoretical efficiencies $\eta_{norm}(\theta)$ at λ = 550 nm and λ = 1650 nm are plotted in Figure 2b,c, respectively. For the coefficients of the diffracted waves, we have used a_n = 0.1, which are estimated by the amount of scattered power which is in the order of the filling factor of the grating taking into account the distribution of the power across four diffracted waves. Note that the toy model represents a coarse approximation of the experimental circumstances and should only be qualitatively compared to experimental results. In each diagram, the black curve represents bare fibers, which all fail to operate beyond angles of 20° ($\eta < 10^{-6}$). In addition to the maximum value at normal incidence (θ = 0°), nanostructure-enhanced fibers exhibit substantially improved incoupling efficiencies particularly at the angles labeled as '-1st' and '-2nd' (refer as local maxima), in which the overall amplitude is close to the order of 10^{-4}. It is notable that different pitches (green line: Λ_1 = 1.9 µm, orange line: Λ_2 = 1.7 µm) could modify the line shape of η and thus lead to an additional degree of freedom to vary the local maxima. For instance, in the case of the short wavelength (λ = 550 nm), the maximum of the -2nd peak for Λ = 1.9 µm (green line in Figure 2b) is located at around 35°, while for the Λ = 1.7 µm the counterpart (orange line) of this maximum approaches 40° (see the vertical dashed lines in Figure 2b).

As shown in ref. [13], the angles of maximal incoupling are related to the lattice constants of the grating and can be approximated by a diffraction order equation, which is given by $\theta = arc\sin(\frac{-n\lambda}{\Lambda})$ [16] and in particular includes the diffraction order n. For the short wavelength (λ = 550 nm), two maxima from two diffraction orders (-1st ($n = -1$) and -2nd ($n = -2$)) induce a substantial improvement in incoupling efficiency, while only the -1st order contributes to the case of the long wavelength (λ = 1650 nm), as already shown in ref. [13]. This effect is clearly visible from the diffraction order equation: the shorter wavelength contributes two values of incidence angle θ (λ = 550 nm, Λ_1 = 1.9 µm, $\theta_{-1\mathrm{st}} \approx 17°$ and $\theta_{-2\mathrm{nd}} \approx 35°$), whereas only one value of θ exists for the identical incidence scenario for the long wavelength (λ = 1650 nm, Λ_1 = 1.9 µm, $\theta_{-1^{st}} \approx 60°$).

Here, we would like to point out that the SMF-28 exhibits higher-order modes at λ = 550 nm, which would demand including the respective modes in the theoretical model. However, our simple analytical model (Figure 2) shows good agreement with experimental data hereafter despite the mode differences. This is partially because the power fraction of fundamental mode dominates, and the higher modes have no major influence on the line shape of the light-incoupling efficiency. Therefore, we believe this simplified model is valuable in revealing the overall characteristics of the light-incoupling process for the configuration investigated here and the errors caused by the multimodeness are sufficiently small.

3. Implementation and Characterization Setup

The array of hexagonal nanodots (lattice constant Λ_1 = 1.9 µm and Λ_2= 1.7 µm, scanning electron microscopy (SEM) images presented in Figure 1b,c) are fabricated on a cleaved end face of an SMF-28 using of modified electron beam lithography (more manufacturing details can be found in ref. [13]). The nanodots have diameters and thicknesses of D = 480 nm and t = 40 nm, resulting in a plasmonic resonance at around λ = 1440 nm with the full width at half maximum (FWHM) of 200 nm (a measured transmission spectrum is provided in ref. [13]).

The coupling efficiencies have been measured at various incidence configurations using a laser source (NKT Photonics SuperK Compact) together with several polarizers (Thorlabs), focusing lenses (Thorlabs) as well as power detectors (Thorlabs). More details to the typical setup are provided in ref. [13]. Note that, due to the detection limit of the powermeter used, several points could not be detected for $\theta > 50°$.

4. Experimental Results

The measured data at λ = 550 nm and λ = 1650 nm are presented in Figure 3a–d, overall showing a strong impact of the nanodot array on incoupling efficiency. In general, for both wavelengths the coupling efficiencies of nanostructure-enhanced fiber have been improved to the maximum of about 10^{-4} at large incidence angles (i.e., from 20° to 85°), while the distributions of local maxima strongly vary depending on wavelength and pitch. In addition to the value maximum at normal incidence ($\theta = 0°$), two less-distinguished local maxima emerge at around $\theta = 20°$ and $\theta = 50°$ for λ = 550 nm (more obvious for the green dots), while only one local maximum (around 50°) was observed for λ = 1650 nm. This agrees with the theoretical model where both -1st and -2nd diffraction orders contribute at shorter wavelengths, while only the -1st order is relevant when the pitch value approximates the operation wavelength. Once again, the local maximum position of all four figures can be predicted by the simple formula $\theta = arc\sin(\frac{-n\lambda}{\Lambda})$, which are indicated as color-coded dashed lines in Figure 3. The local maxima position mismatch between experiments and model predictions can be attributed to beam excitation differences (Gaussian beam in experiment while the plane wave excitation in toy model) and fluctuations of fabricated nanodot grating. Note that the line shape of the incoupling efficiency does not depend strongly on the incidence polarization. Additionally, we have carried out an additional efficiency measurement at 650 nm, resembling the data at 550 nm. Specifically, the overall efficiencies are improved at angles of 30° and 55°, indicating the involvement of both first and second diffraction orders.

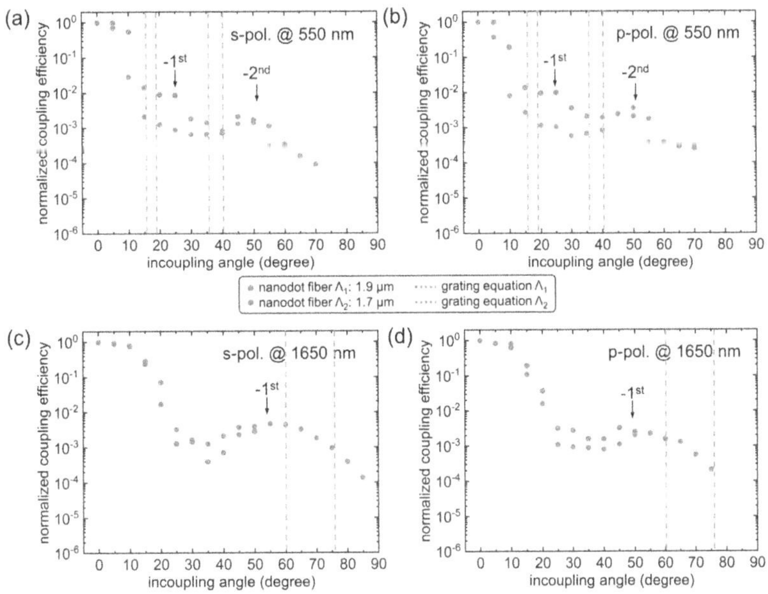

Figure 3. Experimental coupling efficiencies at wavelengths of 550 nm (**a**,**b**) and 1650 nm (**c**,**d**) under various incidence occasions (wavelength, angle, polarization, pitch condition, etc.). The vertical dashed lines situated at different incoupling angles have been obtained from the grating equation (i.e., $\theta = arc\sin(\frac{-n\lambda}{\Lambda})$) and the two colors refer to varied pitch configurations (green: Λ_1 = 1.9 μm, orange: Λ_2= 1.7 μm).

5. Conclusions

Even though optical fibers have been used to great success in a multitude of areas, collecting light under large angles within different spectral regimes remains a key challenge with commonly used fibers due to low numerical aperture. To meet this requirement, we demonstrate in this work that optical fibers functionalized with arrays of metallic nanodots located on the core of the fiber substantially improve light-collection performance at multiple wavelengths, extending the spectral domain of operation towards the visible. We experimentally demonstrate a nanostructure-mediated enhancement of the light-coupling efficiencies at large incident angles (from 30° to 85°) at two selected wavelengths (550 nm to 1650 nm), while nanostructure implementation relies on modified electron beam lithography. The overall improvement lies within the order of 10^{-4}, which can be improved in the future by optimization of disk diameter and thickness, while the the additional maximum in the angle spectrum at visible wavelengths is related to an additional diffraction orders.

Our concept represents a generic approach to increase light-coupling efficiencies and can easily be further improved by using high-performance nanostructures such as metasurfaces [17,18]. Note that commonly used multimode fibers have typical numerical apertures of at NA = 0.5 [19], yielding maximum incoupling angles of θ = 30°. This is substantially smaller than what can be achieved with our concept, emphasizing the importance of nanostructure-mediated incoupling. Therefore, we strongly believe that the concept of light incoupling through nanostructures will have great impacts in various fields of research and applications including bioanalytics (e.g., collection of Raman signals [20]), life science (e.g., in vivo endoscopy [21]) or quantum technology (e.g., collection of light from single emitters [22]). We would like to emphasize that particular applications that require incoupling at large angles would benefit from the presented concept (e.g., in vivo imaging).

Author Contributions: M.A.S. conceived the idea. N.W. conducted the measurements and data analysis. M.Z. and N.W. performed the theory calculations. U.H. fabricated the sample. M.A.S. supervised the research. The manuscript was written by N.W., M.Z. and M.A.S. through contributions of all authors. All authors have read and agreed to the published version of the manuscript.

Funding: This research is partially funded by National Natural Science Foundation of China under grant number 62005224.

Institutional Review Board Statement: Not applicable.

Informed Consent Statement: Not applicable.

Data Availability Statement: Data is contained within the article.

Conflicts of Interest: The authors declare no conflict of interest.

References

1. Ghatak, A.A.; Ghatak, A.; Thyagarajan, K.; Thyagarajan, K. *An Introduction to Fiber Optics*; Cambridge University Press: Cambridge, UK, 1998.
2. Miller, S. *Optical Fiber Telecommunications*; Elsevier: Amsterdam, The Netherlands, 2012.
3. Alexander Schmidt, M.; Argyros, A.; Sorin, F. Hybrid optical fibers-an innovative platform for in-fiber photonic devices. *Adv. Opt. Mater.* **2016**, *4*, 13–36. [CrossRef]
4. Leung, A.; Shankar, P.M.; Mutharasan, R. A review of fiber-optic biosensors. *Sens. Actuators B Chem.* **2007**, *125*, 688–703. [CrossRef]
5. Doherty, B.; Csáki, A.; Thiele, M.; Zeisberger, M.; Schwuchow, A.; Kobelke, J.; Fritzsche, W.; Schmidt, M.A. Nanoparticle functionalised small-core suspended-core fibre–a novel platform for efficient sensing. *Biomed. Opt. Express* **2017**, *8*, 790–799. [CrossRef] [PubMed]
6. Wang, N.; Zeisberger, M.; Hübner, U.; Schmidt, M.A. Nanotrimer enhanced optical fiber tips implemented by electron beam lithography. *Opt. Mater. Express* **2018**, *8*, 2246–2255. [CrossRef]
7. Hong, Y.; Zhao, D.; Wang, J.; Lu, J.; Yao, G.; Liu, D.; Luo, H.; Li, Q.; Qiu, M. Solvent-Free Nanofabrication Based on Ice-Assisted Electron-Beam Lithography. *Nano Lett.* **2020**, *20*, 8841–8846. [CrossRef] [PubMed]
8. Chemnitz, M.; Gebhardt, M.; Gaida, C.; Stutzki, F.; Kobelke, J.; Limpert, J.; Tünnermann, A.; Schmidt, M.A. Hybrid soliton dynamics in liquid-core fibres. *Nat. Commun.* **2017**, *8*, 42. [CrossRef]
9. Sollapur, R.; Kartashov, D.; Zürch, M.; Hoffmann, A.; Grigorova, T.; Sauer, G.; Hartung, A.; Schwuchow, A.; Bierlich, J.; Kobelke, J.; et al. Resonance-enhanced multi-octave supercontinuum generation in antiresonant hollow-core fibers. *Light. Sci. Appl.* **2017**, *6*, e17124. [CrossRef]
10. Klas, R.; Kirsche, A.; Gebhardt, M.; Buldt, J.; Stark, H.; Hädrich, S.; Rothhardt, J.; Limpert, J. Ultra-short-pulse high-average-power Megahertz-repetition-rate coherent extreme-ultraviolet light source. *PhotoniX* **2021**, *2*, 1–8. [CrossRef]
11. Choi, Y.; Yoon, C.; Kim, M.; Yang, T.D.; Fang-Yen, C.; Dasari, R.R.; Lee, K.J.; Choi, W. Scanner-free and wide-field endoscopic imaging by using a single multimode optical fiber. *Phys. Rev. Lett.* **2012**, *109*, 203901. [CrossRef] [PubMed]
12. Bovino, F.A.; Varisco, P.; Colla, A.M.; Castagnoli, G.; Di Giuseppe, G.; Sergienko, A.V. Effective fiber-coupling of entangled photons for quantum communication. *Opt. Commun.* **2003**, *227*, 343–348. [CrossRef]
13. Wang, N.; Zeisberger, M.; Hübner, U.; Schmidt, M.A. Boosting light collection efficiency of optical fibers using metallic nanostructures. *ACS Photonics* **2019**, *6*, 691–698. [CrossRef]
14. Yermakov, O.; Schneidewind, H.; Hübner, U.; Wieduwilt, T.; Zeisberger, M.; Bogdanov, A.; Kivshar, Y.; Schmidt, M.A. Nanostructure-empowered efficient coupling of light into optical fibers at extraordinarily large angles. *ACS Photonics* **2020**, *7*, 2834–2841. [CrossRef]
15. Snyder, A.W.; Love, J. *Optical Waveguide Theory*; Springer Science & Business Media: Berlin/Heidelberg, Germany, 2012.
16. Saleh, B.E.; Teich, M.C. *Fundamentals of Photonics*; John Wiley & Sons: Hoboken, NJ, USA, 2019.
17. Khorasaninejad, M.; Capasso, F. Metalenses: Versatile multifunctional photonic components. *Science* **2017**, *358*. [CrossRef] [PubMed]
18. Zou, X.; Zheng, G.; Yuan, Q.; Zang, W.; Chen, R.; Li, T.; Li, L.; Wang, S.; Wang, Z.; Zhu, S. Imaging based on metalenses. *PhotoniX* **2020**, *1*, 1–24. [CrossRef]

19. Thorlabs. 0.50 NA Step-Index Multimode Fibers. Available online: https://www.thorlabs.com/newgrouppage9.cfm?objectgroup_id=362 (accessed on 16 June 2021).
20. Cordero, E.; Latka, I.; Matthäus, C.; Schie, I.W.; Popp, J. In-vivo Raman spectroscopy: From basics to applications. *J. Biomed. Opt.* **2018**, *23*, 071210. [CrossRef] [PubMed]
21. Pahlevaninezhad, H.; Khorasaninejad, M.; Huang, Y.W.; Shi, Z.; Hariri, L.P.; Adams, D.C.; Ding, V.; Zhu, A.; Qiu, C.W.; Capasso, F.; et al. Nano-optic endoscope for high-resolution optical coherence tomography in vivo. *Nat. Photonics* **2018**, *12*, 540–547. [CrossRef] [PubMed]
22. Li, X.; Scully, R.A.; Shayan, K.; Luo, Y.; Strauf, S. Near-unity light collection efficiency from quantum emitters in boron nitride by coupling to metallo-dielectric antennas. *ACS Nano* **2019**, *13*, 6992–6997. [CrossRef] [PubMed]

photonics

Review

Advances in Multicore Fiber Grating Sensors

Zhiyong Zhao [1], **Yunli Dang** [1,*] **and Ming Tang** [1,2]

[1] Wuhan National Lab for Optoelectronics (WNLO), School of Optical and Electronic Information, Huazhong University of Science and Technology, Wuhan 430074, China; zhiyongzhao@hust.edu.cn (Z.Z.); tangming@mail.hust.edu.cn (M.T.)

[2] Optics Valley Laboratory, Wuhan 430074, China

* Correspondence: yunlidang@hust.edu.cn

Abstract: In recent years, multicore fiber (MCF) has attracted increasing interest for sensing applications, due to its unique fiber structure of multiple parallel cores in a single fiber cladding, which offers a flexible configurable platform to establish diverse functional fiber devices for sensing applications. So far, a variety of discrete fiber sensors using MCF have been developed, among which one of the major categories is the MCF grating sensors. The most distinct characteristic of MCF that differs from the normal single mode fibers is that the off-center cores of a MCF are sensitive to bending, which is caused by the bending induced tangential strain in off-center waveguides through either compression or stretching. The bending sensitivity has been widely developed for bending/curvature sensing or measuring physical parameters that are associated with bending. In this paper, we review the research progress on MCF-based fiber grating sensors. MCF-based diverse fiber grating sensors will be introduced, whose working principles will be discussed, and various types of applications of the MCF grating sensors will be summarized. Finally, the challenges and prospects of MCF grating for sensing applications will be presented.

Keywords: multicore fiber; fiber Bragg grating; long period grating; fiber grating sensors

Citation: Zhao, Z.; Dang, Y.; Tang, M. Advances in Multicore Fiber Grating Sensors. *Photonics* **2022**, *9*, 381. https://doi.org/10.3390/photonics9060381

Received: 30 January 2022
Accepted: 9 March 2022
Published: 26 May 2022

1. Introduction

In recent years, multicore fiber (MCF) technology has been extensively investigated for applications spanning optical fiber transmission, sensing and laser [1–6], etc. Thanks to the promotion of bandwidth requirement in the field of optical fiber communications, the MCF-based space-division multiplexing technique has developed rapidly in the last decade as a promising solution to increase the transmission capacity per fiber [7], which has substantially facilitated the development of MCF manufacturing and the relevant optical components. On the other hand, since multiple spatial cores of the MCF allow for independent parallel optical transmission within a single fiber, the unique MCF also turns out to be a flexible configurable platform that allows to develop diverse functional fiber devices and systems for sensing applications.

Multicore fiber contains more than one core in a single fiber cladding, whose core numbers and their spatial distributions can be flexibly designed in terms of the tolerance of inter-core crosstalk. Some common core arrangements of MCF include hexagonal structure [8], square lattice structure [9], ring structure [10], and linear array structure [11], etc. On the other hand, MCF can be manufactured with homogeneous or heterogeneous fiber cores, depending on the used material compositions of the cores, i.e., the cores of a homogeneous MCF have identical refractive index profiles, while those of the heterogeneous MCF are not the same. Currently, most of the MCFs are weakly-coupled MCFs, whose core-to-core pitch is deliberately set to be large (generally >30 um) in order to avoid high crosstalk between adjacent cores [12]. As a result, the weakly-coupled MCFs can transmit optical signals independently in different spatial cores. In addition, some effective fiber structure designs have been proposed to suppress inter-core crosstalk, including the trench-assisted cladding structure and the air-hole-assisted cladding structure [2,13], etc. However, in contrast, strongly coupled MCFs have also been developed recently, whose cores are much closer to each other in comparison with the weakly coupled MCFs. This kind of MCF supports super-modes transmission [14], which is therefore also considered as a form of multimode fiber. In a MCF-based optical system, one of the most critical optical components is the multiplexer/de-multiplexer (fan-in/fan-out), which is used to achieve optical coupling between each core of the MCF and the single mode fibers (SMFs). The proposed schemes include free space lens coupling [15,16], waveguide coupling [17], and etched or tapered fiber bundle coupling [8,18–21], etc.

Multicore fiber possesses some outstanding advantages, including multiple independent spatial channels, all-solid cores, small size and compact, excellent mechanical robust properties, etc. Specifically, different from the situation in the normal SMFs, where the cores are located in the fiber center, i.e., the strain neutral axis, the off-center cores of MCF are sensitive to bending, since fiber bending will generate tangential strain in off-center positions of the waveguide. The bending sensitivity has been used to develop various bending/curvature sensors, as well as to measure some other physical parameters that are associated with bending, e.g., vibration, acceleration, flow velocity of liquid and gas, etc. MCF sensors can be divided into two categories, i.e., distributed fiber sensors and discrete fiber sensors [22,23]. On one hand, MCF-based distributed fiber sensors have shown unprecedented performance that allows for distributed curvature and 3-D shape sensing [24]. In addition, MCF space-division multiplexed system configurations also turn out to be effective solutions of many critical problems in conventional distributed sensors, e.g., addressing the intrinsic cross-sensitivity issue in Brillouin distributed sensors [25,26], enabling large dynamic range and ultra-high measurement resolution simultaneously [27], achieving multi-parameter sensing [28,29], as well as extending the sensing range and frequency measurement range [30,31], etc. On the other hand, so far, a variety of discrete fiber sensors using MCF have been developed, which can roughly be categorized into two types, including the MCF grating sensors and the MCF interferometric sensors. MCF gratings sensors are one of the major categories of MCF discrete sensors, where fiber gratings are written in the cores of MCF. In addition, the multi-core structure of the MCF provides an excellent platform to fabricate in-fiber spatially integrated interferometers by means of splicing with other fibers or tapering, as well as using the fan-in/out coupler [32–34], etc., as shown in Figures 1 and 2, respectively.

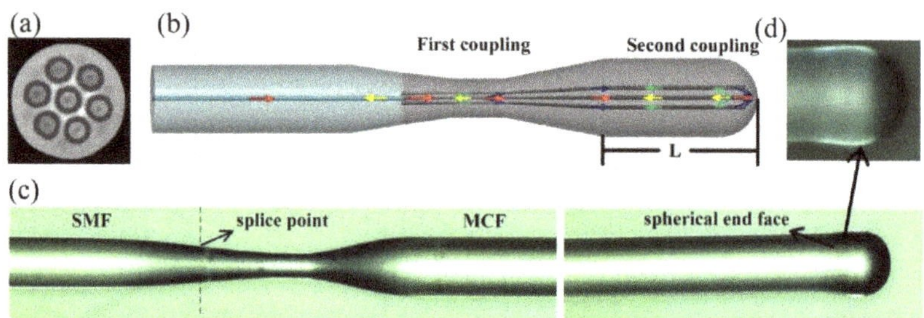

Figure 1. MCF sensor fabricated with tapering. (**a**) The cross-section image of the MCF. (**b**) Schematic diagram and operation principle of the Michelson-type multipath interferometer. (**c**) MCF-MI structure under microscope. (**d**) Spherical end face of MCF under microscope. Reprinted with permission from Ref. [34] © 2016 Optica Publishing Group.

In this paper, we review the research progress of fiber grating sensors that are using multicore fibers. Firstly, the unique bending sensitivity in off-center cores of MCF will be introduced in Section 2. Then, the advances in MCF-based fiber Bragg grating (FBG) sensors, tilted fiber Bragg grating (TFBG) sensors, and long-period grating (LPG) sensors will be summarized thoroughly, which will be presented in Section 3, Section 4, and Section 5, respectively. Section 6 will discuss the challenges and future prospects of MCF grating sensors. Finally, a conclusion will be given in Section 7.

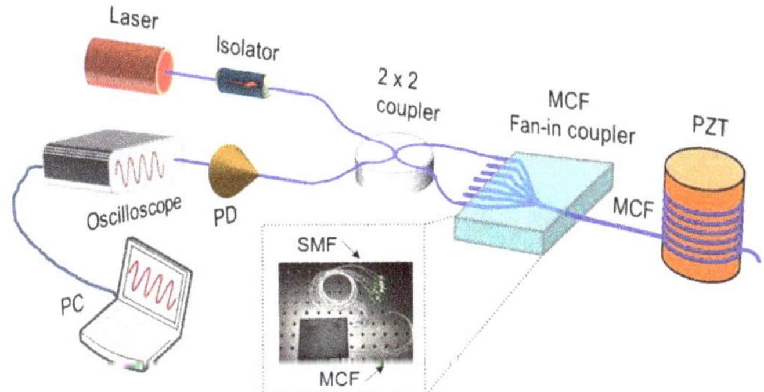

Figure 2. MCF Michelson interferometer sensor fabricated by using the fan-in/out coupler. Reprinted with permission from Ref. [33] © 2018 Optica Publishing Group.

2. Bending Sensitivity in Off-Center Cores of MCF

There has been increasing interest in developing various MCF grating sensors, including MCF-based FBG, TFBG, and LPG sensors, etc. It is well-known that SMF-based fiber grating sensors have been widely used for measuring temperature and strain, etc. The fiber core of SMF is located on the strain neutral axis, so theoretically FBG that is inscribed in the central core of a SMF is not sensitive to bending, as bending will not give rise to strain to the FBG. While in a bent MCF, as shown in Figure 3, the cores on the outer side of the neutral plane will be stretched, and the cores on the inner side of the neutral plane will be compressed, as a result fiber bending will give rise to tangential strain in off-center cores [24,31,35]. The fiber gratings written in the off-center cores of MCF will thus be sensitive to bending, due to fiber bending induced strain. By using the unique bending sensitivity, MCF has been widely used for bending sensing and relevant applications.

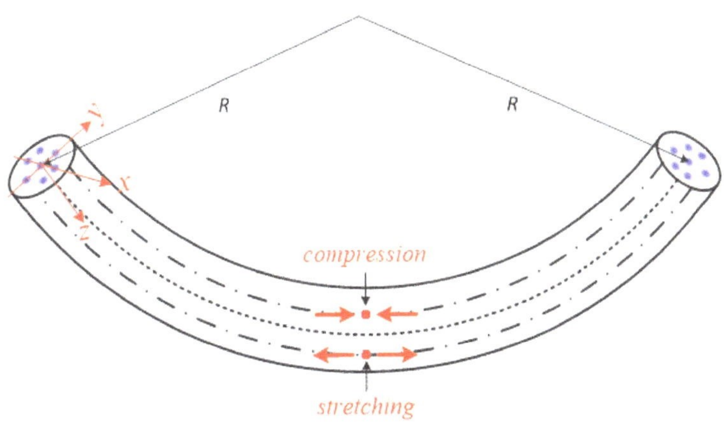

Figure 3. Schematic diagram of a bent MCF with a bending radius of R. The inner side cores will be compressed, while the outer side cores will be stretched. As a result, bending will generate different local tangential strains in distinct spatial cores. Adapted with permission from Ref. [31] © 2018 IEEE.

Let us take a seven-core fiber as an example, as shown in Figure 4 [36]. Fiber bending induced local tangential strain in MCF can be quantitatively determined, as given by [24,35]

$$\varepsilon_i = -\frac{d_i}{R}\cos(\theta_b - \theta_i) \tag{1}$$

where d_i is the distance of core i to the fiber center, R is the bending radius ($\kappa = 1/R$), θ_b is the bending angle which is defined to be the angular offset from the local x-axis to the local fiber bending direction, and θ_i is the angular position of core i, i.e., the angle from local x-axis to a specific core i. It is worth noticing that for the central core $d = 0$, so it can be inferred from Equation (1) that fiber bending will not generate strain in the central core of MCF.

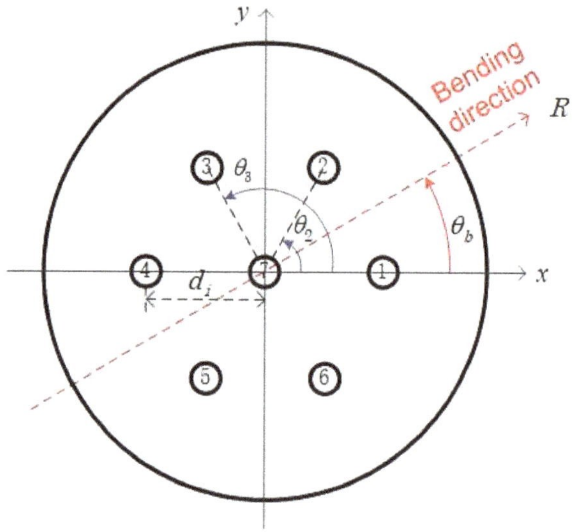

Figure 4. The transversal spatial distribution of cores of a seven-core fiber, showing the definitions of some important geometrical parameters. Reprinted with permission from Ref. [36] © 2016 Optica Publishing Group.

3. MCF-Based Fiber Bragg Grating Sensors

The first demonstration of bending sensing using fiber Bragg gratings in MCF was reported by Gander et al. [37]. Since fiber bending will give rise to local tangential strain in off-center cores, assuming that two cores are in the bending plane, then the strain difference between the two cores is determined by

$$\Delta\varepsilon = \varepsilon_1 - \varepsilon_2 = \frac{2d}{R} \tag{2}$$

Note that this fiber bending induced tangential strain in off-center cores will lead to the shift of Bragg wavelengths of FBGs, as a result the difference in Bragg wavelength between two adjacent gratings written in different cores can be used to retrieve the local curvature, as governed by

$$\Delta\lambda_B = \lambda_B(1 - p_\varepsilon)\Delta\varepsilon \tag{3}$$

where p_ε is the effective photo-elastic coefficient. Equation (3) indicates that the Bragg wavelengths difference is proportional to the strain difference $\Delta\varepsilon$, and thus the curvature. Therefore, curvature can be obtained from the measurement of two FBGs. The benefit of the MCF-based curvature sensor is that it is temperature-independent, because the two FBGs in the MCF have almost identical temperature sensitivities.

By using a pair of FBGs in two cores of the MCF, it is only able to measure a one-axis bending, while in order to measure a two-axis bending, two pairs of orthogonal FBGs in three cores are required [38], e.g., cores 1–3 in Figure 5a, where the difference between the wavelength shifts in each grating pair reveals the differential strain and hence the curvature. For a specific azimuthal placement of the four-core fiber, bending sensitivity of the FBGs has been calibrated by bending the fiber in the horizontal and vertical (i.e., x and y) planes separately. The Bragg wavelength differences between the two core pairs (i.e., cores 1 and 2, cores 2 and 3) are plotted in Figure 5b,c, respectively [38].

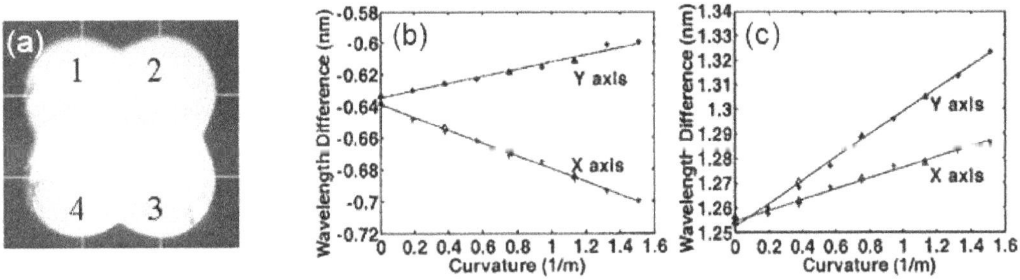

Figure 5. (**a**) The cleaved face of the four-core fiber. (**b**,**c**) are the measured wavelength difference versus applied curvature for x-axis and y-axis bending with a linear least-square fit: (**b**) pairing of cores 1 and 2; (**c**) pairing of cores 2 and 3. Adapted with permission from Ref. [38] © 2003 Optica Publishing Group.

Note that by using the two pairs of FBGs, their wavelength difference can be described by a matrix in terms of curvature κ_x and κ_y, as given by

$$\begin{bmatrix} \Delta\lambda_{12} \\ \Delta\lambda_{23} \end{bmatrix} = \begin{bmatrix} \Delta\lambda_{12}^0 \\ \Delta\lambda_{23}^0 \end{bmatrix} + \begin{bmatrix} p & q \\ r & s \end{bmatrix} \begin{bmatrix} \kappa_x \\ \kappa_y \end{bmatrix} \tag{4}$$

where $\Delta\lambda_{12}^0$ and $\Delta\lambda_{23}^0$ are the constant absolute wavelength difference between FBGs, p, q, r, and s are the calibrated response coefficients of FBGs along the two bending axes. Inverse matrix transformation can then be performed to determine the relative bending in the x and y axes, so two-axis curvature measurement is achieved.

The result indicates that the responses of FBGs under different bending angle are distinct. As a matter of fact, it can be inferred from Equation (1) that the curvature sensitivity (wavelength shift vs. curvature variation) of an outer core should be angular position dependent, which is subject to a cosinoidal function with respect to the angular offset between the fiber core and the bending direction, i.e., $\theta_b - \theta_i$. The offset angle dependence of the curvature sensitivity of the off-center core has been investigated, whose experimental setup is shown in Figure 6a [39]. The setup allows to adjust the curvature of the sensing fiber, as well as to rotate the fiber with about 2° angular accuracy. Here, the curvature κ of the sensing fiber is governed by [39,40]

$$\sin\left(\frac{L\kappa}{2}\right) = \frac{(L - \Delta L)\kappa}{2} \tag{5}$$

where L is the initial length between the two fixed ends of the fiber without bending and ΔL is the movement distance of the adjustable stage.

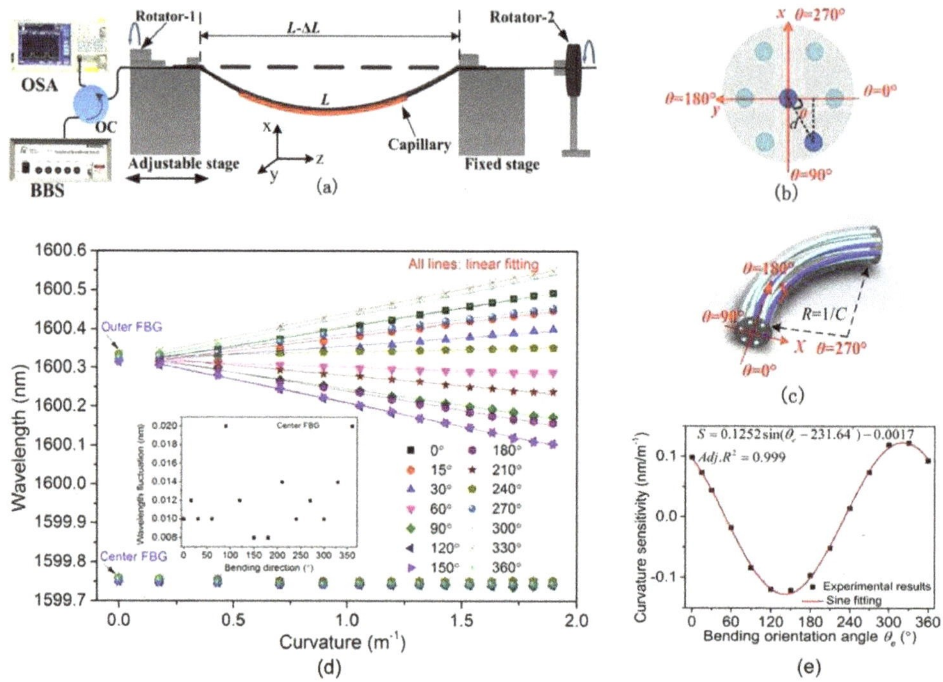

Figure 6. (**a**) Experimental setup of the proposed curvature sensor; (**b**) theoretical definition of the bending orientation angle; (**c**) schematic diagram of a curved fiber when the bending angle is 270°; (**d**) Bragg wavelength shifts versus curvature for the two FBGs at different bending angles. The inset shows the wavelength fluctuations of the central FBG at different bending angles over the curvature range from 0 to 1.896 m^{-1}. (**e**) Relationship between the curvature sensitivity of FBG in the off-center core and the experimental bending orientation angle. Adapted with permission from Ref. [39].

As shown in Figure 6b, a heterogeneous seven-core fiber was used in the experiment, while only one outer core and the central core have been used for characterization. A schematic diagram of the curved fiber with 270° bending angle is presented in Figure 6c. By rotating the fiber, a different bending angle (i.e., the angular offset) with respect to the selected fiber core can be generated, note that the initial state is trivially considered to be 0°; meanwhile for each applied bending angle, the Bragg wavelength shifts under different curvatures were measured, and the result is plotted in Figure 6d. It is observed that for each specific bending angle, the Bragg wavelength shift of FBG in the outer core is linear to the curvature of the sensing fiber, and different bending angle leads to distinct curvature sensitivities. In contrast, as shown in Figure 6d, the Bragg wavelengths of FBG in the central core remain almost unchanged under different curvature, so it turns out that it is not sensitive to bending, because the FBG is located on the strain neutral axis of the fiber. Using the measurement results of Figure 6d, the obtained curvature sensitivities of FBG in the off-center core as a function of the bending angles is presented in Figure 6e. It verifies that the curvature sensitivity of FBG in the off-center core varies with the relative bending angle. Specifically, as indicated by Equation (1), it follows a cosinoidal (or sinusoidal) function with respect to the bending angle, and a period of 360°. Since the FBG in the central core is insensitive to fiber bending, it can be used to compensate the effects of temperature and externally applied axial strain.

The bending angle dependence of the curvature sensitivity of a specific off-center core has been investigated in a twin-core few-mode fiber (TC-FMF) [41]. As shown in Figure 7, each core of the TC-FMF supports LP_{01} and LP_{11} modes. While three resonance peaks were observed in the FBG reflection spectrum of the TC-FMF, which result from the LP_{01} mode resonance, LP_{01}–LP_{11} mode cross-coupling resonance, and LP_{11} mode resonance, respectively. The bending response of one resonance peak has been investigated by rotating the fiber from 0° to 360° with a step of 15°, and the measurement was then repeated for some different curvatures of the sensing fiber. The result is shown in Figure 8a. It indicates that for a specific fiber curvature, the resonance wavelength varies periodically with respect to the bending angle, which matches well with the inference of Equation (1). The curvature sensitivities (wavelength shift vs. curvature variation) can then be calculated from the measurement result, which is shown in polar coordinate in Figure 8b. The result indicates that the curvature sensitivity reaches to the maximum when the bending angle is 0° (or 360°) and 180°, which is because the used sensing fiber core is located in the bending plane and bending caused strain is the maximum in this case, as inferred from Equation (1); while the curvature sensitivity is zero when the bending angle is 90° and 270°, which is because the used sensing fiber core is located in the strain neutral axis plane in this case, as a result bending will not give rise to strain in the fiber core.

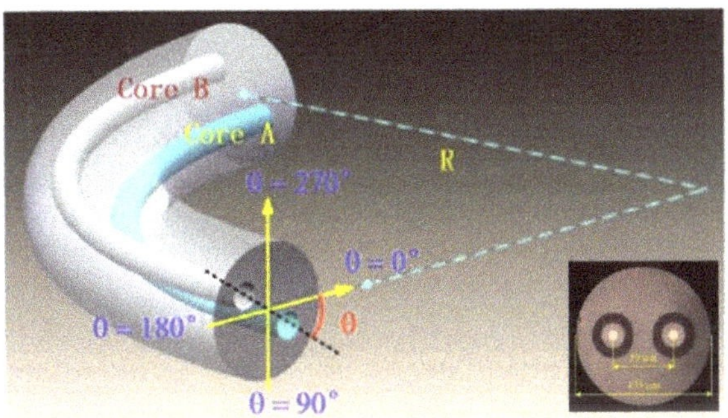

Figure 7. Schematic diagram of a curved twin-core few-mode fiber with bending radius of R, where bend direction angle θ is the angular offset between the fiber bend plane and the axis connecting core A and core B. The inset shows the microscopic image of the TC-FMF. Adapted with permission from Ref. [41] © 2017 IEEE.

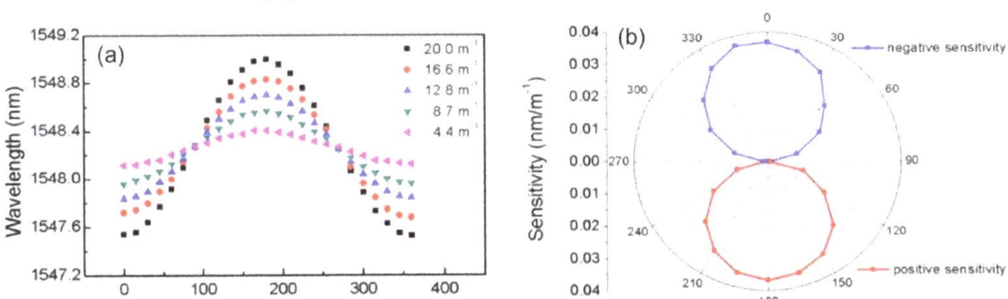

Figure 8. (**a**) Resonant wavelength of one peak in the reflection spectrum of TC-FM FBG as a function of the bend direction angle under different bend curvatures. (**b**) Bend sensitivities of the TC-FM FBG in different bend directions. Adapted with permission from Ref. [41] © 2017 IEEE.

The bending responses of all the three resonance peaks have also been compared. For a given bending direction, different bending radii were applied to the sensing fiber, meanwhile the reflection spectra were recorded, respectively, as shown in Figure 9a. The wavelength shifts of the three peaks as a function of the curvature are shown in Figure 9b. The linearly fitted curvature sensitivities are −36.26, −37.41, and −36.35 pm/m^{-1}, respectively. Therefore, it turns out that there is no large difference between the resonance peaks in terms of the response of fiber bending.

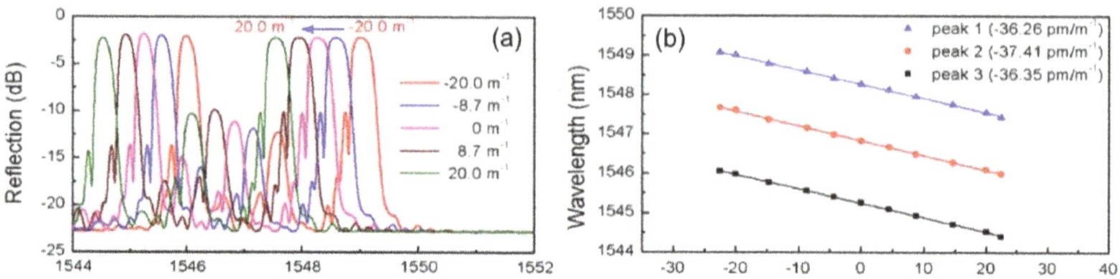

Figure 9. Bend response of the TC-FM FBG with different bend curvatures. (**a**) Reflection spectra evolution of the TC-FM FBG with different bend curvatures. (**b**) Wavelength shift of the three resonance peaks in the reflection spectra of the TC-FM FBG as a function of bend curvature. Adapted with permission from Ref. [41] © 2017 IEEE.

The curvature sensitivities of FBGs in all the off-center cores of a seven-core fiber have also been characterized [42], as shown in Figure 10. For a given curved MCF with bending angle θ_b, as each outer core has different angular position θ_i, i.e., their relative angle offsets to the bending direction are distinct, the outer cores have different curvature sensitivities. While because the six outer cores are arranged hexagonally, the curvature sensitivities are actually the same with 60° offset in the polar coordinate between two adjacent outer cores.

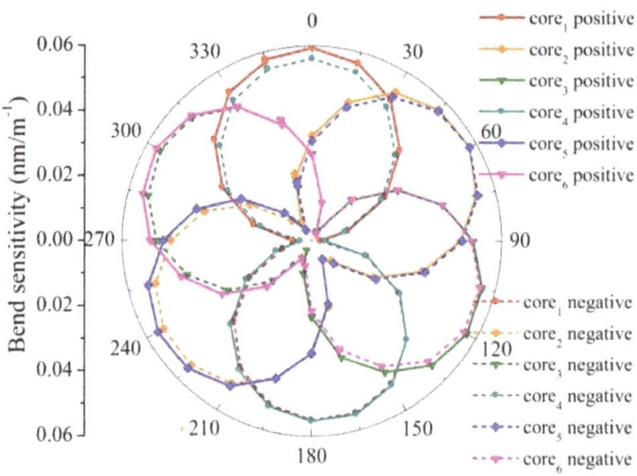

Figure 10. Bend sensitivities for the six outer-core FBGs in the seven-core fiber plotted for various bend directions (from 0° to 360°). Reprinted with permission from Ref. [42] © 2018 Optica Publishing Group.

In addition to the silica fiber, polymer optical fiber (POF) has also been investigated for bending sensing [43]. As shown in Figure 11a, the POF has off-center cores, in which Bragg grating is inscribed. Using the experiment setup shown in Figure 11b, bending was then applied to the sensing fiber along four directions, i.e., 0°, 90°, 180°, and 270°, and the measured wavelength shifts as a function of the curvature is shown in Figure 11c. The fiber core is compressed along the 0° bending direction, and it is stretched along the 180° bending direction. While fiber bending should not give rise to strain in the fiber core along 90° and 270° bending directions, since the fiber core is in the neutral plane, and the wavelength shift of FBG along these two bending directions is assumed to be caused by misalignment of the fiber axis or a slight twist of the POF during measurement. The benefit of polymer fiber is that it has a smaller Young's modulus and its failure critical strain is much larger in comparison with the silica fiber, so it has potential in the bending sensing applications that have very small bending radius.

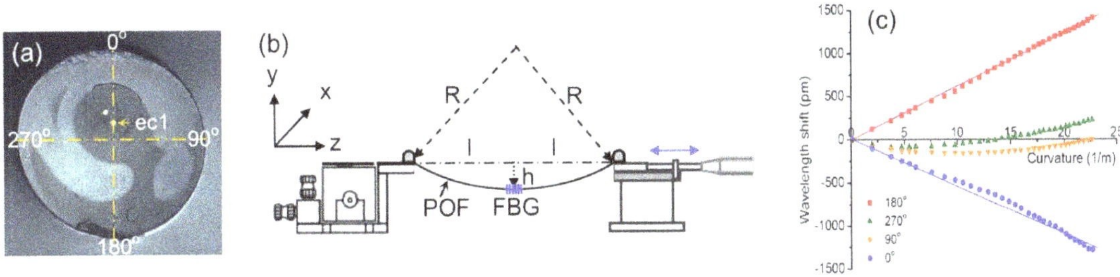

Figure 11. (**a**) Microscope image of the polymer fiber cross-section marked with the orientations; (**b**) schematic of the experimental setup for characterizing the FBG-based sensor. R is the bend radius and h is the displacement; (**c**) wavelength shifts of the eccentric polymeric FBG against curvatures at four different orientations. The symbols show the experimental data and lines depict the best fit lines. Adapted with permission from Ref. [43] © 2010 IEEE.

It is realized that the most prominent advantage of FBG in MCF is that it allows for bending measurement, which is a unique characteristic that the FBG does not possess in the core of a normal SMF. As a matter of fact, FBG in MCF has also enabled the measurement of parameters that are associated with bending, e.g., transverse loading, acceleration, vibration, flow velocity of liquid and gas, etc.

For example, it is observed that when transverse loading is applied to an MCF, the reflection spectrum of FBG in the MCF will split due to transverse stress induced birefringence [44], as shown in Figure 12a, and it turns out that the splitting is a function of the applied loading, as shown in Figure 12b. It can be observed that the peak splitting of FBG is also angular position dependent for a given loading, because transverse stress is the largest around the cladding edge along the diameter that is aligned with the loading axis i.e., cores 1 and 3 in Figure 12b, and it is the lowest in the cores that are far from the loading axis, i.e., cores 2 and 4 in Figure 12b.

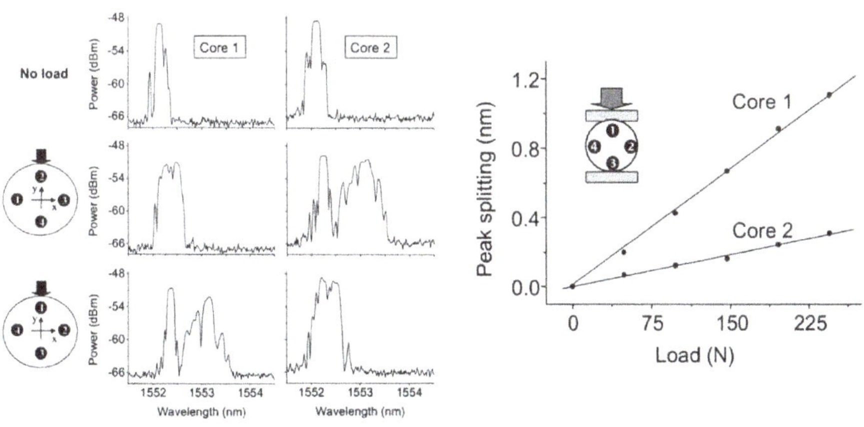

Figure 12. Birefringence-induced peak splitting due to increasing loading. Adapted with permission from Ref. [44] © 2004 Optica Publishing Group.

In fact, the orientation of transverse loading can also be retrieved from the measurement of FBG peak splitting in MCF. Because transverse loading induced birefringence is different at different angular positions, as a result the peak splitting of FBGs in different fiber cores is distinct as well, as is shown in Figure 12b. Therefore, the load angle relative to the cores can be found from the ratio of the Bragg peak splitting between pairs of cores [45].

This has been demonstrated by using FBGs that were inscribed at the same axial position in all cores of a four-core fiber, whose cores are arranged at the vertices of a square. Specifically, a parameter α that defines the ratio of peaking splitting coefficients (i.e., the slope of peaking splitting versus loading at a specific angle) of two adjacent cores is used to deduce the orientation of loading, as given by

$$\alpha = \frac{d\Delta\lambda_i}{dF} / \frac{d\Delta\lambda_j}{dF} \tag{6}$$

where $\Delta\lambda_i$ and $\Delta\lambda_j$ are the peak splitting of FBGs in two neighboring cores, F is the applied loading. The dependence of splitting coefficients ratio on loading angle is presented in Figure 13, where both the theoretical curve and experimental data are shown, and measurement was carried out by using two cores with 90° separation, i.e., cores 1 and 2 in the inset of Figure 13.

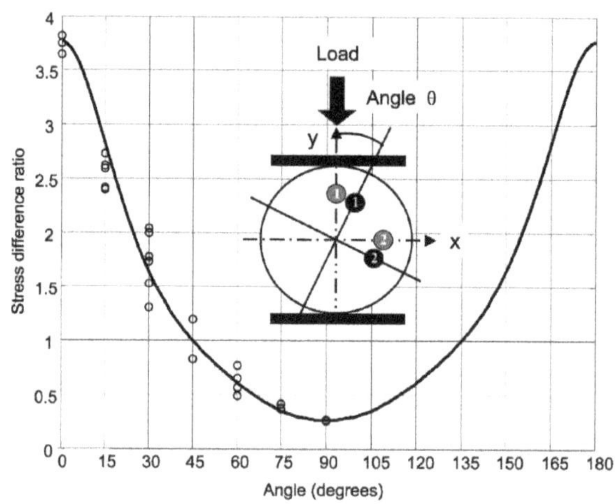

Figure 13. Theoretical curve (solid curve) and experimental data (points) for the stress difference ratio of two cores at 90° separation. Reprinted with permission from Ref. [45] © 2005 Optica Publishing Group.

Making use of the fiber bending induced differential strain between the FBGs in a four-core fiber, pitch and roll sensing has also been demonstrated [46]. As shown in Figure 14a, the four-core fiber was supported at one end and a small mass was attached to the other end of the fiber, so a cantilever is formed. The free end of the fiber trends to be curved due to gravity, thus, in this case the fiber bending induced differential strains between opposite grating pairs are determined by the orientation of fiber in pitch (in the vertical plane) and roll (azimuth) with respect to gravity. Specifically, the roll angle can be acquired from the ratio of the strain differences between two orthogonal core pairs, e.g., core pairs (1, 2) and (2, 3), as shown in the inset of Figure 14a. Moreover, the pitch angle can be achieved from the magnitude of the differential strain between pairs [46]. The measurement result of pitch and roll is shown in Figure 14b.

FBGs in a four-core fiber have also been used for dynamic two-axis curvature measurements by making use of the bending induced differential strain in off-center cores of the MCF [47]. For this application, in order to achieve dynamical strain measurement, arrayed waveguide grating (AWG) was used to interrogate the Bragg wavelength of the FBG by monitoring the optical power of different AWG channels, and then the ratio of the power in these channels was used to infer the grating wavelength, which permits strain measurement dynamically at kilohertz frequency.

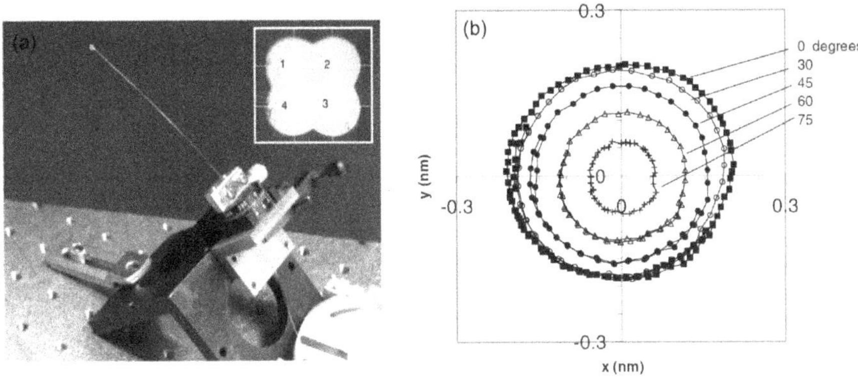

Figure 14. (**a**) Sensing fiber mounted in rotary stage; in this case the pitch angle is set at 45°. The inset shows the used four-core fiber in this experiment; (**b**) experimental results from pitch (θ) and roll (φ) measurements. $x = \lambda 1 - \lambda 2$, $y = \lambda 2 - \lambda 3$ are the differential Bragg wavelength shifts (in nm with offset removed) between orthogonal pairs (1, 2) and (2, 3) of MCF gratings. Data are plotted for $\theta = 0°$, 30°, 45° (φ range 0–540°) and $\theta = 60°$, 75° (φ range 0–360°). Adapted with permission from Ref. [46].

Note that the strain difference between two gratings in the MCF can be used to determine the radius of curvature R in the plane of the cores, as governed by

$$R = \frac{d}{\varepsilon_1 - \varepsilon_2} \tag{7}$$

where d is the distance between the cores and ε_n is the strain in the nth core. Therefore, the curvature in two dimensions can be obtained by using two orthogonal FBG pairs. As shown in Figure 15a, FBGs were written co-located in the cores of the MCF, and three cores have been used in the experiment. The MCF containing the FBGs was inserted into a stainless steel tube, and a cantilever was formed by clamping the tube at one end. A high speed camera was used to capture the displacement of the end of the cantilever during cantilever vibration at 28 Hz frequency; so that it can be used to compare with the result that was measured by the FBGs. Figure 15b presents the measured deflections by both the FBGs in MCF and the camera along x-axis and y-axis directions, showing good reliability.

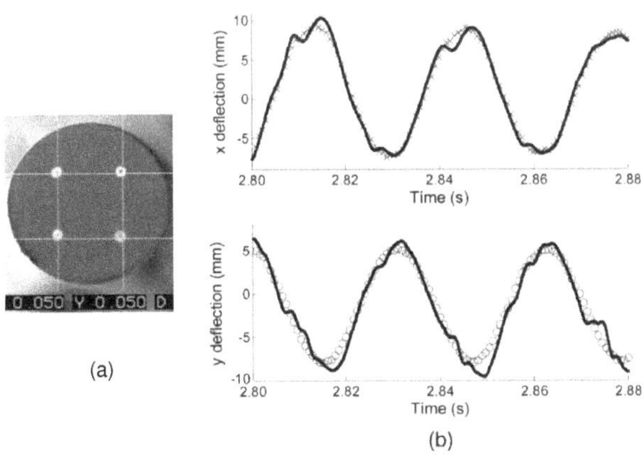

Figure 15. (**a**) Cross-section of the MCF. (**b**) Upper plot, x-deflection measured by MCF FBGs (solid curve) and camera (x). Lower plot, y-deflection measured by sensor (solid curve) and camera (o). Adapted with permission from Ref. [47] © 2006 Optica Publishing Group.

Based on the measurement of differential strain using FBGs in a four-core fiber, displacement sensing of the tunnel has also been demonstrated [48]. As shown in Figure 16, in order to measure displacement of the tunnel, the orientation calibrated MCF was mounted on two tunnel sections, as shown in Figure 16b, where FBGs are inscribed co-located in the cores of MCF. Figure 16c shows the schematic diagram of the displacement sensor. As explained previously, measurements of two orthogonal FBG pairs can be used to determine the two-axis curvature, and then the bending plane as well. For small deflection, once the curvature at a known position along the fiber cantilever is obtained, it can be used to determine the translational displacement δ, as governed by [48]

$$w''(x) = 12\frac{\delta}{L^2}(\frac{1}{2} - \frac{x}{L}) \quad (8)$$

where $w(x)$ is the displacement of the cantilever at a point x along the cantilever of length L, and $w''(x)$ is the second derivative of w.

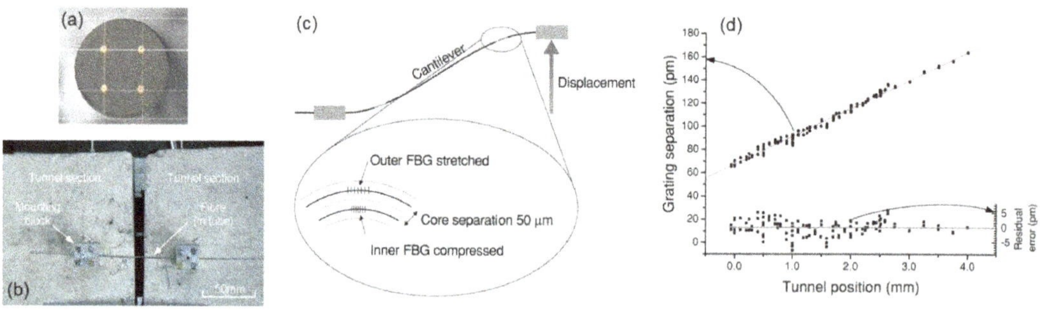

Figure 16. (**a**) Cross-section of MCF used in this experiment; (**b**) cantilever containing MCF attached to tunnel lining; (**c**) FBG pairs act as curvature sensors—the curvature is determined from the strain difference between the gratings; (**d**) experimental results from translation tests. Adapted with permission from Ref. [48].

The measurement of differential response between FBGs in a four-core fiber has also been developed for acceleration sensing [49]. The accelerometer consists of a short fiber cantilever with FBGs inscribed in the fiber cores near the clamping point, as shown in Figure 17a, where the inset shows the used four-core fiber. An epoxy mass is attached to the fiber end of the cantilever, as shown in Figure 17b, which helps to increases the sensitivity of the sensor.

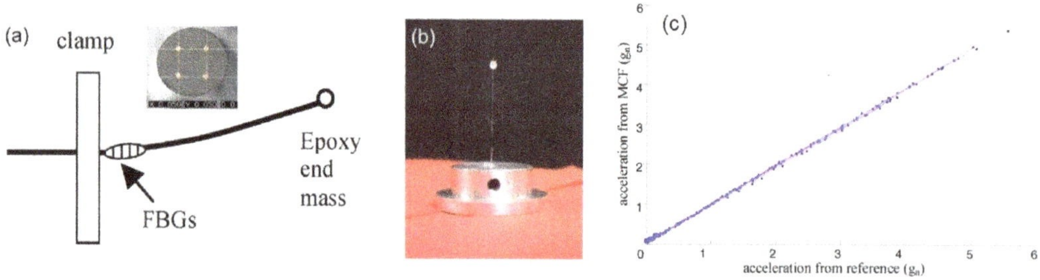

Figure 17. (**a**) Schematic diagram of the MCF cantilever-based accelerometer, the inset shows the four-core fiber used in the experiment; (**b**) the packaged acceleration sensor; (**c**) acceleration response of MCF accelerometer between 30 and 200 Hz. Adapted with permission from Ref. [49].

Using the same principle that has been explained previously, the measured differential strain between two adjacent FBG is used to determine the curvature ($1/R$) in the plane of the two cores, and by using two orthogonal pairs of cores, two-axis curvature measurement can be achieved. Then, acceleration a can be calculated once curvature is determined, as governed by [49]

$$a = \frac{EI}{m(x-l)} \cdot \frac{1}{R} \tag{9}$$

where E is the Young's modulus, I is the second moment of area of the fiber, m is the end mass, x is the distance of the measurement position from the clamp, and l is the length of the fiber cantilever. The measurement result of acceleration is shown in Figure 17c. On the other hand, it is worth mentioning that the frequency response of the sensor is also related to the cantilever length l and the fiber mass M, as given by [49,50]

$$f \propto \sqrt{\frac{3EI}{Ml^3}} \tag{10}$$

Therefore, the frequency response of the MCF-based accelerometer can be adjusted by changing the fiber length and the end mass that is attached to the fiber end. In addition, FBGs in a twin-core fiber have also been used for an acceleration sensor [51], but since two cores (one core is located in the fiber center) can only determine the curvature in a one-dimensional plane, it can be used as a one-axis accelerometer. While, using several FBGs simultaneously in multiple cores of seven-core fibers, two-dimensional accelerometers have been demonstrated, which are capable of measuring the vibration azimuthal angle as well as the acceleration simultaneously [52,53].

What is more, based on the same principle of the MCF accelerometer, MCF and multifiber inclinometers have also been developed [54,55]. As shown in Figure 18, the inclinometers are made of MCF or multifiber bundle cantilevers, in which FBGs are inscribed in the fiber cores, and a phase generated carrier demodulation technique has been used to extract the differential phases between the FBGs [56,57]. The differential strain measurements of FBGs in three cores (e.g., cores A–C in Figure 18b) can be used to retrieve the bending orientation and bending radius [58], which is then used to calculate the inclination through a transform equation that is derived from the calibration procedure [54]. In addition, fiber Bragg gratings in a seven-core fiber have also been developed in a two-dimensional inclinometer, which is able to measure the azimuthal angle and the inclination angle simultaneously [59].

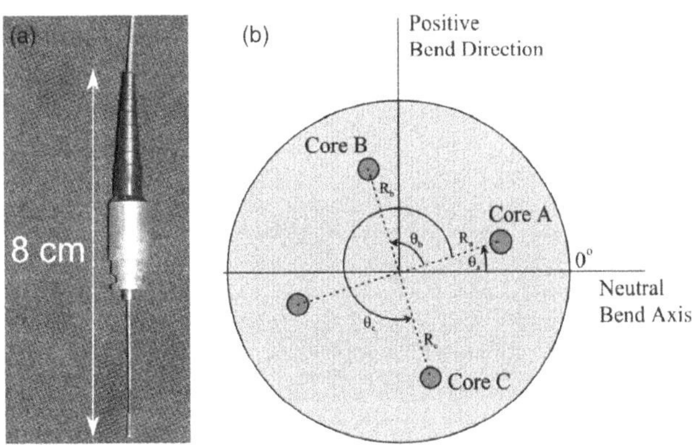

Figure 18. (**a**) The packaged MCF inclinometer. (**b**) Core geometry of the MCF. Bend direction denotes the direction of positive radii of curvature (end of the fiber points down). Neutral bend axis defines the plane of bending. The cores are labeled with their respective core radii and angular offsets from 0°. Figure reproduced from Ref. [54] © 2009 SPIE.

Utilizing the bending sensitivity of off-center cores of the MCF, parametrical fiber three-dimensional (3D) shape sensing has been widely investigated in recent years [60]. The technology relies on the measurement of fiber bending induced strains in multiple cores originally through axially co-located discrete FBGs arrays, then the obtained differential strains are used to retrieve the local bending angles and bending radii along the fiber length, and by using these parameters eventually the overall shape of the MCF can be parametrically reconstructed by solving the Frenet–Serret equations [35,61]. Figure 19 shows an example of MCF enabled shape sensing by using this technology [62]. In order to ensure high shape reconstruction accuracy, the sensing fiber needs to contain high density FBGs arrays, e.g., typically 1 cm spacing [62]. In this case, the interrogation of the Bragg wavelengths of the FBGs is unable to be accomplished by the traditional optical time-domain reflectometry (OTDR) technique due to its bad spatial resolution, i.e., in the order of meters. On the other hand, the wavelength-division multiplexed FBG arrays scheme suffers from finite bandwidth of the light source, which results in very limited multiplexing numbers of FBGs. Therefore, normally the high density FBGs arrays in MCF are measured by using the optical frequency-domain reflectometry (OFDR) technique [62,63], since it can offer up to micrometers spatial resolution.

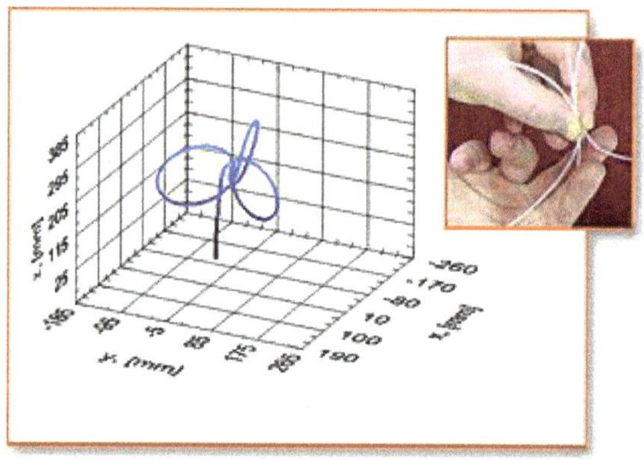

Figure 19. An example of multicore fiber enabled shape measurement. Figure reproduced from Ref. [62] © 2005 SPIE.

It is worth mentioning that when FBG arrays are used to measure the local strains in cores of the MCF, the spatial resolution of strain measurements is determined by the interval of two adjacent FBG groups. As mentioned previously, the interval is typically 1 cm, which might be still insufficient. The sparse discrete sets of strains that are used for shape reconstruction might lead to accumulated large angular error, which rapidly reduces the accuracy of the retrieved shape. Therefore, it turns out that localization of strain with higher spatial resolution is preferred. This can be done by measuring the intrinsic Rayleigh backscattering signals through OFDR, then strain can be retrieved by calculating the cross-correlation of two sets of Rayleigh measurements [64–66]. What is more, in order to increase the signal-to-noise ratio (SNR) of measurement, continuous weak fiber Bragg gratings can be inscribed in the cores of MCF along the whole fiber length [67–69].

As a matter of fact, the sensing fiber may also have an arbitrary twist along the fiber length, so in addition to bending, fiber twist will also give rise to additional strain in the fiber cores, which might cause error in shape reconstruction of the sensing fiber [70,71]. Therefore, fiber twist induced strain needs to be separated from the overall strain measurement in MCF. In order to be able to measure fiber twist, a kind of helical MCF with permanent twisted off-center cores and a straight central core has been developed [72–77]. As an example, Figure 20 presents the schematic diagram of a helical seven core fiber [77], in which continuous weak fiber Bragg gratings have been written in the cores of MCF along the whole fiber length during the fiber drawing process, so it allows for a fiber twist effect compensated strain measurement with high SNR, which will be very helpful for shape sensing. In fact, in addition to fiber twist, the error of core position as well as inaccurate placement of the FBGs within each cross-section of the MCF will also eventually decrease the accuracy of shape reconstruction [78,79]. Moreover, it is found that the parameters of FBG length, fiber geometry, photo-elastic coefficient, etc., also have an impact on the shape reconstruction accuracy [80,81].

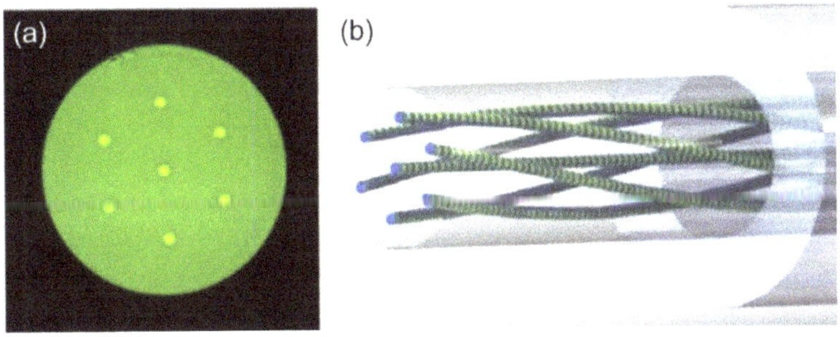

Figure 20. (**a**) End-view image of a seven-core fiber with coating removed; (**b**) twisted multicore fiber schematic showing UV transparent coating (right). Bare glass region (left) shown to highlight twisted multicore continuous gratings. Adapted with permission from Ref. [77] © 2017 IEEE.

In fact, it is realized that precise fully 3D shape sensing using MCF is complicated and difficult, because the angular error will accumulate along the fiber length during shape reconstruction, so the technology is highly dependent on the accuracy of strain measurement and the localization resolution of strain, while MCF enabled curvature sensing turns out to be easier and reliable. So far, multipoint two-dimensional curvature sensing using discrete FBG arrays in MCF [82,83], as well as distributed curvature sensing using the Brillouin optical time domain analysis (BOTDA) technique in MCF have been demonstrated with high accuracy [24]. As for the application fields, MCF enabled curvature and shape sensors have shown great potential in many industrial applications [60], e.g., for real time sensing of a wing's shape in flight, and for medical instruments such as a catheter or robotics that can be integrated into minimally invasive surgery systems [84–87], etc.

Here, it is worth mentioning that writing gratings in MCF turns out to be more complicated than in SMF, due to the geometric difference in spatial position between cores of the MCF, which adds some technical challenges in fabrication, e.g., due to the lens effect of the fiber, the grating strength may have variations between cores, and it may also lead to different transmission/reflection profiles of the FBGs [39,62,76,88–91]. Currently, the most widely used way to write gratings in MCF is to use UV light irradiation, and this method will normally bring in gratings in all cores of the MCF. However, in order to achieve selective FBG inscription in any specific core of a MCF, a femtosecond laser writing method can be used [41,92–94]. In addition, the small spot direct ultraviolet writing technique has also been used to achieve FBG inscription into individual cores of a seven-core fiber [95].

In addition to the widely developed applications for bending and shape sensing, FBGs in MCF have also been used for measuring temperature, strain, and refractive index [96–99], etc. Efforts have also been made for the demodulation of FBG in MCF. For example, a broadband source with a sinusoidal spectrum has been used to interrogate the wavelength of FBGs in MCF [100], which transforms the wavelength shift of FBGs into the variation in reflected power, thus, providing a fast wavelength measurement method. The microwave photonics filtering (MPF) technique has also been used to measure the wavelengths of FBGs in MCF, which retrieves the curvature of fiber by monitoring the notch frequency shift of the MPF, offering a high resolution curvature measurement solution [101]. On the other hand, it is worth mentioning that similar functionality to MCF has been achieved by using a fiber bundle that is composed of multiple single mode fibers, in which FBGs have been inscribed, and they have been used to measure inclinometer [55], curvature and temperature [102], as well as 2D flow velocity [103], etc.

4. MCF-Based Tilted Fiber Bragg Grating Sensors

In addition to the conventional FBG, tilted fiber Bragg grating (TFBG) has also been written in MCF. As shown in Figure 21, inter-core crosstalk induced by TFBG has been observed in a seven-core fiber [104], including forward transmission crosstalk and backward reflection crosstalk. As an example, Figure 21b shows the measured transmission spectrum, crosstalk, and back crosstalk between core 1 and core 3. The experimental result also indicates that the magnitude and direction of fiber bending, as well as the external refractive index have an impact on the inter-core crosstalk, which might be used for sensing applications. For example, the inter-core crosstalk caused by TFBG in MCF has been used for temperature and surrounding refractive index sensing [105]. The result shows that for refractive index sensing, the higher the tilted angle the higher the sensor sensitivity.

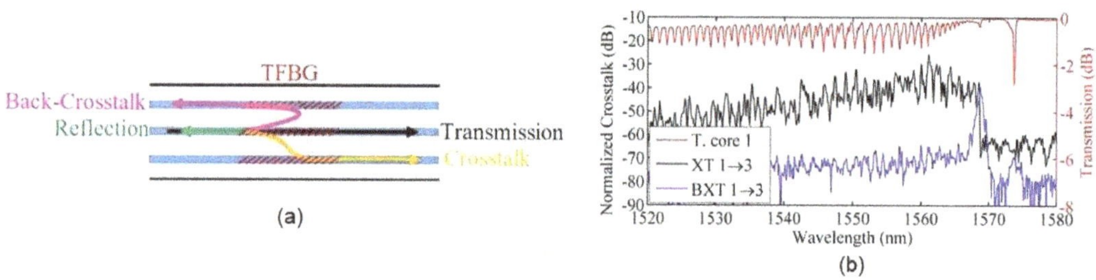

Figure 21. (**a**) Schematic diagram of the MCF with TFBG inscribed, showing inter-core crosstalk induced by TFBG; (**b**) transmission spectra of inscribed TFBG, crosstalk, and back crosstalk between cores 1 and 3. Adapted with permission from Ref. [104] © 2017 Optica Publishing Group.

Tilted fiber Bragg gratings in 4-core fibers have also been used to develop refractometers [106]. Thanks to the reduction in the core-surrounding medium distance in the multicore fiber, higher sensitivity can be achieved in comparison with their counterparts in SMF. Particularly, a plasmonic sensor-based refractometer was demonstrated using gold-coated TFBG in the 4-core fiber, which brought a 10fold sensitivity increase, showing great potential for biosensing applications. However, it is worth mentioning that so far the number of investigations on the MCF-TFBG sensor is still very low, which is much less than that of the MCF based FBG sensors and LPG sensors. It is believed that MCF will enable more TFBG based sensing applications in the future.

5. MCF-Based Long Period Grating Sensors

Long period gratings (LPGs) have also been inscribed in MCFs [107–111]. Similar to the characteristic of FBG in MCF, LPG in MCF is also bending sensitive, owing to the reason that fiber bending will give rise to tangential strain in off-center positions of the fiber. The bending dependence of LPGs in a seven-core fiber has been characterized, as shown in Figure 22 [108]. The seven-core fiber contains six outer cores that are arranged in a hexagonal pattern and a central core. The resonant wavelength shift of each LPG as a function of curvature is presented in Figure 22a, where the fiber is placed at the straight position. The bending angle dependence of the resonant wavelength under a constant curvature of 1.25 m^{-1} has also been measured, as shown in Figure 22b, which shows sinusoidal response curves, owing to the angular dependence of the strain caused by fiber bending. It is also worth mentioning that LPG in MCF is also found to be sensitive to torsion [108].

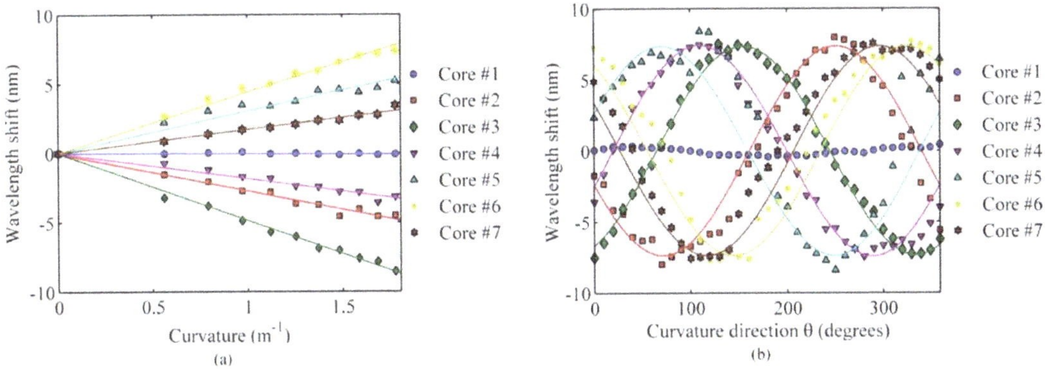

Figure 22. (**a**) Wavelength shift with the curvature magnitude. (**b**) Wavelength shift with curvature direction at a constant curvature of 1.25 m^{-1}. Adapted with permission from Ref. [108] © 2018 IEEE.

LPGs in a three-core fiber have also been used to develop a two-dimensional microbend sensor [109]. As shown in Figure 23, the three-core fiber consists of one central core and two outer cores with isosceles triangle arrangement and 120° angular offset. LPGs with different periods have been written in the two outer cores, therefore, different resonant wavelengths are obtained, as shown in Figure 23. The resonant wavelength shift of the two LPGs as a function of curvature under different bending angles has been measured, as shown in Figure 23c [109]. It is then proposed that the calibrated result can be used to estimate a coordinate system whose x- and y-axis are the resonant wavelengths of the two LPGs, respectively, as shown in Figure 23d. Therefore, the coordinate can be used to infer the bending angle. Quantitatively, this is achieved by calculating a matrix. As has been mentioned previously, since the curvature sensitivity s_i (obtained from Figure 23c) of LPG in the MCF follows a sinusoidal response curve, which can be fitted by

$$s_i = y_i + A_i \sin(\theta - \theta_i),\ i = 1,\ 2 \tag{11}$$

where θ is the bending angle, and the other parameters can be determined by experiments. Then, curvature C and bending angle θ can be obtained by calculating the matrix

$$\begin{bmatrix} \Delta\lambda_1 \\ \Delta\lambda_2 \end{bmatrix} = C \cdot \begin{bmatrix} y_1 + A_1 \sin(\theta - \theta_1) \\ y_2 + A_2 \sin(\theta - \theta_2) \end{bmatrix} \tag{12}$$

where $\Delta\lambda_1$ and $\Delta\lambda_2$ are the resonant wavelength shifts of the two LPGs.

LPGs in a seven-core fiber have also been used for external refractive index sensing, where the surface of MCF is partially coated by a thin film SnO_2 layer [110]. In addition to the widely used UV irradiation method, LPGs have also been generated in a heterogeneous seven-core fiber through electrical arc discharges by using a commercial fiber fusion splicer [111]. The LPGs in heterogeneous cores have been used for strain and temperature discriminative sensing.

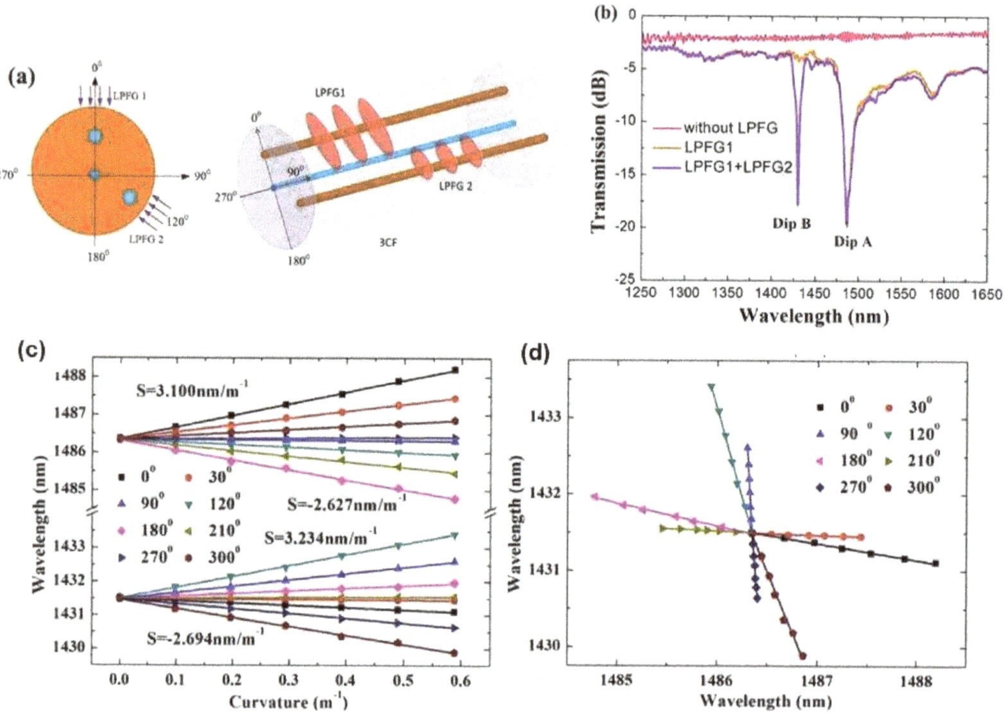

Figure 23. (**a**) Schematic diagram of the LPFGs in the three-core fiber; (**b**) measured transmission spectra of the LPFG1 and LPFG2; (**c**) measured resonance wavelength as a function of curvature for different orientations; (**d**) bending vector sensing coordinate system. Adapted with permission from Ref. [109] © 2017 Optica Publishing Group.

6. Discussion

Thanks to the unique structure of multicore fiber, which contains multiple cores in a single fiber cladding, multicore fiber turns out to be an excellent platform with unprecedented advantages that allows to develop diverse functional lab-on-fiber devices, with much better flexibility and diversity than the normal single mode fiber. So far, multicore fibers have been used to fabricate various in-fiber integrated devices and sensors, e.g., a variety of MCF gratings and interferometers. Particularly, owing to the bending sensitivity in off-center cores, multicore fibers have opened up a new field of sensing applications, i.e., bending and 3D shape sensing. A summary of various multicore fiber sensors is presented, as shown in Table 1.

Despite the exciting developments and bright future of multicore fiber sensors, there are still some critical issues that need to be addressed. Firstly, it remains a technical challenge to manipulate an individual core in the multicore fiber due to the restriction imposed by the geometric structure and lens effect of the fiber, e.g., it is difficult to write fiber gratings or fabricate interferometers with different properties in distinct cores in co-located positions of a multicore fiber. In order to enable lab-on-fiber application, high-precision core-selective fabrication technology is required in the future. Secondly, though MCF has been widely developed for bending and shape sensing, it requires calibration of the original azimuthal angle that is used for reference. Unfortunately, high precision position calibration is difficult due to the small size of the fiber. The angular deflection of the original fiber placement turns out to be one of the major factors that cause error in bending sensing. In addition, another problem is the low bending sensitivity of fiber gratings, as a result it is difficult to retrieve small curvature change, especially when the bending radius is large, which also leads to large error in shape reconstruction. Therefore, it is necessary to improve the sensitivity of bending measurement. This may be done by using a MCF that has a larger distance from the off-center core to the fiber center, as inferred from Equation (1), i.e., the longer the distance, the higher the bending sensitivity. In addition, instead of using fiber gratings, other high sensitivity sensing techniques can also be employed in bending and shape measurement, e.g., phase-sensitive optical time-domain reflectometry (φ-OTDR) [112], etc. However, it should be pointed out that the meter-scale spatial resolution of φ-OTDR is not comparable with that of FBG. Finally, it is worth mentioning that, comparing with the traditional widely used curvature sensing scheme employing optical fiber modal interferometers [113], MCF grating-based curvature sensors have the merits of being compact, robust, and having excellent mechanical strength, but their curvature sensitivity may not be comparable with some of the modal interferometer-based curvature sensors.

Table 1. Summary of different multicore fiber sensors.

Type of Sensor	Application	Principle/Feature	Refs.
MCF-based FBG sensor	Bending/curvature sensing	Fiber bending causes differential response in different cores	[37–39,41–43, 82,83,100,101]
	Transverse loading sensing	Transverse stress causes birefringence change	[44,45]
	Vibration sensing	Fiber bending causes output power change of FBGs	[47]
	Displacement sensing	Displacement causes fiber curvature change	[48]
	Acceleration sensing	Fiber bending causes differential strain in different cores	[49–53]
	Inclinometer sensor	Fiber bending causes differential strain in different cores	[54,55,59]
	3-D shape sensing	Differential strain in different cores	[35,60–62,64,68, 71,73,74,77–81,84–87]
	Temperature, strain sensing	Temperature and strain sensitivity	[96,97]
MCF-based TFBG sensor	Refractive index sensing	Using etched multicore fiber	[98,99]
	Temperature, curvature and refractive index sensing	Inter-core crosstalk, surface plasmon resonance	[104–106]
MCF-based LPG sensor	Bending sensing	Fiber bending causes wavelength shift of LPGs	[107–109]
	Refractive index sensing	Fiber surface is partially coated with a thin film SnO_2 layer	[110]
	Temperature, strain sensing	Temperature and strain sensitivity	[111]
MCF-based distributed fiber sensor	Curvature and 3-D shape sensing	Measuring the differential strain using BOTDA technique	[24]
	Temperature and strain discriminative sensing	Hybrid Raman and Brillouin sensor, or using heterogeneous multicore fiber	[25,26]
	High performance distributed fiber sensing	Large dynamic range and ultra-high measurement resolution simultaneously	[27]
	Multi-parameter sensing	Using a hybrid sensing system, e.g., DTS and DAS, simultaneously	[28,29]
	Distributed vibration sensing	MCF interferometer enables wide frequency measurement range	[30,31]

7. Conclusions

In this review, a comprehensive survey of research progress on multicore fiber grating sensors was presented. Several key aspects of the achievements using the multicore fiber in fiber grating sensing systems are summarized, including multicore fiber-based fiber Bragg grating sensors, tilted fiber Bragg grating sensors, and long period grating sensors. In addition to the traditional sensing applications, e.g., measuring temperature, stain, and refractive index, etc., the unique bending sensitivity in off-center cores of MCF has been widely developed for bending/curvature sensing or measuring physical parameters that are associated with bending, e.g., transverse loading, acceleration, vibration, flow velocity of liquid and gas, and 3D shape, etc. Thanks to the structure of multiple parallel cores in MCF, it shows unprecedented advantages in allowing to develop diverse functional lab-on-fiber devices, and it is believed that multicore fibers will enable more exciting applications in the future.

Author Contributions: Conceptualization, Z.Z.; writing—original draft preparation, Z.Z. and Y.D.; writing—review and editing, Z.Z. and M.T. All authors have read and agreed to the published version of the manuscript.

Funding: This research was funded by the National Key Research and Development Program of China (2021YFB2800902), National Natural Science Foundation of China (61931010, 62105111), Hubei Province Key Research and Development Program (2021BAA008), Fundamental Research Funds for the Central Universities (HUST: 2021XXJS026), and Natural Science Foundation of Hubei Province (2021CFB049).

Data Availability Statement: Not applicable.

Conflicts of Interest: The authors declare no conflict of interest.

References

1. Winzer, P.J. Making spatial multiplexing a reality. *Nat. Photonics* **2014**, *8*, 345–348. [CrossRef]
2. Van Uden, R.G.H.; Correa, R.A.; Lopez, E.A.; Huijskens, F.M.; Xia, C.; Li, G.; Schülzgen, A.; Waardt, H.; Koonen, A.M.J.; Okonkwo, C.M. Ultra-high-density spatial division multiplexing with a few-mode multicore fibre. *Nat. Photonics* **2014**, *8*, 865–870. [CrossRef]
3. Tang, M. Multicore Fibers. In *Handbook of Optical Fibers*; Peng, G.D., Ed.; Springer: Singapore, 2019. [CrossRef]
4. Ortiz, A.M.; Sáez, R.L. Multi-core optical fibers: Theory, applications and opportunities. In *Selected Topics on Optical Fiber Technologies and Applications*; IntechOpen: London, UK, 2017; pp. 63–102.
5. Awaji, Y.; Saitoh, K.; Matsuo, S. *Optical Fiber Telecommunications VIB: Chapter 13. Transmission Systems Using Multicore Fibers*; Elsevier: Amsterdam, The Netherlands, 2013.
6. Glas, P.; Naumann, M.; Schirrmacher, A.; Pertsch, T. The multicore fiber–a novel design for a diode pumped fiber laser. *Opt. Commun.* **1998**, *151*, 187–195. [CrossRef]
7. Mizuno, T.; Takara, H.; Sano, A.; Miyamoto, Y. Dense space-division multiplexed transmission systems using multi-core and multi-mode fiber. *J. Lightwave Technol.* **2015**, *34*, 582–592. [CrossRef]
8. Zhu, B.; Taunay, T.F.; Yan, M.F.; Fini, J.M.; Fishteyn, M.; Monberg, E.M.; Dimarcello, F.V. Seven-core multicore fiber transmissions for passive optical network. *Opt. Express* **2010**, *18*, 11117–11122. [CrossRef] [PubMed]
9. Matsui, T.; Sakamoto, T.; Goto, Y.; Saito, K.; Nakajima, K.; Yamamoto, F.; Kurashima, T. Design of 125 μm cladding multi-core fiber with full-band compatibility to conventional single-mode fiber. In Proceedings of the 2015 European Conference on Optical Communication (ECOC), Valencia, Spain, 27 September–1 October 2015; pp. 1–3.
10. Takara, H.; Sano, A.; Kobayashi, T.; Kubota, H.; Kawakami, H.; Matsuura, A.; Miyamoto, Y.; Abe, Y.; Ono, H.; Shikama, K.; et al. 1.01-Pb/s (12 SDM/222 WDM/456 Gb/s) crosstalk-managed transmission with 91.4-b/s/Hz aggregate spectral efficiency. In Proceedings of the European Conference and Exhibition on Optical Communication, Amsterdam, The Netherlands, 16–20 September 2012; p. Th-3.
11. Li, M.J.; Hoover, B.; Nazarov, V.N.; Butler, D.L. Multicore fiber for optical interconnect applications. In Proceedings of the 2012 17th Opto-Electronics and Communications Conference, Busan, Korea, 2–6 July 2012; pp. 564–565.
12. Saitoh, K.; Matsuo, S. Multicore fiber technology. *J. Lightwave Technol.* **2016**, *34*, 55–66. [CrossRef]
13. Takenaga, K.; Arakawa, Y.; Tanigawa, S.; Guan, N.; Matsuo, S.; Saitoh, K.; Koshiba, M. Reduction of crosstalk by trench-assisted multi-core fiber. In Proceedings of the Optical Fiber Communication Conference 2011, Los Angeles, CA, USA, 6–10 March 2011; Optical Society of America: Washington, DC, USA, 2011; p. OWJ4.
14. Xia, C.; Bai, N.; Ozdur, I.; Zhou, X.; Li, G. Supermodes for optical transmission. *Opt. Express* **2011**, *19*, 16653–16664. [CrossRef]
15. Klaus, W.; Sakaguchi, J.; Puttnam, B.J.; Awaji, Y.; Wada, N.; Kobayashi, T.; Watanabe, M. Free-space coupling optics for multicore fibers. *IEEE Photonics Technol. Lett.* **2012**, *24*, 1902–1905. [CrossRef]
16. Tottori, Y.; Kobayashi, T.; Watanabe, M. Low loss optical connection module for seven-core multicore fiber and seven single-mode fibers. *IEEE Photonics Technol. Lett.* **2012**, *24*, 1926–1928. [CrossRef]
17. Thomson, R.R.; Bookey, H.T.; Psaila, N.D.; Fender, A.; Campbell, S.; Macpherson, W.N.; Barton, J.S.; Reid, D.T.; Kar, A.K. Ultrafast-laser inscription of a three dimensional fan-out device for multicore fiber coupling applications. *Opt. Express* **2007**, *15*, 11691–11697. [CrossRef]
18. Watanabe, K.; Saito, T.; Imamura, K.; Shiino, M. Development of fiber bundle type fan-out for multicore fiber. In Proceedings of the 2012 17th Opto-Electronics and Communications Conference, Busan, Korea, 2–6 July 2012; pp. 475–476.
19. Noordegraaf, D.; Skovgaard, P.M.W.; Nielsen, M.D.; Bland-Hawthorn, J. Efficient multi-mode to single-mode coupling in a photonic lantern. *Opt. Express* **2009**, *17*, 1988–1994. [CrossRef] [PubMed]
20. Li, B.; Feng, Z.; Tang, M.; Xu, Z.; Fu, S.; Wu, Q.; Deng, L.; Tong, W.; Liu, S.; Shum, P.P. Experimental demonstration of large capacity WSDM optical access network with multicore fibers and advanced modulation formats. *Opt. Express* **2015**, *23*, 10997–11006. [CrossRef] [PubMed]

21. Tang, M.; Zhao, Z.; Gan, L.; Wu, H.; Wang, R.; Li, B.; Fu, S.; Liu, S.; Liu, D.; Wei, H.; et al. Spatial-division multiplexed optical sensing using MCF and FMF. In Proceedings of the Specialty Optical Fibers 2016, Vancouver, BC, Canada, 18–20 July 2016; Optical Society of America: Washington, DC, USA, 2016; p. SoM2G.3.
22. Zhao, Z.; Tang, M.; Lu, C. Distributed multicore fiber sensors. *Opto-Electron. Adv.* **2020**, *3*, 02190024. [CrossRef]
23. Zhao, Z.; Tang, M. Distributed fiber sensing using SDM fibers. In Proceedings of the Optoelectronics and Communications Conference 2021, Hong Kong, China, 3–7 July 2021; Optical Society of America: Washington, DC, USA, 2021; p. W4D.1.
24. Zhao, Z.; Soto, M.A.; Tang, M.; Thévenaz, L. Distributed shape sensing using Brillouin scattering in multi-core fibers. *Opt. Express* **2016**, *24*, 25211–25223. [CrossRef] [PubMed]
25. Zhao, Z.; Dang, Y.; Tang, M.; Duan, L.; Wang, M.; Wu, H.; Fu, S.; Tong, W.; Shum, P.P.; Liu, D. Spatial-division multiplexed hybrid Raman and Brillouin optical time-domain reflectometry based on multi-core fiber. *Opt. Express* **2016**, *24*, 25111–25118. [CrossRef]
26. Zhao, Z.; Dang, Y.; Tang, M.; Li, B.; Gan, L.; Fu, S.; Wei, H.; Tong, W.; Shum, P.; Liu, D. Spatial-division multiplexed Brillouin distributed sensing based on a heterogeneous multicore fiber. *Opt. Lett.* **2017**, *42*, 171–174. [CrossRef]
27. Dang, Y.; Zhao, Z.; Tang, M.; Zhao, C.; Gan, L.; Fu, S.; Liu, T.; Tong, W.; Shum, P.P.; Liu, D. Towards large dynamic range and ultrahigh measurement resolution in distributed fiber sensing based on multicore fiber. *Opt. Express* **2017**, *25*, 20183–20193. [CrossRef]
28. Zhao, Z.; Dang, Y.; Tang, M.; Wang, L.; Gan, L.; Fu, S.; Yang, C.; Tong, W.; Lu, C. Enabling simultaneous DAS and DTS through space-division multiplexing based on multicore fiber. *J. Lightwave Technol.* **2018**, *36*, 5707–5713. [CrossRef]
29. Dang, Y.; Zhao, Z.; Wang, X.; Liao, R.; Lu, C. Simultaneous distributed vibration and temperature sensing using multicore fiber. *IEEE Access* **2019**, *7*, 151818–151826. [CrossRef]
30. Zhao, Z.; Shen, L.; Dang, Y.; Lu, C.; Tang, M. Enabling long range distributed vibration sensing using multicore fiber interferometers. *Opt. Lett.* **2021**, *46*, 3685–3688. [CrossRef]
31. Zhao, Z.; Tang, M.; Wang, L.; Guo, N.; Tam, H.Y.; Lu, C. Distributed vibration sensor based on space-division multiplexed reflectometer and interferometer in multicore fiber. *J. Lightwave Technol.* **2018**, *36*, 5764–5772. [CrossRef]
32. Zhao, Z.; Tang, M.; Fu, S.; Liu, S.; Wei, H.; Cheng, Y.; Tong, W.; Shum, P.; Liu, D. All-solid multi-core fiber-based multipath Mach–Zehnder interferometer for temperature sensing. *Appl. Phys. B* **2013**, *112*, 491–497. [CrossRef]
33. Zhao, Z.; Liu, Z.; Tang, M.; Fu, S.; Wang, L.; Guo, N.; Jin, C.; Tam, H.Y.; Lu, C. Robust in-fiber spatial interferometer using multicore fiber for vibration detection. *Opt. Express* **2018**, *26*, 29629–29637. [CrossRef] [PubMed]
34. Duan, L.; Zhang, P.; Tang, M.; Wang, R.; Zhao, Z.; Fu, S.; Gan, L.; Zhu, B.; Tong, W.; Liu, D.; et al. Heterogeneous all-solid multicore fiber based multipath Michelson interferometer for high temperature sensing. *Opt. Express* **2016**, *24*, 20210–20218. [CrossRef]
35. Moore, J.P.; Rogge, M.D. Shape sensing using multi-core fiber optic cable and parametric curve solutions. *Opt. Express* **2012**, *20*, 2967–2973. [CrossRef]
36. Zhao, Z.; Soto, M.A.; Tang, M.; Thévenaz, L. Curvature and shape distributed sensing using Brillouin scattering in multi-core fibers. In Proceedings of the Optical Sensors 2016, Vancouver, BC, Canada, 18–20 July 2016; Optical Society of America: Washington, DC, USA, 2016; p. SeM4D.4.
37. Gander, M.J.; MacPherson, W.N.; McBride, R.; Jones, J.D.; Zhang, L.; Bennion, I.; Blanchard, P.M.; JBurnett, J.G.; Greenaway, A.H. Bend measurement using Bragg gratings in multicore fibre. *Electron. Lett.* **2000**, *36*, 120–121. [CrossRef]
38. Flockhart, G.M.H.; MacPherson, W.N.; Barton, J.S.; Jones, J.D.C.; Zhang, L.; Bennion, I. Two-axis bend measurement with Bragg gratings in multicore optical fiber. *Opt. Lett.* **2003**, *28*, 387–389. [CrossRef]
39. Zhang, H.; Wu, Z.; Shum, P.P.; Wang, R.; Dinh, X.Q.; Fu, S.; Tong, W.; Tang, M. Fiber Bragg gratings in heterogeneous multicore fiber for directional bending sensing. *J. Opt.* **2016**, *18*, 085705. [CrossRef]
40. Zhou, W.; Zhou, Y.; Dong, X.; Shao, L.; Cheng, J.; Albert, J. Fiber-optic curvature sensor based on cladding-mode Bragg grating excited by fiber multimode interferometer. *IEEE Photonics J.* **2012**, *4*, 1051–1057. [CrossRef]
41. Yang, K.; He, J.; Liao, C.; Wang, Y.; Liu, S.; Guo, K.; Zhou, J.; Li, Z.; Tan, Z.; Wang, Y. Femtosecond laser inscription of fiber Bragg grating in twin-core few-mode fiber for directional bend sensing. *J. Lightwave Technol.* **2017**, *35*, 4670–4676. [CrossRef]
42. Hou, M.; Yang, K.; He, J.; Xu, X.; Ju, S.; Guo, K.; Wang, Y. Two-dimensional vector bending sensor based on seven-core fiber Bragg gratings. *Opt. Express* **2018**, *26*, 23770–23781. [CrossRef] [PubMed]
43. Chen, X.; Zhang, C.; Webb, D.J.; Kalli, K.; Peng, G.D. Highly sensitive bend sensor based on Bragg grating in eccentric core polymer fiber. *IEEE Photonics Technol. Lett.* **2010**, *22*, 850–852. [CrossRef]
44. Silva-Lopez, M.; Li, C.; MacPherson, W.N.; Moore, A.J.; Barton, J.S.; Jones, J.D.; Zhao, D.; Zhang, L.; Bennion, I. Differential birefringence in Bragg gratings in multicore fiber under transverse stress. *Opt. Lett.* **2004**, *29*, 2225–2227. [CrossRef]
45. Silva-Lopez, M.; MacPherson, W.N.; Li, C.; Moore, A.J.; Barton, J.S.; Jones, J.D.; Zhao, D.; Zhang, L.; Bennion, I. Transverse load and orientation measurement with multicore fiber Bragg gratings. *Appl. Opt.* **2005**, *44*, 6890–6897. [CrossRef] [PubMed]
46. MacPherson, W.N.; Flockhart, G.M.; Maier, R.R.; Barton, J.S.; Jones, J.D.; Zhao, D.; Zhang, L.; Bennion, I. Pitch and roll sensing using fibre Bragg gratings in multicore fibre. *Meas. Sci. Technol.* **2014**, *15*, 1642–1646. [CrossRef]
47. Fender, A.; Rigg, E.J.; Maier, R.R.; MacPherson, W.N.; Barton, J.S.; Moore, A.J.; Jones, J.D.; Zhao, D.; Zhang, L.; Bennion, I.; et al. Dynamic two-axis curvature measurement using multicore fiber Bragg gratings interrogated by arrayed waveguide gratings. *Appl. Opt.* **2006**, *45*, 9041–9048. [CrossRef]

48. MacPherson, W.N.; Silva-Lopez, M.; Barton, J.S.; Moore, A.J.; Jones, J.D.C.; Zhao, D.; Zhang, L.; Bennion, I.; Metje, N.; Rogers, C.D.F.; et al. Tunnel monitoring using multicore fiber displacement sensor. *Meas. Sci. Technol.* **2006**, *17*, 1180–1185. [CrossRef]
49. Fender, A.; MacPherson, W.N.; Maier, R.; Barton, J.S.; George, D.S.; Howden, R.I.; Smith, G.W.; Jones, B.J.; McCulloch, S.; Chen, X.; et al. Two-axis temperature-insensitive accelerometer based on multicore fiber Bragg gratings. *IEEE Sens. J.* **2008**, *8*, 1292–1298. [CrossRef]
50. Fender, A.; MacPherson, W.N.; Maier, R.R.J.; Barton, J.S.; George, D.S.; Howden, R.I.; Smith, G.W.; Jones, B.J.; McCulloch, S.; Chen, X.; et al. Two-axis accelerometer based on multicore fibre Bragg gratings. In Proceedings of the Third European Workshop on Optical Fibre Sensors, Napoli, Italy, 4–6 July 2007; International Society for Optics and Photonics: Bellingham, WA, USA, 2007; Volume 6619, p. 66190Q.
51. He, X.; Dong, X. Accelerometer Based on the FBG Inscribed in a Twin-Core Fiber. In Proceedings of the Asia Communications and Photonics Conference 2013, Beijing China, 12–15 November 2013; Optical Society of America: Washington, DC, USA, 2013; p. AF2I.16.
52. Cui, J.; Liu, Z.; Gunawardena, D.S.; Zhao, Z.; Tam, H.Y. Two-dimensional vector accelerometer based on Bragg gratings inscribed in a multi-core fiber. *Opt. Express* **2019**, *27*, 20848–20856. [CrossRef]
53. Zhou, R.; Chen, F.; Li, S.; Wang, R.; Qiao, X. Three-Dimensional Vector Accelerometer Using a Multicore Fiber Inscribed with Three FBGs. *J. Lightwave Technol.* **2021**, *39*, 3244–3250. [CrossRef]
54. Miller, G.A.; Askins, C.G.; Cranch, G.A. Interferometric interrogation of a multicore fiber, two-axis inclinometer. In Proceedings of the 20th International Conference on Optical Fibre Sensors, Edinburgh, UK, 5–9 October 2009; International Society for Optics and Photonics: Bellingham, WA, USA, 2009; Volume 7503, p. 75032R.
55. Miller, G.A. Fabrication of a multifiber optical inclinometer. *IEEE Photon. Technol. Lett.* **2015**, *27*, 1289–1292. [CrossRef]
56. Flockhart, G.M.H.; Cranch, G.A.; Kirkendall, C.K. Differential phase tracking applied to Bragg gratings in multicore fiber for high-accuracy curvature measurement. In Proceedings of the Smart Structures and Materials 2006: Smart Sensor Monitoring Systems and Applications, San Diego, CA, USA, 27 February–1 March 2006; International Society for Optics and Photonics: Bellingham, WA, USA, 2006; Volume 6167, p. 616701.
57. Flockhart, G.M.H.; Cranch, G.A.; Kirkendall, C.K. Differential phase tracking applied to Bragg gratings in multi-core fibre for high accuracy curvature measurement. *Electron. Lett.* **2006**, *42*, 390–391. [CrossRef]
58. Askins, C.G.; Miller, G.A.; Friebele, E.J. Bend and twist sensing in a multiple-core optical fiber. In Proceedings of the 2008 Conference on Optical Fiber Communication/National Fiber Optic Engineers Conference, San Diego, CA, USA, 24–28 February 2008; pp. 1–3.
59. Cui, J.; Gunawardena, D.S.; Liu, Z.; Zhao, Z.; Tam, H.Y. All-fiber two-dimensional inclinometer based on Bragg gratings inscribed in a seven-core multi-core fiber. *J. Lightwave Technol.* **2020**, *38*, 2516–2522. [CrossRef]
60. Amanzadeh, M.; Aminossadati, S.M.; Kizil, M.S.; Rakic, A.D. Recent developments in fibre optic shape sensing. *Measurement* **2018**, *128*, 119–137. [CrossRef]
61. Moore, J.P. Shape sensing using multi-core fiber. In Proceedings of the 2015 Optical Fiber Communications Conference and Exhibition (OFC), Los Angeles, CA, USA, 22–26 March 2015; pp. 1–3.
62. Duncan, R. *Sensing Shape: Fiber-Bragg-Grating Sensor Arrays Monitor Shape at a High Resolution*; SPIE Newsroom: Bellingham, WA, USA, 2005.
63. Soller, B.J.; Gifford, D.K.; Wolfe, M.S.; Froggatt, M.E. High resolution optical frequency domain reflectometry for characterization of components and assemblies. *Opt. Express* **2005**, *13*, 666–674. [CrossRef]
64. Duncan, R.G.; Froggatt, M.E.; Kreger, S.T.; Seeley, R.J.; Gifford, D.K.; Sang, A.K.; Wolfe, M.S. High-accuracy fiber-optic shape sensing. In Proceedings of the Sensor Systems and Networks: Phenomena, Technology, and Applications for NDE and Health Monitoring 2007, San Diego, CA, USA, 19–21 March 2007; International Society for Optics and Photonics: Bellingham, WA, USA, 2007; Volume 6530, p. 65301S.
65. Froggatt, M.; Moore, J. High spatial resolution distributed strain measurement in optical fiber using Rayleigh scatter. *Appl. Opt.* **1998**, *37*, 1735–1740. [CrossRef]
66. Kreger, S.T.; Gifford, D.K.; Froggatt, M.E.; Soller, B.J.; Wolfe, M.S. High resolution distributed strain or temperature measurements in single-and multi-mode fiber using swept-wavelength interferometry. In Proceedings of the Optical Fiber Sensors 2006, Cancun, Mexico, 23–27 October 2006; Optical Society of America: Washington, DC, USA, 2006; p. ThE42.
67. Guo, H.; Tang, J.; Li, X.; Zheng, Y.; Yu, H.; Yu, H. On-line writing identical and weak fiber Bragg grating arrays. *Chin. Opt. Lett.* **2013**, *11*, 030602.
68. Westbrook, P.S.; Feder, K.S.; Kremp, T.; Taunay, T.F.; Monberg, E.; Kelliher, J.; Ortiz, R.; Bradley, K.; Abedin, K.S.; Au, D.; et al. Integrated optical fiber shape sensor modules based on twisted multicore fiber grating arrays. In Proceedings of the Optical Fibers and Sensors for Medical Diagnostics and Treatment Applications XIV, San Francisco, CA, USA, 1–2 February 2014; International Society for Optics and Photonics: Bellingham, WA, USA, 2014; Volume 8938, p. 89380H.
69. Westbrook, S.P.; Kremp, T.; Feder, K.S.; Ko, W.; Monberg, E.M.; Wu, H.; Simoff, D.A.; Ortiz, R.M. Improving distributed sensing with continuous gratings in single and multi-core fibers. In Proceedings of the Optical Fiber Communication Conference 2018, San Diego, CA, USA, 11–15 March 2018; Optical Society of America: Washington, DC, USA, 2018; p. W1K.1.
70. Floris, I.; Sales, S.; Calderón, P.A.; Adam, J.M. Measurement uncertainty of multicore optical fiber sensors used to sense curvature and bending direction. *Measurement* **2019**, *132*, 35–46. [CrossRef]

71. Khan, F.; Barrera, D.; Sales, S.; Misra, S. Curvature, twist and pose measurements using fiber Bragg gratings in multi-core fiber: A comparative study between helical and straight core fibers. *Sens. Actuators A Phys.* **2021**, *317*, 112442. [CrossRef]
72. Askins, C.G.; Miller, G.A.; Friebele, E.J. Bend and twist sensing in a multi-core optical fiber. In Proceedings of the LEOS 2008-21st Annual Meeting of the IEEE Lasers and Electro-Optics Society, Newport Beach, CA, USA, 9–13 November 2008; pp. 109–110.
73. Froggatt, M.; Klein, J.; Gifford, D. Shape sensing of multiple core optical fiber. In Proceedings of the Applied Industrial Optics: Spectroscopy, Imaging and Metrology, AIO 2011, Toronto, ON, Canada, 10–14 July 2011; Optical Society of America: Washington, DC, USA, 2011; p. AIMB2.
74. Lally, E.M.; Reaves, M.; Horrell, E.; Klute, S.; Froggatt, M.E. Fiber optic shape sensing for monitoring of flexible structures. In Proceedings of the Sensors and Smart Structures Technologies for Civil, Mechanical, and Aerospace Systems 2012, San Diego, CA, USA, 12–15 March 2012; International Society for Optics and Photonics: Bellingham, WA, USA, 2012; Volume 8345, p. 83452Y.
75. Cooper, L.J.; Webb, A.S.; Gillooly, A.; Hill, M.; Read, T.; Maton, P.; Hankey, J.; Bergonzo, A. Design and performance of multicore fiber optimized towards communications and sensing applications. In Proceedings of the Optical Components and Materials XII, San Francisco, CA, USA, 9–11 February 2015; International Society for Optics and Photonics: Bellingham, WA, USA, 2015; Volume 9359, p. 93590H.
76. Westbrook, P.S.; Feder, K.S.; Kremp, T.; Taunay, T.F.; Monberg, E.; Ortiz, R.; Puc, G. Improved uniformity of sensor gratings in offset core fibers with fiber lensing mitigation. In Proceedings of the Optical Fibers and Sensors for Medical Diagnostics and Treatment Applications XVI, San Francisco, CA, USA, 13–14 February 2016; International Society for Optics and Photonics: Bellingham, WA, USA, 2016; Volume 9702, p. 97020J.
77. Westbrook, P.S.; Kremp, T.; Feder, K.S.; Ko, W.; Monberg, E.M.; Wu, H.; Simoff, D.A.; Taunay, T.T.; Ortiz, R.M. Continuous multicore optical fiber grating arrays for distributed sensing applications. *J. Lightw. Technol.* **2017**, *35*, 1248–1252. [CrossRef]
78. Floris, I.; Calderón, P.A.; Sales, S.; Adam, J.M. Effects of core position uncertainty on optical shape sensor accuracy. *Measurement* **2019**, *139*, 21–33. [CrossRef]
79. Pauer, H.; Ledermann, C.; Tuschmann, W.; Woern, H. Non-linear compensation of production inaccuracies and material drift by adjusting the sensor data fusion algorithms for shape sensing based on FBG-optical fibers. In Proceedings of the 2014 International Conference on Multisensor Fusion and Information Integration for Intelligent Systems (MFI), Beijing, China, 28–29 September 2014; pp. 1–5.
80. Floris, I.; Madrigal, J.; Sales, S.; Adam, J.M.; Calderón, P.A. Experimental study of the influence of FBG length on optical shape sensor performance. *Opt. Lasers Eng.* **2020**, *126*, 105878. [CrossRef]
81. Idrisov, R.; Floris, I.; Rothhardt, M.; Bartelt, H. Characterization and calibration of shape sensors based on multicore optical fibre. *Opt. Fiber Technol.* **2021**, *61*, 102319. [CrossRef]
82. Barrera, D.; Gasulla, I.; Sales, S. Multipoint two-dimensional curvature optical fiber sensor based on a nontwisted homogeneous fourcore. *J. Lightw. Technol.* **2015**, *33*, 2445–2450. [CrossRef]
83. Sonoda, N.; Takagi, R.; Saito, I.; Abe, T.; Zhao, S.; Tanaka, Y. Multipoint bending measurement using multicore fiber Bragg grating and two-photon absorption process in Si-APD. *IEEE Sens. J.* **2021**, *21*, 25736–25742. [CrossRef]
84. Moon, H.; Jeong, J.; Kang, S.; Kim, K.; Song, Y.-W.; Kim, J. Fiber Bragg-grating-based ultrathin shape sensors displaying single-channel sweeping for minimally invasive surgery. *Opt. Lasers Eng.* **2014**, *59*, 50–55. [CrossRef]
85. Khan, F.; Denasi, A.; Barrera, D.; Madrigal, J.; Sales, S.; Misra, S. Multi-core optical fibers with Bragg gratings as shape sensor for flexible medical instruments. *IEEE Sens. J.* **2019**, *19*, 5878–5884. [CrossRef]
86. Khan, F.; Donder, A.; Galvan, S.; Baena, F.R.; Misra, S. Pose measurement of flexible medical instruments using fiber bragg gratings in multi-core fiber. *IEEE Sens. J.* **2020**, *20*, 10955–10962. [CrossRef]
87. Al-Ahmad, O.; Ourak, M.; Van Roosbroeck, J.; Vlekken, J.; Vander Poorten, E. Improved fbg-based shape sensing methods for vascular catheterization treatment. *IEEE Robot. Autom. Lett.* **2020**, *5*, 4687–4694. [CrossRef]
88. Lindleyet, E.; Min, S.S.; Leon-Saval, S.; Cvetojevic, N.; Lawrence, J.; Ellis, S.; Bland-Hawthorn, J. Demonstration of uniform multicore fiber Bragg gratings. *Opt. Express* **2014**, *22*, 31575–31581. [CrossRef]
89. Stępień, K.; Slowikowski, M.; Tenderenda, T.; Murawski, M.; Szymanski, M.; Szostkiewicz, L.; Becker, M.; Rothhardt, M.; Bartelt, H.; Mergo, P.; et al. Fiber Bragg gratings in hole-assisted multicore fiber for space division multiplexing. *Opt. Lett.* **2014**, *39*, 3571–3574. [CrossRef] [PubMed]
90. Ishihara, H.; Uemura, H.; Sasaki, Y.; Omichi, K.; Fujisawa, T.; Saitoh, K. Grating inscription to few-mode multi-core optical fiber. In Proceedings of the Asia-Pacific Optical Sensors Conference 2016, Shanghai, China, 11–14 October 2016; Optical Society of America: Washington, DC, USA, 2016; p. Th4A.13.
91. Idrisov, R.; Becker, M.; Rothhardt, M.; Bierlich, J.; Bartelt, H. Optimisation of fibre Bragg gratings inscription in multicore fibres. In Proceedings of the Optical Fiber Sensors 2018, Lausanne, Switzerland, 24–28 September 2018; Optical Society of America: Washington, DC, USA, 2018; p. WF64.
92. Donko, A.; Beresna, M.; Jung, Y.; Hayes, J.; Richardson, D.J.; Brambilla, G. Point-by-point femtosecond laser micro-processing of independent core-specific fiber Bragg gratings in a multi-core fiber. *Opt. Express* **2018**, *26*, 2039–2044. [CrossRef] [PubMed]
93. Lee, K.K.C.; Mariampillai, A.; Haque, M.; Standish, B.A.; Yang, V.X.; Herman, P.R. Temperature-compensated fiber-optic 3D shape sensor based on femtosecond laser direct-written Bragg grating waveguides. *Opt. Express* **2013**, *21*, 24076–24086. [CrossRef]
94. Wolf, A.; Dostovalov, A.; Bronnikov, K.; Babin, S. Arrays of fiber Bragg gratings selectively inscribed in different cores of 7-core spun optical fiber by IR femtosecond laser pulses. *Opt. Express* **2019**, *27*, 13978–13990. [CrossRef]

95. Jantzen, S.L.; Bannerman, R.H.S.; Jantzen, A.; Mennea, P.L.; Smith, D.H.; Gates, J.C.; Boyd, L.J.; Smith, P.G.R.; Holmes, C. Individual inscription of spectrally multiplexed Bragg gratings in optical multicore fibers using small spot direct UV writing. *Opt. Express* **2020**, *28*, 21300–21309. [CrossRef]
96. Suo, R.; Lousteau, J.; Li, H.; Jiang, X.; Zhou, K.; Zhang, L.; MacPherson, W.N.; Bookey, H.T.; Barton, J.S.; Kar, A.K.; et al. Fiber Bragg gratings inscribed using 800 nm femtosecond laser and a phase mask in single and multi-core mid-IR glass fibers. *Opt. Express* **2009**, *17*, 7540–7548. [CrossRef] [PubMed]
97. Zhang, Y.; Zhang, W.; Zhang, Y.; Wang, S.; Yu, L.; Yan, Y. Simultaneous measurement of curvature and temperature based on LP11 mode Bragg grating in seven-core fiber. *Meas. Sci. Technol.* **2017**, *28*, 055101. [CrossRef]
98. Kilic, S.G.; Zhu, Y.; Sheng, Q.; Inci, M.N.; Han, M. Refractometer with etched chirped fiber Bragg grating Fabry–Perot interferometer in multicore fiber. *IEEE Photon. Technol. Lett.* **2019**, *31*, 575–578. [CrossRef]
99. Hu, W.; Li, C.; Cheng, S.; Mumtaz, F.; Du, C.; Yang, M. Etched multicore fiber Bragg gratings for refractive index sensing with temperature in-line compensation. *OSA Contin.* **2020**, *3*, 1058–1067. [CrossRef]
100. Zheng, D.; Madrigal, J.; Chen, H.; Barrera, D.; Sales, S. Multicore fiber-Bragg-grating-based directional curvature sensor interrogated by a broadband source with a sinusoidal spectrum. *Opt. Lett.* **2017**, *42*, 3710–3713. [CrossRef]
101. Zheng, D.; Madrigal, J.; Barrera, D.; Sales, S.; Capmany, J. Microwave photonic filtering for interrogating FBG-based multicore fiber curvature sensor. *IEEE Photon. Technol. Lett.* **2017**, *29*, 1707–1710. [CrossRef]
102. Araújo, F.M.; Ferreira, L.A.; Santos, J.L. Simultaneous determination of curvature, plane of curvature and temperature by use of a miniaturized sensing head based on fibre Bragg gratings. *Appl. Opt.* **2002**, *41*, 2401–2407. [CrossRef] [PubMed]
103. Wolf, B.J.; Morton, J.A.S.; MacPherson, W.B.N.; van Netten, S.M. Bio-inspired all -optical artificial neuromast for 2D flow sensing. *Bioinspir. Biomim.* **2018**, *13*, 026013. [CrossRef] [PubMed]
104. Barrera, D.; Madrigal, J.; Sales, S. Tilted fiber Bragg gratings in multicore optical fibers for optical sensing. *Opt. Lett.* **2017**, *42*, 1460–1463. [CrossRef] [PubMed]
105. Madrigal, J.; Barrera, D.; Sales, S. Refractive index and temperature sensor based on TFBGs in multicore fiber. In Proceedings of the Optical Fiber Sensors 2018, Lausanne, Switzerland, 24–28 September 2018; Optical Society of America: Washington, DC, USA, 2018; p. ThE11.
106. Ortega-Gomez, A.; Loyez, M.; Lobry, M.; Chah, K.; Zubia, J.; Villatoro, J.; Caucheteur, C. Plasmonic sensors based on tilted Bragg gratings in multicore optical fibers. *Opt. Express* **2021**, *29*, 18469–18480. [CrossRef] [PubMed]
107. Saffari, P.; Allsop, T.; Adebayo, A.; Webb, D.; Haynes, R.; Roth, M.M. Long period grating in multicore optical fiber: An ultra-sensitive vector bending sensor for low curvatures. *Opt. Lett.* **2014**, *39*, 3508–3511. [CrossRef]
108. Barrera, D.; Madrigal, J.; Sales, S. Long period gratings in multicore optical fibers for directional curvature sensor implementation. *J. Lightw. Technol.* **2018**, *36*, 1063–1068. [CrossRef]
109. Wang, S.; Zhang, W.; Chen, L.; Zhang, Y.; Geng, P.; Zhang, Y.; Yan, T.; Yu, L.; Hu, W.; Li, Y. Two-dimensional microbend sensor based on long period fiber gratings in an isosceles triangle arrangement three-core fiber. *Opt. Lett.* **2017**, *42*, 4938–4941. [CrossRef]
110. Barrera, S.; Goicoechea, J.; Madrigal, J.; González-Larequi, M.; Arregui, F.J.; Sales, S. Partially coated long period fiber Bragg gratings in multicore optical fibers. In Proceedings of the Optical Sensors 2018, Zurich, Switzerland, 2–5 July 2018; Optical Society of America: Washington, DC, USA, 2018; p. SeM4E.5.
111. Wang, R.; Tang, M.; Fu, S.; Feng, Z.; Tong, W.; Liu, D. Spatially arrayed long period gratings in multicore fiber by programmable electrical arc discharge. *IEEE Photon. J.* **2017**, *9*, 1–10. [CrossRef]
112. Szostkiewicz, Ł.; Soto, M.A.; Yang, Z.; Dominguez-Lopez, A.; Parola, I.; Markiewicz, K.; Pytel, A.; Kołakowska, A.; Napierała, M.; Nasiłowski, T.; et al. High-resolution distributed shape sensing using phase-sensitive optical time-domain reflectometry and multicore fiber. *Opt. Express* **2019**, *27*, 20763–20773. [CrossRef]
113. Wang, Q.; Liu, Y. Review of optical fiber bending/curvature sensor. *Measurement* **2018**, *130*, 161–176. [CrossRef]

photonics

MDPI

Review

Characteristics of Critical-Wavelength-Existed Fiber-Optic Mach–Zehnder Interferometers and Their Sensing Applications

Chenxu Lu [1], Xiaopeng Dong [2,*] and Chi Wu [1]

[1] Southern Marine Science and Engineering Guangdong Laboratory (Guangzhou), Guangzhou 511458, China; chenxulu@gmlab.ac.cn (C.L.); qi.wu@gmlab.ac.cn (C.W.)

[2] Department of Electronic Engineering, School of Electronic Science and Engineering, Xiamen University, Xiamen 361005, China

* Correspondence: xpd@xmu.edu.cn

Abstract: In this paper, we review the characteristics of critical wavelength (CWL)-existed fiber-optic Mach–Zehnder interferometers (MZIs), including special few-mode fibers and microfibers, and their sensing applications in physical, chemical, and marine fields. Owing to the existence of CWL in the transmission spectra, the in-line MZIs show some specific characteristics. The closer the peak/dip wavelength to the CWL, the larger the wavelength shift or the related sensitivity when the interferometer is under testing. Meanwhile, CWL shifts monotonically with the variations in measurands, such as temperature (in the air or seawater), axial strain, water pressure, surrounding refractive index, etc., when they are applied to the sensing fibers. These characteristics of the CWL-existed in-line MZIs make them appealing solutions for fabricating various interferometric sensors, with the advantages of large measurement range, high sensitivity, multiparameter sensing, etc. Theoretical and experimental studies on the properties of the CWL-existed in-line MZIs are reviewed and discussed in this paper.

Keywords: fiber-optic sensors; few-mode fiber; microfiber; Mach–Zehnder interferometer; critical wavelength

Citation: Lu, C.; Dong, X.; Wu, C. Characteristics of Critical-Wavelength-Existed Fiber-Optic Mach–Zehnder Interferometers and Their Sensing Applications. *Photonics* **2022**, *9*, 378. https://doi.org/10.3390/photonics9060378

Received: 7 April 2022
Accepted: 12 May 2022
Published: 26 May 2022

1. Introduction

Fiber-optic in-line Mach–Zehnder interferometers (MZIs) have the advantages of high sensitivity, compact size, flexible structure, immunity to electromagnetic interference, and durability in harsh and corrosive environments and, therefore, have been intensively studied and applied in physical, chemical, and marine sensing fields, to monitor a considerable variety of essential parameters [1–5]. An in-line MZI can be constructed by splicing a piece of few-mode fiber (FMF) between two pieces of single-mode fibers (SMF)—termed SFS structure [6–15]—or replacing the FMF with the multimode fiber (MMF) [16–21], two-core fiber [22], thin-core fiber [23], no-core fiber [24], photonic crystal fiber [25] and few-mode microfiber [26–36], where offset fusion splicing [37], fiber taper [38], and collapsed splicing [39] are often used in the fabrication.

A typical in-line MZI shows periodic interference fringes. The interference peaks/dips in the periodically changed transmission spectrum exhibit the same behavior, i.e., shifting in the same direction, with the same sensitivity, when monitoring sensing parameters. In this case, when such a sensor system is being applied to detect any particular parameter through monitoring the wavelength variations of a specific interference peak/dip or intensity variations at a particular wavelength, some critical issues will limit the veracity of the sensor performance: (1) multivalue outputs and limited measurement range induced by overlapping of adjacent peaks/dips; (2) limitations in multiparameter applications caused by the uniformity of the interference fringes and their wavelength sensitivity.

However, in the transmission spectra of some in-line MZIs, a critical wavelength (CWL) may appear in the operating wavelength range. At the CWL, the wavelength response period seen is infinite in the transmission spectrum. This interference fringe, which shows the least dependence on wavelength, is defined as the achromatic fringe [26]. In fact, the CWL is the wavelength at which the two interference modes' group velocities equal each other. Therefore, the propagation constant difference in the two interference modes changes nonlinearly with wavelength and turns around at CWL. In this way, the CWL is also called the dispersion turning point (DTP).

The CWL-existed in-line MZIs exhibit intriguing characteristics that researchers have thoroughly studied and applied in a wide range of sensing fields: (1) The peak lying closest to CWL at each side has its maximum spacing from the adjacent peak. Thus, compared with periodically changed interference fringes, the CWL is easy to identify in the transmission spectrum. Moreover, the CWL is exclusive at the operating wavelength range, which becomes an excellent candidate for measuring various parameters, especially in large-range measurement. (2) The interference peak/dip closest to the CWL has the maximum wavelength sensitivity among all the interference peaks/dips. (3) The interference peaks/dips on each side of the CWL shift to opposite directions under sensing parameter variations. The characteristics of the CWL and the interference peaks/dips have been proved with competitive advantages in large-range measurement [7,8,30], high wavelength sensitivities of interference peaks/dips [10,28–33,35,36], and co-located multiparameter sensing [9,11]. With these advantages, the CWL-existed in-line MZIs have been reported in sensing parameters of axial strain, curvature, displacement, surrounding refractive index, temperature, relative humidity, static pressure, chemical solution concentration, etc., according to the particular circumstances. Difference fibers such as liquid-core fiber, MMF, FMF, and few-mode microfibers are employed in these CWL-existed in-line MZIs.

In this paper, we provide a simple but overall review of the developments and recent progress on CWL-existed in-line MZIs and their sensing applications. In Section 2, the generation of the CWL is explained. In Section 3, we review the wavelength sensitivity characteristics of the interference peaks/dips lying near the CWL and their applications in sensors. The large-range measurement sensors employing CWL detection are reviewed in Section 4. Lastly, in Section 5, some conclusions and prospects are given.

2. Theoretical Analysis of the Transmission Characteristics of the CWL-Existed in-Line MZIs

CWL has been experimentally observed in different kinds of in-line MZIs employing MMFs [17–20], FMFs [6–15], liquid-core fibers [40], tapered few-mode microfibers [25–35], highly birefringent (HB) fibers [41] or birefringent side-hole fibers [42]. The mode interferences in these CWL-existed in-line MZIs are mainly between the fundamental mode and the first circularly symmetric high-order mode or the first antisymmetric high-order mode and between two orthogonal (x and y) polarization modes. The schematic diagram of the sensor system is shown in Figure 1. Compared with mode interference, where the polarization-dependent light is included, the in-line MZI with two circularly symmetric modes interference provides merits of stability and polarization independence [43]. As an example, in this section, we use an SFS structure reported by our group in references [6–14], in which the LP_{01} and LP_{02} mode interferences were employed, to explain the transmission characteristics of the CWL-existed MZIs theoretically.

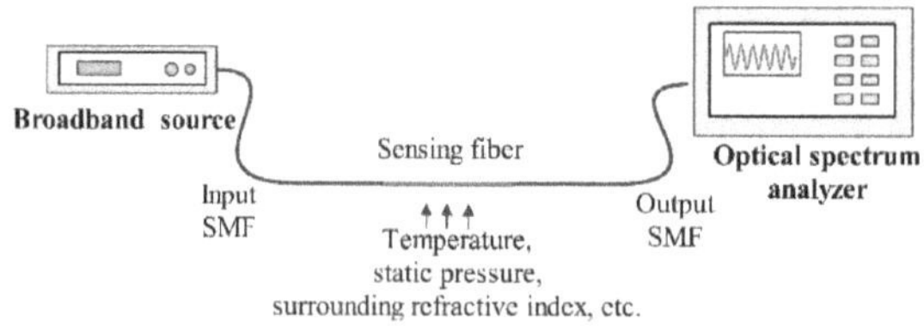

Figure 1. Schematic of the critical-wavelength-existed in-line Mach–Zehnder interferometer measurement system.

Figure 2 shows the cross-section structure of the FMF and the schematic diagram of the SFS structure. The relative refractive index difference is defined as $\Delta n_{co/cli} = (n_{co/cli} - n_0)/n_0$, where the $n_{co/cli}$ and n_0 are the refractive indices of the core/i^{th} inner cladding and pure silica of the FMF, respectively. FMF is specially designed with five layers and a W-shaped refractive index difference profile. Therefore, at the operational wavelength planned, the FMF supports only the LP_{01} and LP_{02} modes propagating in the fiber under well-centered splicing [44].

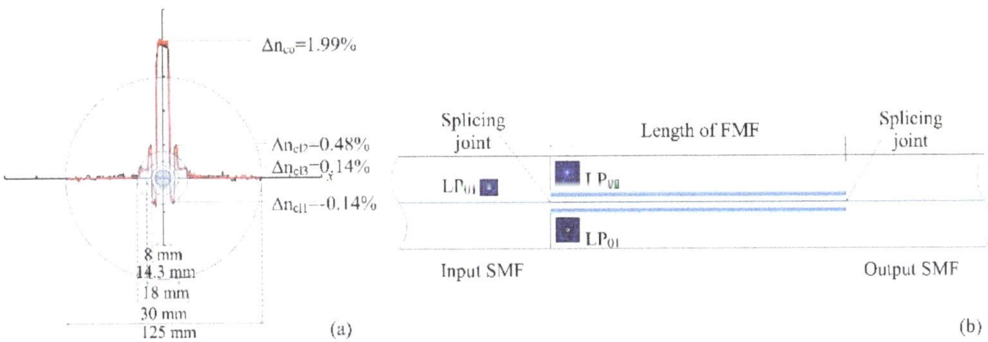

Figure 2. (**a**) Geometrical structure and relative refractive index difference profile of the few-mode fiber; (**b**) the SMF–FMF–SMF (SFS) structure. The insets are simulated mode field distributions of the LP_{01} mode propagating in SMF and the LP_{01} and LP_{02} modes propagating in FMF at 1550 nm wavelength using the finite element model. SMF: single-mode fiber; FMF: few-mode fiber.

The LP_{01} and LP_{02} modes propagating in the FMF are excited by the fundamental mode LP_{01} in input SMF. The interference between the LP_{01} and LP_{02} modes is transmitted by output SMF. The ratios of optical power transferred to the LP_{01} and LP_{02} modes in the FMF from input SMF are $t_{01} = P_{01}/P_{in}$ and $t_{02} = P_{02}/P_{in}$, and calculated as 81% and 11.5%, respectively. The transmission, T, through the SFS structure with L_{FMF}-length FMF is given by [15] as follows:

$$T = \frac{P_{out}}{P_{in}} = t_{01}^2 + t_{02}^2 + 2t_{01}t_{02}cos(\varphi(\lambda)) \quad (1)$$

where $\varphi(\lambda) = \Delta\beta \cdot L_{FMF}$ is the phase difference between LP_{01} and LP_{02} modes. $\Delta\beta$ is the propagation constant difference between LP_{01} and LP_{02} modes.

With the parameters of FMF given in Figure 1, a finite element model is established to simulate the relationship between the $\Delta\beta$ and wavelength, which is depicted as a dashed line in Figure 3. The calculated transmission spectrum combining Equation (1) and the simulated value of $\Delta\beta$ is shown as a solid line in Figure 3, where the physical length of the employed FMF is 50 cm. Because the dispersion of $\Delta\beta$ exhibits nonlinear behavior and has a maximum, corresponding to the value of CWL in the transmission spectrum at the operational wavelength, the wavelength response period at CWL is infinite. Moreover, the periods of interference fringes closest to CWL from both sides reach a maximum. As shown in Figure 3, numbering from the peak that lies closest to CWL from each side, the peaks that are located on the higher wavelength side of CWL (i.e., within the region from CWL to 1580 nm) are denoted as $P_{H,1}$, $P_{H,2}, \ldots$, respectively. Peaks located on the lower wavelength side of CWL (i.e., within the region from 1530 nm to CWL) are denoted as $P_{L,1}$, $P_{L,2}, \ldots$, respectively. The interference dips are also numbered in the same way.

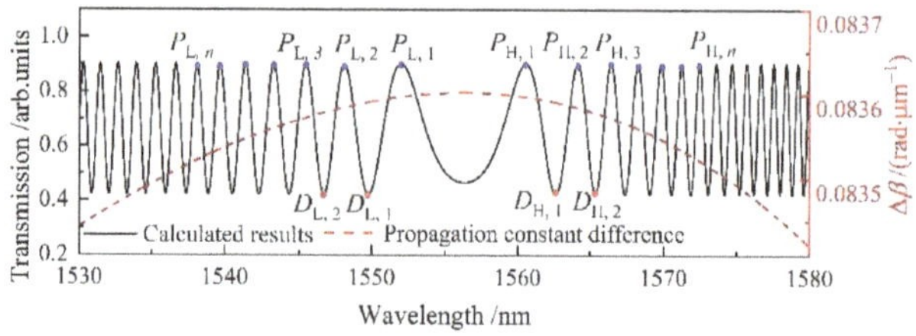

Figure 3. The simulated curve of the propagation constant difference $\Delta\beta$ of LP_{01} and LP_{02} modes propagating in the few-mode fiber, and the transmission spectrum of the SFS structure ($L_{FMF} = 50$ cm) under the temperature of 25 °C versus wavelength.

With the achromatic fringes in the transmission spectrum, multiple sensing indicators are provided in the sensors employing CWL-existed in-line MZIs, including monitoring wavelength shifts in CWL [6–8,18,40], $P_{H,1}/D_{H,1}$, $P_{L,1}/D_{L,1}$ [9–11,15,17,28–36], and intensity variations [12] at a specific wavelength or wavelength ranges. Studies in CWL-existed in-line MZIs show that utilizing CWL shifts or other monitoring parameters listed above, the CWL-existed in-line MZIs can provide competitive advantages over conventional periodically changed interferometers in the applications of large-range measurement, co-located multiparameter sensing, and high-sensitivity sensing. Moreover, the theoretical study of wavelength sensitivities of interference fringes offers a direction for mode interferometers to maximize their peak wavelength sensitivities.

3. Characteristics of Interference Peaks/Dips in the CWL-Existed in-Line MZIs and Their Applications

In a CWL-existed transmission spectrum, the interference peaks/dips, especially those lying closest to CWL on both sides, exhibit essential characteristics different from the typical periodically changing fringes. These differences were experimentally observed by Salik et al. [15] and Zhou et al. [17]. However, theoretical analysis was not presented. The wavelength sensitivities of the interference fringes under temperature, axial strain, static pressure, and surrounding refractive index (SRI) variations were further theoretically and experimentally studied in Refs. [10,11,28–30,32,36], which are reviewed in this section.

3.1. *Characteristics of Interference Peaks/Dips in the CWL-Existed SFS Structures*

For the CWL-existed SFS structure employing the FMF shown in Figure 2, because the geometrical structure and relative refractive index difference profile of the FMF are determined, the temperature, axial strain, and static pressure wavelength sensitivities of the interference peaks/dips are governed by the wavelength separations between the measured wavelengths and the CWL. The temperature (T), axial strain (ε), and static pressure (P) wavelength sensitivities can be written as follows [11,13]:

$$\frac{\Delta\lambda}{\Delta \mathrm{T}} = -\left(\frac{\partial(\Delta\beta)}{\partial \mathrm{T}} + \Delta\beta\cdot\alpha\right)\left(\frac{\partial(\Delta\beta)}{\partial\lambda}\right)^{-1} \quad (2)$$

$$\frac{\Delta\lambda}{\Delta\varepsilon} = -\left(\Delta\beta + \frac{k_0^2}{2\beta_1\beta_2}\left(\gamma\left(n_0^4\Delta\beta + \left(n_{co}^4 - n_0^4\right)\left(\beta_1 b_1 - \beta_2 b_2\right)\right) + V\delta\left(\beta_1\frac{\partial b_1}{\partial \mathrm{V}} - \beta_2\frac{\partial b_2}{\partial \mathrm{V}}\right)\right)\right)\left(\frac{\partial(\Delta\beta)}{\partial\lambda}\right)^{-1} \quad (3)$$

$$\frac{\Delta\lambda}{\Delta P} = -\left(\frac{-(1-2\nu)}{E}\Delta\beta + \frac{k_0(1-2\nu)}{2E}\left(2\rho_{12} + \rho_{11}\right)\left(n_1^3 - n_2^3\right) - \frac{(1-2\nu)V^3}{2r_{co}^2 E}\left(\frac{1}{\beta_1}\frac{\partial b_1}{\partial \mathrm{V}} - \frac{1}{\beta_2}\frac{\partial b_2}{\partial \mathrm{V}}\right)\right)\left(\frac{\partial(\Delta\beta)}{\partial\lambda}\right)^{-1} \quad (4)$$

where β_{01} and β_{02} are the propagation constants of the LP_{01} and LP_{02} modes, respectively. $\alpha = (1/L_{FMF})(\partial L_{FMF}/\partial T) = 5.1 \times 10^{-7}/°C$ is the thermal expansion coefficient for pure silica [20]. $\gamma = \rho_{12} - \nu(\rho_{11} + \rho_{12})$ and $\delta = \nu(n_{co}^2 - n_0^2) + (\gamma/2)(n_{co}^4 - n_0^4)$. $\rho_{11} = 0.12$, $\rho_{12} = 0.27$ are strain-optic coefficients for fused silica, respectively [45]. $\nu = 0.17$ and $E = 78$ GPa is the Poisson ratio and Young's modulus for pure silica, respectively [46]. $k_0 = 2\pi/\lambda$ is the free space wave vector. r_{co} is the core radius of the FMF. n_1 and n_2 is the effective refractive index of the LP_{01} and LP_{02} modes propagating in the FMF, respectively. $V = k_0 r_{co}\sqrt{n_{co}^2 - n_0^2}$ is the normalized frequency of the FMF. β_i and b_i (i = 1, 2) is the propagation constant and normalized propagation constant of LP_{0i} mode, respectively, and can be represented as

$$\beta_i^2 = k_0^2\left(n_0^2 + b_i\left(n_{co}^2 - n_0^2\right)\right) \tag{5}$$

The wavelength sensitivity curves under temperature, axial strain, and static pressure are calculated and depicted in solid curves in Figure 4a–c, together with the experimental results. From Equations (2)–(5) and Figure 4, the wavelength sensitivities of interference fringes in the spectrum of the SFS structure with CWL can be summarized as follows:

(1) The interference peaks/dips lying on the different sides of the CWL shift to opposite directions under temperature, axial strain, and static pressure variations. For example, when temperature increases, there is a blueshift for peaks on the lower wavelength side of the CWL (i.e., within the region from 1500 nm to the CWL), while for peaks on the longer wavelength side of the CWL (i.e., within the region from the CWL to 1580 nm), there is a redshift. The experimental transmission spectra of the SFS structure under temperature variations are shown in Figure 4d, which agrees well with the calculation results.

(2) The wavelength sensitivity of the interference peak is only related to the wavelength spacing between the measured wavelength and the CWL. The L_{FMF} used in the SFS sensor and the sensing parameters pre-applied to the FMF section may change the wavelength separation between the peak wavelength and the CWL. Thus, the variation in wavelength separation will affect the corresponding wavelength sensitivity of the measured peaks, as shown in Figure 4a–c. The experimental results show close agreement with the simulated curves.

(3) As the peak wavelength value approaches the CWL value, i.e., $\lambda/\lambda_{CWL} \to 1$, the wavelength sensitivity increases significantly, as shown in Figure 4a–c, from both the theoretical and experimental points of view. However, the mathematically derived infinitely high sensitivity at the CWL has no practical meaning because $\partial(\Delta\beta)/\partial\lambda = 0$ at CWL from the theoretical analysis, and an infinite length of FMF is needed in experiments to minimize the wavelength spacing between the wavelength of $P_{L,1}/P_{H,1}$, and the CWL, which is unpractical in applications.

Because the wavelength sensitivities of interference peaks and CWL are different under sensing parameter variations, the SFS structure can be applied in multiparameter sensing by simultaneously measuring the shift in CWL and interference peaks. Simultaneous measurements of temperature and axial strain [9] have been studied by detecting wavelength shifts in $P_{H,1}$ and $P_{L,1}$, taking advantage of their highest wavelength sensitivities among all of the peaks on the right and left sides of the CWL, respectively.

To further improve the sensitivity performance, a cascaded CWL-existed SFS structure was proposed by our group, which has been applied in static water pressure and temperature sensing [11]. Figure 5 shows the schematic diagram and transmission spectra under different static water pressures of the cascaded SFS structure. Vernier effect was adopted to magnify the wavelength sensitivities of interference peaks. The free spectrum range of the envelope in the superimposed transmission spectra varies with wavelength and reaches the maximum when approaching the envelope critical wavelength (CWL_E). Therefore, the envelop peaks, $P_{EL,1}$ and $P_{EL,2}$, are easy to identify. A maximum static water pressure sensitivity of 4.072 nm/MPa and temperature sensitivity of 1.753 nm/°C is achieved using this sensing approach, which is more than 7 times larger in terms of the wavelength sensitivity than those reported in sensors using a single CWL-existed SFS structure [14]. This cascaded CWL-existed SFS static-pressure sensor can be applied in marine pressure sensing with temperature compensation by simultaneously measuring the envelope peaks $P_{EL,1}$ and $P_{EL,2}$.

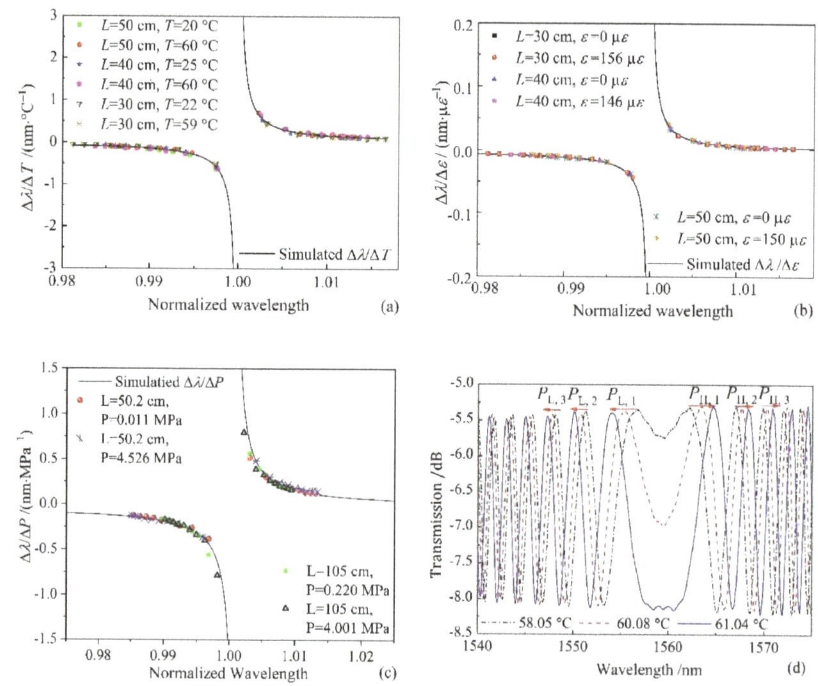

Figure 4. Characteristics of interference fringes in the transmission spectra of the CWL-existed SFS structure: (**a**) temperature wavelength sensitivities of interference fringes; (**b**) axial strain wavelength sensitivities of interference fringes; (**c**) static pressure wavelength sensitivities of interference fringes; (**d**) transmission spectra of the SFS structure (L_{FMF} = 50 cm) under temperature variations.

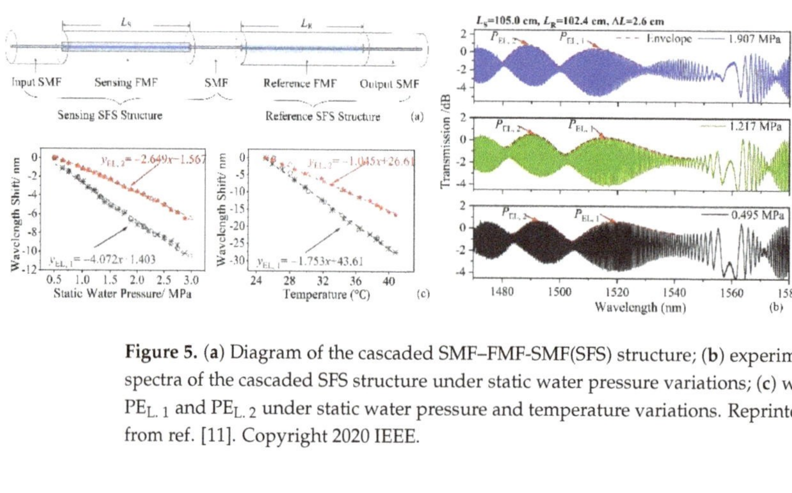

Figure 5. (**a**) Diagram of the cascaded SMF–FMF-SMF(SFS) structure; (**b**) experimental transmission spectra of the cascaded SFS structure under static water pressure variations; (**c**) wavelength shifts in $PE_{L,1}$ and $PE_{L,2}$ under static water pressure and temperature variations. Reprinted with permission from ref. [11]. Copyright 2020 IEEE.

3.2. Characteristics of Interference Peaks/Dips in the CWL-Existed Few-Mode Microfiber in-Line MZIs

The few-mode microfiber interferometer fabricated by tapering a normal-sized optical fiber to the diameter of several micrometers is widely applied in SRI sensing using the evanescent waves propagating in the microfiber section. Wei et al. [28] proposed a few-mode microfiber interferometer designed at the CWL. They explored the SRI sensing properties of tapered microfiber interferometers in a gaseous medium, where the SRI changes around 1. The nonadiabatic few-mode microfiber is tapered from a standard SMF. The fundamental HE_{11} mode and the high-order HE_{12} mode are excited and propagate in the uniform waist region, as shown in Figure 6.

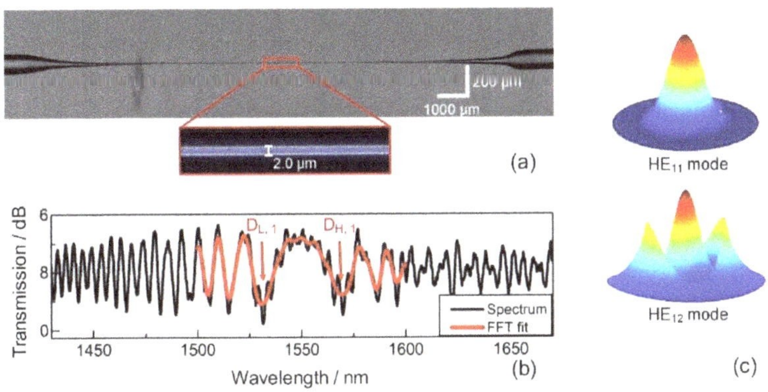

Figure 6. (**a**) The microscopic view of the tapered optical microfiber; (**b**) the origin (black) and the fast Fourier transform (FFT)-filtered (red) transmission spectra of the tapered few-mode microfiber surrounded by air; (**c**) the simulated mode profiles of HE_{11} and HE_{12} modes. Adapted with permission from [28].

For a particular interference peak/dip with the peak/dip wavelength of λ_N, the SRI wavelength sensitivity is written as follows [28,47]:

$$\frac{\Delta\lambda_N}{\Delta SRI} = \frac{\lambda_N}{G}\frac{\partial\left(\Delta n_{eff}\right)}{\partial n} \tag{6}$$

where $G = n_g^{HE_{11}} - n_g^{HE_{12}}$ is the difference between the group effective index of HE_{11} and HE_{12} modes. The group effective index can be calculated by $n_g = n_{eff} - \lambda_N \partial\left(n_{eff}\right)/\partial\lambda_N$. Therefore, with known material and diameter parameters of the uniform microfiber section, the term $\Delta\lambda_N/\Delta SRI$ increases dramatically as the value of G approaches zero, which corresponds to the CWL point. Moreover, the interference peaks/dips lying on each side of the CWL shift to opposite wavelength directions under SRI variations. These conclusions agree well with the wavelength sensitivity study of CWL-existed SFS structure. The calculated wavelength sensitivity curves of SRI under different microfiber diameters are shown in Figure 7. By measuring the peak wavelength separation of $D_{L,1}$ and $D_{H,1}$, a maximum sensitivity of −69,984.3 nm/RIU ("−" denotes blue wavelength shifts) around the air refractive index is obtained in experiments. The temperature influence on the gas RI sensor is also reported, and the measured temperature sensitivities of $D_{L,1}$ and $D_{H,1}$ are −2.134 nm/°C and 1.899 nm/°C, respectively. Thus, the cross-sensitivity of the gas refractive index sensor is 5.763×10^{-5} RIU/°C.

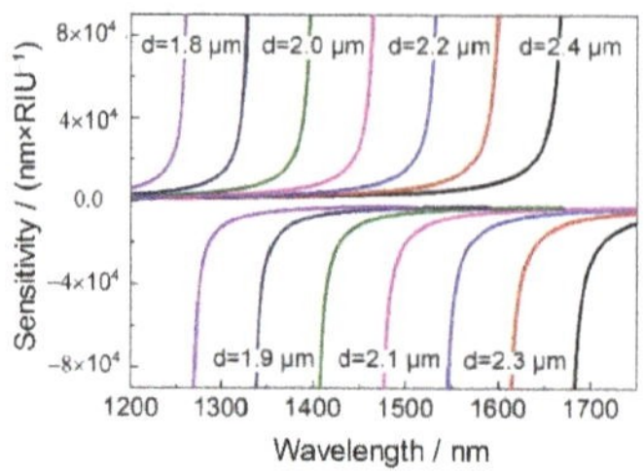

Figure 7. The surrounding refractive index sensitivities of interference fringes around CWLs at different waist diameters of microfibers. Adapted with permission from Ref. [28].

With the merits of excellent sensitivity performances and simple fabrication processes, few-mode microfibers have been further studied and applied in various SRI-related sensing parameters. Guan et al. [29] thoroughly investigated the sensitivity of a two-mode microfiber MZI as a function of SRI. Since Equation (7) can be rewritten as

$$\frac{\Delta\lambda_N}{\Delta n_{SRI}} = \lambda_N \frac{1}{\Gamma}\left(\frac{1}{\Delta n_{eff}}\frac{\partial\left(\Delta n_{eff}\right)}{\partial n_{SRI}}\right) \tag{7}$$

where $\Gamma = 1 - \lambda_N/\Delta n_{eff} \cdot \Delta n_{eff}/\partial\lambda$ is a dispersion factor that characterizes the effect of index difference variations with wavelength, $\Delta n_{eff}/\partial\lambda$. Since the whole term in the bracket was negative [48], the sign of Γ is crucial in determining the *SRI* sensitivity sign. CWL exists at the point where the sign of the *SRI* sensitivity changes. The simulated dispersion factors Γ and $\Delta\lambda_N/\Delta n_{SRI}$ as functions of the fiber diameter is shown in Figure 8. The numerical simulation revealed that two CWLs exist as the microfiber diameter decreases. Experiments were conducted to study the wavelength sensitivities of interference fringes around both CWLs. At the second CWL, an ultrahigh *SRI* sensitivity of approximately 95,836 nm/RIU with the microfiber diameter of 1.87 μm was achieved around the *SRI* of air. Wang et al. [33] applied the few-mode microfiber MZI in marine fields by measuring sodium nitrate in seawater. They achieved the wavelength sensitivity of 5.98 pm/ppm using $D_{L,1}$, corresponding to the *SRI* sensitivity of 50,396.09 nm/RIU. CWL was also observed in Z-shaped few-mode microfibers between the interference of HE_{11} and HE_{21} modes [34]. The SRI sensitivity of 1.46×10^5 nm/RIU was experimentally achieved around the refractive index of water.

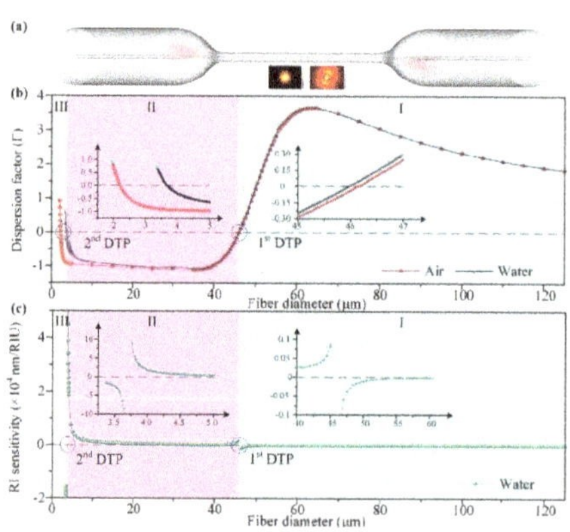

Figure 8. (**a**) Schematic diagram of the taper-based in-line MZI; the inset presents calculated intensity profiles of the interference two modes, HE_{11} and HE_{12}, respectively; (**b**) simulated dispersion factor Γ and (**c**) RI sensitivity as functions of the fiber diameter. Adapted with permission from Ref. [29].

Zhao et al. [32] demonstrated an ultrasensitive sensor for SRI and temperature measurement by employing a CWL-existed two-mode microfiber. An ultrahigh SRI sensitivity of 24,209 nm/RIU around 1.3320 and a high-temperature sensitivity of −2.47 nm/°C were experimentally achieved by a microfiber with a waist diameter of ~4.8 μm. Meanwhile, the CWL-existed microfiber was coated with a polydimethylsiloxane (PDMS) layer to increase the temperature sensitivity to about 8.33 nm/°C. To further increase the measurement range of the interference dips, they [30] proposed a fiber Bragg grating (FBG) cascaded microfiber in-line MZI operating at the CWL and applied it in seawater temperature measurement, as shown in Figure 9a. The FBG resonant dip, located in the achromatic fringe region, works as a rough measurement tool to identify the temperature measurement ranges. Meanwhile, the interference dip closest to the CWL with the highest sensitivity among all the interference dips was employed to measure the water temperature once the rough range was determined. The two-mode microfiber MZI was packaged in PDMS to increase temperature sensitivity. The transmission spectrum of the FBG-cascaded two-mode microfiber MZI in PDMS package is shown in Figure 9b. With the aid of FBG, the temperature measurement range could be enlarged from ~2 °C to ~36 °C, with a temperature sensitivity of ~38 nm/°C, as shown in Figure 9c.

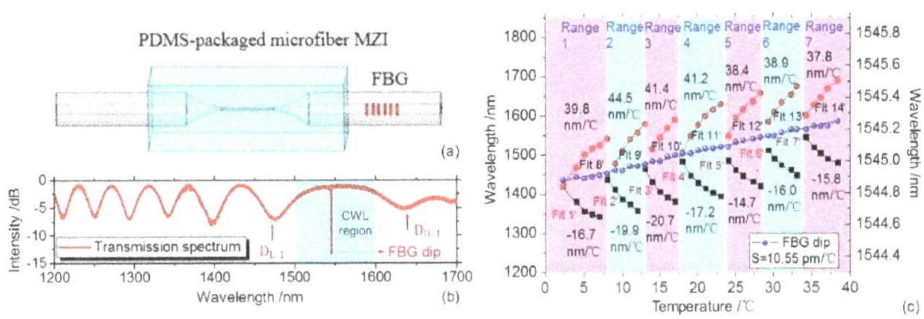

Figure 9. (**a**) Schematic of the polydimethylsiloxane (PDMS)-packaged two-mode microfiber for temperature measurement; (**b**) the transmission spectrum of the PDMS-packaged two-mode microfiber MZI cascaded with FBG; (**c**) cooperation of the FBG resonant dip and the interference dips ($D_{L,1}$ and $D_{H,1}$) lying on each side of the CWL in temperature measurement. Reprinted with permission from Ref. [30]. Copyright 2020 Elsevier.

Apart from using nonadiabatic tapers to fabricate the few-mode microfibers by tapering a standard SMF, other special fibers were employed to increase the sensor performance and redshift the cutoff wavelength of the high-order mode in the microfiber. In 2020, Wang et al. [31] proposed a CWL-existed in-line MZI by tapering a ring-core two-mode fiber. Quasi-adiabatic transition tapers were fabricated to provide HE_{11} and HE_{21} modes with continuous mode evolutions to the microfiber section. The two modes launched into the fiber were collected at the fiber output. A CWL was observed when the tapered two-mode microfiber was immersed in a liquid environment. The SRI sensitivities of 79,283.7 nm/RIU and −48,772 nm/RIU were reported, respectively, by detecting the right and left interference dip closest to CWL, i.e., $D_{L,1}$ and $D_{H,1}$. Thus, a maximum sensitivity of 128,055.7 nm/RIU was achieved by measuring the wavelength separation variations between these two interference dips. This CWL-based two-mode microfiber was further applied in an immunoglobulin G (IgG) level measurement. The limits of detection of 10 fg/mL of IgG in phosphate-buffered saline solution were achieved in the experiments. Luo et al. [36] reported a refractive index sensor fabricated by tapering a single stress-applying fiber (SSAF), where a stress-applying part was introduced into the fiber cladding to engineer the cladding material index. The SSAF-based two-mode microfiber could significantly reduce the sensitivity of the group effective refractive index difference between the two interference modes—HE_{11} and TE_{01}. Therefore, the operation bandwidth of the two-mode microfiber can be broadened to 500 nm, compared with that of conventional

SMF-based two-mode microfiber with the same diameter operating at CWL. By detecting the wavelength separation changes between $D_{L,1}$ and $D_{H,1}$, a maximum sensitivity of 30,563 nm/RIU was reported in the refractive index measurement range of 1.3212~1.3216.

The interference fringes in the transmission spectrum of a CWL-existed few-mode microfiber structure can also be applied in vibration measurement. Luo et al. [35] demonstrated a highly sensitive vibration sensor using a CWL-existed two-mode microfiber MZI and packaging it in a rectangular through-hole cantilever beam. By detecting the interference dip near the CWL, the maximum axial-strain sensitivity reached −45.55 pm/με. This structure was also applied in acceleration sensing experiments, where the acceleration sensitivity of the sensor was 0.764 V/g at 45 Hz.

The sensors employing the CWL-based in-line MZI structures and measuring the wavelength shifts in the interference peaks/dips closest to the CWL on both sides are listed in Table 1. Compared with SFS structures, the few-mode microfibers exhibit higher sensitivity in temperature, axial strain, acceleration, and surrounding refractive index measurements. However, the taper structures and ultrasmall waists of the microfibers may cause structural fragility. Therefore, the microfibers and the taper regions should be packaged carefully to guarantee robustness in applications. During experiments, the noise induced by temperature fluctuations and axial strain in the packaging process should also be considered. The CWL-based in-line MZIs have advantages in that the CWL and interference fringes near the CWL exhibit different wavelength sensitivities under the same measurement and variation. Thus, the temperature- and strain-induced noises can be compensated by simultaneously detecting two or three peaks/dips.

Table 1. Performance comparisons of different CWL-existed in-line MZI sensors.

Structure	Sensor Application	Maximum Sensitivity	Refs.
Few-mode fiber	Curvature	−0.745 nm/m^{-1}	[6]
Few-mode fiber	Axial Strain	43 pm/με	[10]
Few-mode fiber	Temperature	704 pm/°C	[10]
Cascaded few-mode fiber	temperature	1.753 nm/°C	[11]
Cascaded few-mode fiber	Static pressure	4.072 nm/MPa	[11]
Few − mode microfiber with a diameter of 2 μm	Refractive index around 1	−69,984.3 nm/RIU	[28]
Few − mode microfiber with a diameter of 1.87 μm	Refractive index around 1	95,836 nm/RIU	[29]
Few − mode microfiber with a diameter of 3.57 μm	Sodium nitrate in seawater	5.98 pm/ppm	[33]
Z-shaped few-mode microfiber	Refractive index around 1.333	1.46 × 105 nm/RIU	[34]
Few-mode microfiber coated with PDMS	Temperature	38 nm/°C	[30]
Few-mode microfiber tapered by a ring-core two-mode fiber	Refractive index around 1.333	128,055.7 nm/RIU	[31]
Few-mode microfiber tapered by a single stress-applying fiber	Refractive index around 1.3212	30,563 nm/RIU	[36]
Few − mode microfiber with a diameter of 2.2 μm	Axial strain	−45.55 pm/με	[35]
Few − mode microfiber with a diameter of 2.2 μm	Acceleration	764 mV/g	[35]

4. Characteristics of the Achromatic Fringe in CWL-Existed in-Line MZIs and Their Applications

In the transmission spectrum of a CWL-existed in-line MZI, achromatic fringes are different from the periodically changing interference fringes and, therefore, are easy to identify. Because the achromatic fringe shifts monotonously with sensing parameters, CWL detection sensors can be applied in large-range measurements. This section reviews a demodulation method of the CWL and sensing applications employing CWL detection.

4.1. Detection Scheme of CWL in the SFS Sensing Structure

The demodulation method of the CWL in the transmission spectrum and its advantages, compared with the traditional interference peak detection, is explained in Ref. [8], using an etched SFS structure in SRI detection. Figure 10 shows the simulated transmission spectra, the shift in CWL, and the wavelength shift in peak/dip located closest to the CWL on each side under different SRIs. The peaks/dips in both sides of CWL will move toward the CWL with increasing SRI. If SRI is measured by monitoring the wavelength shift in peaks/dips lying around the CWL, the peaks/dips will end at CWL with only a small measurement range of SRI, as shown in Figure 10. Therefore, traditional interference peak/dip wavelength detection methods limit the measurement range significantly.

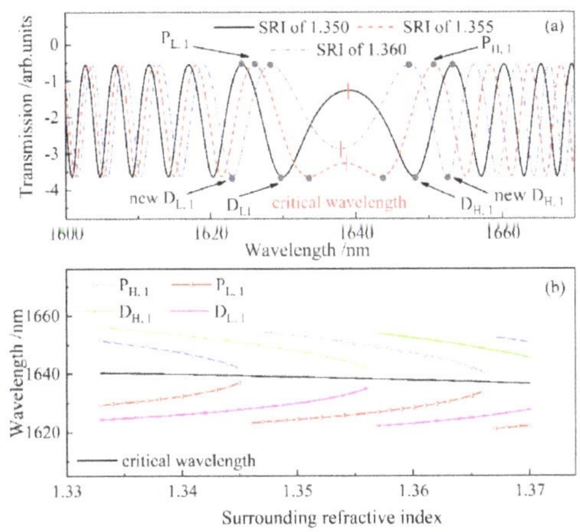

Figure 10. The simulated results of the etched SFS structure (d_{FMF} = 21.3 μm, L_{FMF} = 20 cm): (**a**) the transmission spectra under the SRIs of 1.350, 1.355, and 1.360; (**b**) wavelength shifts in the critical wavelength and interference peaks/dips on each side of the critical wavelength.

Since the peaks and dips in the transmission spectrum are nearly symmetrically distributed on both sides of CWL, to avoid difficulties in identifying CWL from achromatic fringes, we can obtain CWL from Equation (8) by averaging the wavelength of $P_{H,1}$, $D_{H,1}$, $P_{L,1}$, and $D_{L,1}$, written as

$$\lambda_{CWL} = \frac{\lambda_{P_{H,1}} + \lambda_{D_{H,1}} + \lambda_{P_{L,1}} + \lambda_{D_{L,1}}}{4} \tag{8}$$

The shift in *CWL* with *SRI* is calculated from Equation (8) above, and the results are given in Figure 10b. Compared with the change in first peak or dip wavelengths simultaneously plotted in Figure 10b, the *CWL* changes monotonically with SRI in a large measurement range, with unambiguous demodulation of SRIs. Thus, the *CWL* detection sensors can effectively solve the overlapping spectrum problem and provide large sensing ranges in applications.

4.2. Sensing Applications of in-Line MZIs Based on CWL Detection

The advantages of the CWL detection method in large-range measurement are first experimentally observed in temperature detection. Tripathi et al. [18] spliced a section of MMF between two sections of SMF (SMS) and observed a CWL in the transmission spectrum at 1318 nm. The experiments showed that the CWL shifted monotonously with temperature from 25 °C to 1100 °C. Martincek et al. [40] employed a liquid-core optical fiber, which consisted of a fused silica cladding and a core with medicinal oil, to increase the temperature sensitivity of the CWL. The maximum temperature sensitivity of the measured CWL was −56 nm/°C, around 24 °C. The transmission spectra of the liquid-core fiber with temperature variations are shown in Figure 11.

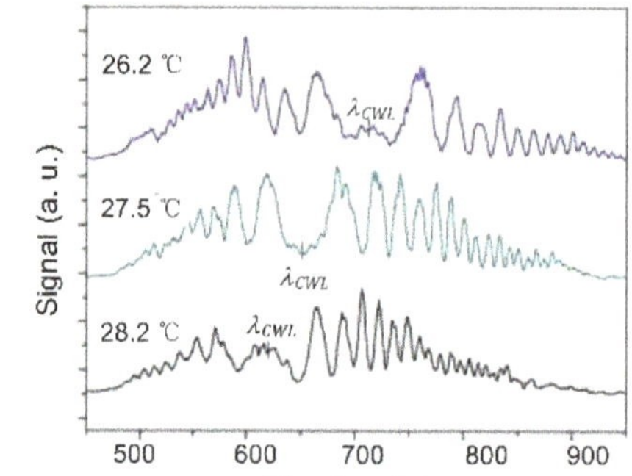

Figure 11. Shift in the CWL in MZI employing the liquid-core fiber shown on measured spectra for temperatures of 28.2, 27.5, and 26.2 °C. Reprinted with permission from Ref. [40]. Copyright 2011 Elsevier.

Our group theoretically studied the characteristics of the CWL in the transmission spectrum of an SFS structure and applied the SFS structure in the large-range measurement of SRI [8], curvature [6], and displacement measurements [7]; the sensing ranges achieved 1.316~1.44 at 1550 nm, 0~125 m^{-1}, and 0~350 mm, respectively, as shown in Figure 12.

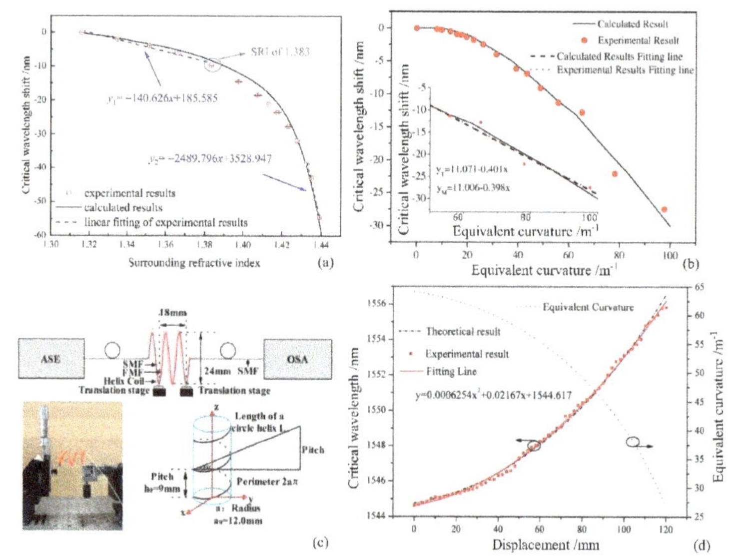

Figure 12. The wavelength shifts in CWL and sensor structures employing the few-mode fiber in the SFS structure; (**a**) experimental (marks with error bars) and calculated results of CWL shift with SRI variations. Reprinted with permission from ref. [8]. Copyright 2017 IEEE. (**b**) calculated and experimental results of CWL shift with equivalent curvature. Reprinted with permission from ref. [6]. Copyright 2016 IEEE.; (**c**) schematic diagram of the experiment setup of the displacement sensor employing SFS structure; (**d**) the experimental and the theoretical results of the CWL changes in the sensor and the theoretical equivalent curvature of FMF in the helix coil (dashed line) versus displacement. Reprinted with permission from ref. [7]. Copyright 2016 IEEE.

5. Conclusions

In this paper, we reviewed the transmission characteristics of CWL-existed in-line MZIs. The detection of the CWL in the transmission spectrum provides a solution to the overlapping problem in conventional MZIs with periodically changing interference fringes. Moreover, the interference peaks/dips on each side of the CWL shift in opposite directions. Their wavelength sensitivities increase significantly when peak wavelengths approach the CWL, which provides a solution for fiber-optic MZIs to enhance their sensitives by designing the operational wavelength near the CWL. Therefore, taking full advantage of the CWL-existed transmission spectrum, these in-line MZIs can be developed into several different sensors, including large-range measurement sensors, multiparameter sensors, and high-sensitivity sensors, to satisfy the requirements in practical applications.

Author Contributions: Principal writing—C.L.; modification and editing—X.D., C.W. and C.L.; project administration, X.D. and C.W.; funding acquisition, X.D. and C.W. All authors have read and agreed to the published version of the manuscript.

Funding: This research was funded, in part, by the PI Project of Southern Marine Science and Engineering Guangdong Laboratory (Guangzhou) under grant number GML2021GD0808; in part, by the National Natural Science Foundation of China, grant number 61775186; and in part by the Marine and Fisheries Bureau of Xiamen, grant number 16CZB025SF03.

Institutional Review Board Statement: Not applicable.

Informed Consent Statement: Not applicable.

Conflicts of Interest: The authors declare no conflict of interest.

References

1. Li, X.; Chen, N.; Zhou, X.; Gong, P.; Wang, S.; Zhang, Y.; Zhao, Y. A review of specialty fiber biosensors based on interferometer configuration. *J. Biophotonics* **2021**, *14*, e202100068. [CrossRef] [PubMed]
2. Bhardwaj, V.; Kishor, K.; Sharma, A.C. Tapered optical fiber geometries and sensing applications based on Mach-Zehnder interferometer: A review. *Opt. Fiber Technol.* **2020**, *58*, 102302. [CrossRef]
3. Zhao, Y.; Zhao, H.; Lv, R.-Q.; Zhao, J. Review of optical fiber Mach–Zehnder interferometers with micro-cavity fabricated by femtosecond laser and sensing applications. *Opt. Lasers Eng.* **2019**, *117*, 7–20. [CrossRef]
4. Min, R.; Liu, Z.; Pereira, L.; Yang, C.; Sui, Q.; Marques, C. Optical fiber sensing for marine environment and marine structural health monitoring: A review. *Opt. Laser Technol.* **2021**, *140*, 107082. [CrossRef]
5. Wang, L.; Yj, W.; Song, S.; Li, F. Overview of fibre optic sensing technology in the field of physical ocean observation. *Front. Phys.* **2021**, *9*, 745487. [CrossRef]
6. Su, J.; Dong, X.; Lu, C. Characteristics of few mode fiber under bending. *IEEE J. Sel. Top. Quantum Electron.* **2016**, *22*, 139–145. [CrossRef]
7. Su, J.; Dong, X.; Lu, C. Property of bent few-mode fiber and its application in displacement sensor. *IEEE Photonics Technol. Lett.* **2016**, *28*, 1387–1390. [CrossRef]
8. Lu, C.; Dong, X.; Su, J. Detection of refractive index change from the critical wavelength of an etched few mode fiber. *J. Lightwave Technol.* **2017**, *35*, 2593–2597. [CrossRef]
9. Lu, C.; Su, J.; Dong, X.; Sun, T.; Grattan, K.T.V. Simultaneous measurement of strain and temperature with a few-mode fiber-based sensor. *J. Lightwave Technol.* **2018**, *36*, 2796–2802. [CrossRef]
10. Lu, C.; Su, J.; Dong, X.; Lu, L.; Sun, T.; Grattan, K.T.V. Studies on temperature and strain sensitivities of a few-mode critical wavelength fiber optic sensor. *IEEE Sens. J.* **2019**, *19*, 1794–1801. [CrossRef]
11. Lu, C.; Dong, X.; Lu, L.; Guan, Y.; Ding, S. Label free all-fiber static pressure sensor based on Vernier effect with temperature compensation. *IEEE Sens. J.* **2020**, *20*, 4726–4731. [CrossRef]
12. Su, J.; Dong, X.; Lu, C. Intensity detection scheme of sensors based on the modal interference effect of few mode fiber. *Measurement* **2016**, *79*, 182–187. [CrossRef]
13. Lei, X.; Dong, X.; Lu, C. Sensitive humidity sensor based on a special dual-mode fiber. *IEEE Sens. J.* **2019**, *19*, 2587–2591. [CrossRef]
14. Lei, X.; Dong, X.; Lu, C.; Sun, T.; Grattan, K.T.V. Underwater pressure and temperature sensor based on a special dual-mode optical fiber. *IEEE Access* **2020**, *8*, 146463–146471. [CrossRef]
15. Salik, E.; Medrano, M.; Cohoon, G.; Miller, J.; Boyter, C.; Koh, J. SMS fiber sensor utilizing a few-mode fibere exhibits critical wavelengthbehavior. *IEEE Photonics Technol. Lett.* **2012**, *24*, 593–595. [CrossRef]
16. Wu, Q.; Semenova, Y.; Wang, P.; Farrell, G. High sensitivity SMS fiber structure based refractometer—Analysis and experiment. *Opt. Express* **2011**, *19*, 7937–7944. [CrossRef]

17. Zhou, J.; He, B.; Gu, X. Transmission spectrum characteristics for a single-mode-multimode-single-mode fiber filter. *IEEE Photonics Technol. Lett.* **2014**, *26*, 2185–2188. [CrossRef]
18. Tripathi, S.M.; Kumar, A.; Marin, E.; Meunier, J. Critical Wavelength in the Transmission Spectrum of SMS Fiber Structure Employing GeO2-Doped Multimode Fiber. *IEEE Photonics Technol. Lett.* **2010**, *22*, 799–801. [CrossRef]
19. Kumar, M.; Kumar, A.; Tripathi, S.M. A comparison of temperature sensing characteristics of SMS structures using step and graded index multimode fibers. *Opt. Commun.* **2014**, *312*, 222–226. [CrossRef]
20. Tripathi, S.M.; Kumar, A.; Varshney, R.K.; Kumar, Y.B.P.; Marin, E.; Meunier, J.P. Strain and temperature sensing characteristics of single-mode–multimode–single-mode structures. *J. Lightwave Technol.* **2009**, *27*, 2348–2356. [CrossRef]
21. Gonthier, F.; Lacroix, S.; Ladouceur, F.; Black, R.J.; Bures, J. Circularly symmetric modal interferometers: Equalization wavelength and equivalent step determination for matched-cladding fibers. *Fiber Integr. Opt.* **1989**, *8*, 217–225. [CrossRef]
22. Zhou, A.; Li, G.; Zhang, Y.; Wang, Y.; Guan, C.; Yang, J.; Yuan, L. Asymmetrical twin-core fiber based michelson interferometer for refractive index sensing. *J. Lightwave Technol.* **2011**, *29*, 2985–2991. [CrossRef]
23. Gu, B.; Yin, M.-J.; Zhang, A.P.; Qian, J.-W.; He, S. Low-cost high-performance fiber-optic pH sensor based on thin-core fiber modal interferometer. *Opt. Express* **2009**, *17*, 22296–22302. [CrossRef] [PubMed]
24. Yi, D.; Huo, Z.; Geng, Y.; Li, X.; Hong, X. PDMS-coated no-core fiber interferometer with enhanced sensitivity for temperature monitoring applications. *Opt. Fiber Technol.* **2020**, *57*, 102185. [CrossRef]
25. Wong, W.C.; Chan, C.C.; Chen, L.H.; Li, T.; Lee, K.X.; Leong, K.C. Polyvinyl alcohol coated photonic crystal optical fiber sensor for humidity measurement. *Sens. Actuators B Chem.* **2012**, *174*, 563–569. [CrossRef]
26. Lacroix, S.; Gonthier, F.; Black, R.J.; Bures, J. Tapered-fiber interferometric wavelength response: The achromatic fringe. *Opt. Lett.* **1988**, *13*, 395–397. [CrossRef]
27. Luo, H.; Sun, Q.; Li, X.; Yan, Z.; Li, Y.; Liu, D.; Zhang, L. Refractive index sensitivity characteristics near the dispersion turning point of the multimode microfiber-based Mach–Zehnder interferometer. *Opt. Lett.* **2015**, *40*, 5042–5045. [CrossRef]
28. Zhang, N.M.Y.; Li, K.; Zhang, N.; Zheng, Y.; Zhang, T.; Qi, M.; Shum, P.; Wei, L. Highly sensitive gas refractometers based on optical microfiber modal interferometers operating at dispersion turning point. *Opt. Express* **2018**, *26*, 29148–29158. [CrossRef]
29. Sun, L.P.; Huang, T.; Yuan, Z.; Lin, W.; Xiao, P.; Yang, M.; Ma, J.; Ran, Y.; Jin, L.; Li, J.; et al. Ultra-high sensitivity of dual dispersion turning point taper-based Mach-Zehnder interferometer. *Opt. Express* **2019**, *27*, 23103–23111. [CrossRef]
30. Xia, F.; Zhao, Y.; Zheng, H.K.; Li, L.K.; Tong, R.J. Ultra-sensitive seawater temperature sensor using an FBG-cascaded microfiber MZI operating at dispersion turning point. *Opt. Laser Technol.* **2020**, *132*, 106458. [CrossRef]
31. Sun, B.; Wang, Y. High-sensitivity detection of IgG operating near the dispersion turning point in tapered two-mode fibers. *Micromachines* **2020**, *11*, 270. [CrossRef] [PubMed]
32. Xia, F.; Zhao, Y.; Peng, Y. In-line microfiber MZI operating at two sides of the dispersion turning point for ultrasensitive RI and temperature measurement. *Sens. Actuators A Phys.* **2020**, *301*, 111754. [CrossRef]
33. Yang, L.; Wang, J.; Wang, S.; Liao, Y.; Li, Y. A new method to improve the sensitivity of nitrate concentration measurement in seawater based on dispersion turning point. *Optik* **2020**, *205*, 164202. [CrossRef]
34. Zhou, W.; Wei, Y.; Wang, Y.; Li, K.; Yu, H.; Wu, Y. Ultrasensitive interferometers based on zigzag-shaped tapered optical microfibers operating at the dispersion turning point. *Opt. Express* **2021**, *29*, 36926–36935. [CrossRef] [PubMed]
35. Liu, K.; Fan, J.; Luo, B.; Zou, X.; Wu, D.; Zou, X.; Shi, S.; Guo, Y.; Zhao, M. Highly sensitive vibration sensor based on the dispersion turning point microfiber Mach-Zehnder interferometer. *Opt. Express* **2021**, *29*, 32983–32995. [CrossRef] [PubMed]
36. Xu, S.; Chang, W.; Luo, Y.; Ni, W.; Zheng, Y.; Wei, L.; Xu, Z.; Lian, Z.; Zhang, Y.a.; Huang, Y.; et al. Ultrasensitive broadband refractometer based on single stress-applying fiber at dispersion turning point. *J. Lightwave Technol.* **2021**, *39*, 2528–2535. [CrossRef]
37. Mascotte, E.H.; Hernandez, J.M.S.; Chavez, R.I.M.; Vazquez, D.J.; Guzman, A.C.; Ayala, J.M.E.; Chavez, A.D.G.; Laguna, R.R. A core-offset Mach Zehnder interferometer based on a non-zero dispersion-shifted fiber and its torsion sensing application. *Sensors* **2016**, *16*, 856. [CrossRef]
38. Ahsani, V.; Ahmed, F.; Jun, M.B.G.; Bradley, C. Tapered fiber-optic Mach-Zehnder interferometer for ultra-high sensitivity measurement of refractive index. *Sensors* **2019**, *19*, 1652. [CrossRef]
39. Gong, H.; Chan, C.C.; Zhang, Y.; Wong, W.; Dong, X. Temperature sensor based on modal interference in hollow-core photonic bandgap fiber with collapse splicing. *IEEE Sens. J.* **2012**, *12*, 1421–1424. [CrossRef]
40. Martincek, I.; Pudis, D.; Kacik, D.; Schuster, K. Investigation of intermodal interference of LP01 and LP11 modes in the liquid-core optical fiber for temperature measurements. *Optik* **2011**, *122*, 707–710. [CrossRef]
41. Hlubina, P.; Kadulova, M.; Ciprian, D.; Mergo, P. Temperature sensing using the spectral interference of polarization modes in a highly birefringent fiber. *Opt. Lasers Eng.* **2015**, *70*, 51–56. [CrossRef]
42. Hlubina, P.; Olszewski, J.; Martynkien, T.; Mergo, P.; Makara, M.; Poturaj, K.; Urbanczyk, W. Spectral-domain measurement of strain sensitivity of a two-mode birefringent side-hole fiber. *Sensors* **2012**, *12*, 12070–12081. [CrossRef]
43. Wei, C.; Lin, G.; Dong, X.; Tao, S. A tunable polarization-independent comb filter based on high-order mode fiber. *J. Opt.* **2013**, *15*, 055403. [CrossRef]
44. Vengsarkar, A.M.; Walker, K.L. Article Comprising a Dispersion-Compensating Optical Waveguide. U.S. Patent 5,448,674, 5 September 1995.

45. Kumar, A.; Goel, N.K.; Varshney, R.K. Studies on a few-mode fiber-optic strain sensor based on LP01-LP02 mode interference. *J. Lightwave Technol.* **2001**, *19*, 358. [CrossRef]
46. Hocker, G. Fiber-optic sensing of pressure and temperature. *Appl. Opt.* **1979**, *18*, 1445–1448. [CrossRef]
47. Shu, X.; Zhang, L.; Bennion, I. Sensitivity characteristics near the dispersion turning points of long-period fiber gratings in B/Ge codoped fiber. *Opt. Lett.* **2001**, *26*, 1755–1757. [CrossRef]
48. Tan, Y.; Sun, L.P.; Jin, L.; Li, J.; Guan, B.O. Microfiber Mach-Zehnder interferometer based on long period grating for sensing applications. *Opt. Express* **2013**, *21*, 154–164. [CrossRef]

photonics

MDPI

Review

Review of Specialty Fiber Based Brillouin Optical Time Domain Analysis Technology

Dora Juan Juan Hu [1,*], Georges Humbert [2], Hui Dong [1], Hailiang Zhang [1], Jianzhong Hao [1] and Qizhen Sun [3]

[1] Institute for Infocomm Research, Agency for Science, Technology and Research, Singapore 138632, Singapore; hdong@i2r.a-star.edu.sg (H.D.); Zhang_Hailiang@i2r.a-star.edu.sg (H.Z.); haoemily@i2r.a-star.edu.sg (J.H.)
[2] XLIM Research Institute, UMR CNRS/University of Limoges, 123 Av. Albert Thomas, 87060 Limoges, France; georges.humbert@xlim.fr
[3] School of Optical and Electronic Information & National Engineering Laboratory for Next Generation Internet Access System (NGIA) & Wuhan National Laboratory for Optoelectronics (WNLO), Huazhong University of Science and Technology, Wuhan 430074, China; qzsun@mail.hust.edu.cn
* Correspondence: jjhu@i2r.a-star.edu.sg

Abstract: Specialty fibers have introduced new functionalities and opportunities in distributed fiber sensing applications. Particularly, Brillouin optical time domain analysis (BOTDA) systems have leveraged the unique features of specialty fibers to achieve performance enhancement in various sensing applications. This paper provides an overview of recent developments of the specialty fibers based BOTDA technologies and their sensing applications. The specialty fibers based BOTDA systems are categorized and reviewed based on the new features or performance enhancements. The prospects of using specialty fibers for BOTDA systems are discussed.

Keywords: specialty fibers; Brillouin optical time domain analysis; BOTDA; distributed fiber sensor; photonic crystal fiber; strain and temperature sensing

Citation: Hu, D.J.J.; Humbert, G.; Dong, H.; Zhang, H.; Hao, J.; Sun, Q. Review of Specialty Fiber Based Brillouin Optical Time Domain Analysis Technology. *Photonics* **2021**, *8*, 421. https://doi.org/10.3390/photonics8100421

Received: 23 July 2021
Accepted: 26 September 2021
Published: 30 September 2021

1. Introduction

Novel specialty optical fibers with engineered materials or specially designed structures have enabled new opportunities to explore novel functionalities and applications. One of the successful applications is in the Brillouin optical time domain analysis (BOTDA) systems, which has emerged as a fast-developing distributed fiber sensor technology for measuring temperature and strain in the past two decades. The capability of distributed strain and temperature measurement and the benefits of a fiber optic sensor system have made the BOTDA systems a favorable solution for long term structural health monitoring (SHM) and assessment of infrastructure. The correlation of strain data and structural health assessment have been investigated and implemented for practical SHM applications, such as damage detection of operating tunnels-based cross-section curvature [1], blade structural fatigue damage of wind turbine [2], structural crack identification and early detection [3], etc. A few examples of field significant applications in the real word are presented in [4], including SHM monitoring in civil infrastructures, such as bridges and railways, geotechnical structures monitoring and pipeline monitoring, etc. Design considerations of BOTDA systems could be from the interrogator to address the technology challenges and meet the performance requirements in spatial resolution, sensing range and dynamic measurement, which are well summarized in several reviews [5–9]. Alternatively, the design can be from the perspectives of fibers or fiber cables. In particular, the exploration of different stimulated Brillouin scattering (SBS) materials and their applications are thoroughly reviewed in [10]. The perspectives of fiber manufacturers on fiber materials and their effects on the Brillouin scattering properties and system designs are presented in [11]. The research and application regarding optical fiber cables for Brillouin distributed sensing are reviewed in [12], showing successful exploitation of the Brillouin sensing technique using appropriate fiber cable designs. Moreover, post-processing using image processing techniques [13] or machine learning techniques have been reported to improve sensor performance without hardware modification [14].

This paper intends to provide a comprehensive review on the different sensing features and performance enhancements provided by the specialty fibers that are used in BOTDA systems. This paper will cover new developments that were reported after the previous papers, which presented specialty fibers in Brillouin based sensing systems [11,15]. The principle of BOTDA technology is introduced in Section 2, followed by a discussion of the specialty fiber based BOTDA with enhanced sensing capability and performance in Section 3. Firstly, the sensor performance enhancements and breakthroughs are introduced, such as latest breakthroughs in Brillouin amplifications in gas filled hollow-core fibers, sensitivity enhancement and suppression of polarization noise. Secondly, the discriminative sensing of temperature and strain using several techniques are reviewed. Thirdly, multiplexed sensing systems of BOTDA and other distributed fiber sensor (DFS) technologies are reviewed. Fourthly, distributed sensing of other parameters, such as curvatures, pressure and salinity, etc., are reviewed. Lastly, the prospects of using specialty fibers for BOTDA systems are discussed in Section 4.

2. Principle of BOTDA Technology

Typical BOTDA systems are based on Brillouin interaction between a pulsed pump and a counter-propagating continuous wave (CW) probe (Figure 1). The probe wave is amplified through SBS process when the frequency difference between the pulsed and the CW light is turned to the Brillouin frequency of the fiber [16]. The probe power is measured as a function of time, which is translated to a distance-dependent information to derive local Brillouin gain. The local Brillouin frequency shift (BFS) at each point of the fiber is derived from reconstruction of the Brillouin gain spectrum (BGS) by a frequency sweep with every frequency detuning around the Brillouin frequency. Both temperature and strain contribute to BFS variations.

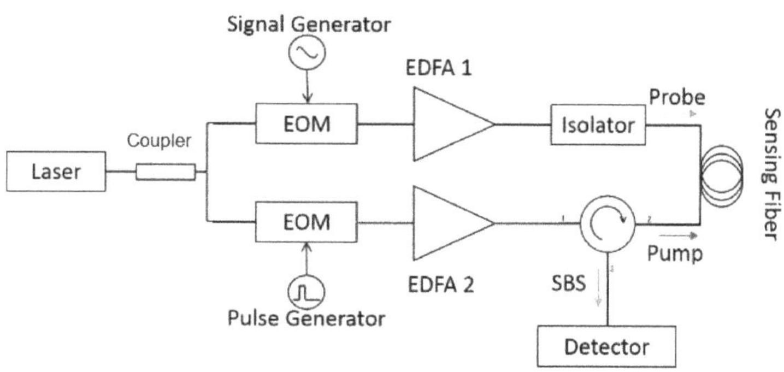

Figure 1. Basic layout of a BOTDA system.

Specialty fibers such as PCFs and hollow-core fibers contain holey structures, which provide an ideal platform to integrate with high gain SBS medium to boost the overall Brillouin amplification. Peculiar Brillouin scattering features of specialty fibers are leveraged to improve the sensor performance of BOTDA systems, such as polarization noise suppression, enhancement of sensitivity, etc. The temperature-compensated distributed strain measurement, or discriminating the response from temperature and strain, is a key challenge in the BOTDA solutions. To overcome this challenge, various specialty fibers have been introduced to the BOTDA solution, such as photonic crystal fibers (PCFs), few mode fibers (FMFs) and multicore fibers etc., which are reviewed in this paper. In addition, multiplexed distributed sensing that consists of the BOTDA technology and other DFS technology, e.g., Raman optical time domain reflectometry (R-OTDR) or phase-sensitive optical time domain reflectometry (φ-OTDR), are reported based on multi-core fibers via spatial multiplexing. The aforementioned opportunities have generated substantial interest and driven the advancement of the specialty fiber based BOTDA systems with favorable features and performance enhancements as reviewed in the next section.

3. Reviews on Specialty Fiber Based BOTDA Technology

Through the design of optical fiber structures and engineering of the optical fiber materials, specialty fibers have offered abundant opportunities in bringing new features, functionalities in fiber devices and enabling new applications. In BOTDA systems, the Brillouin gain spectrum can be customized and optimized through the selection of fiber materials and the design of optical and acoustic waveguide properties in specialty fibers [11]. In this section, a comprehensive review of the specialty fiber based BOTDA with enhanced sensing capability and performance is presented.

3.1. Sensor Performance Enhancements and Breakthroughs

PCF provides a flexible platform to engineer the optical and acoustic waveguide properties to achieve enhanced or reduced Brillouin gain. Brillouin gain in PCFs can be significantly increased by using high Brillouin gain materials as fiber materials or infiltrating high Brillouin gain materials, such as gases, into the holey structure of the PCF. The Brillouin scattering properties and the BGS can be influenced by fiber configurations, fiber material designs and the dopants into the silica glass [11]. The Brillouin gain coefficients of a chalcogenide glass microstructured single mode fiber (SMF) was 100 times larger than fused silica [17]. More recently, a gas-filled hollow-core PCF has demonstrated highly nonlinear Brillouin gain, which scales with the square of the gas pressure and achieved a strain-free high performance distributed temperature sensing [18,19]. On the contrary, solid core PCFs have demonstrated reduced Brillouin gain and a three-fold increase of the Brillouin threshold power with a uniform all-silica fiber. They could be exploited for suppression of Brillouin scattering [20]. Anti-resonant hollow-core fibers exhibited at least three (five if evacuated) orders of magnitude weaker Brillouin scattering compared with conventional SMFs, paving the way to explore ultra-low noise applications for quantum information processing and optical communication [21].

In addition to fiber design and material engineering, the fiber sensing cable design and fiber coatings play a significant role in optimizing the Brillouin distributed fiber sensor performance [12]. For example, SMF with single-layer and double-layer polymer coatings have been used to demonstrate enhanced hydrostatic pressure sensitivities through pressure-induced strain measurement based on Lame formula. The pressure sensitivity is enhanced by increasing the outer coating radius or decreasing the outer coating Young's modulus and Poisson's ratio. The maximum pressure sensitivity of the double-coated SMF is about five times higher than the single-coated SMF [22]. Similarly, enhanced temperature sensitivity can be achieved by putting optical fibers into suitable coatings [23,24].

Polarization effects on the SBS in PCFs have demonstrated interesting properties. In small core single mode PCFs, a higher SBS threshold of 3 dB is observed due to unintentional birefringence as a result of the asymmetry in the fiber structure geometry as well as the large index contrast. In addition, strong polarization dependance of Brillouin gain has been reported. In particular, the SBS has a strong dependence on pump polarization. The minimum and maximum SBS are 3 dB different and 90° apart. For lower powers below threshold, the transmission presents a 180° polarization periodicity due to polarization dependent losses (PDL). For higher powers above threshold, the transmission shows a 90° polarization periodicity such as SBS as PDL becomes negligible [25]. A large-controlled birefringence could be introduced in these fibers to suppress SBS effect for potential applications that are limited by SBS.

3.2. Discriminative Sensing of Temperature and Strain

Discriminative sensing of temperature and strain is an important and desirable capability in BOTDA systems because the optical fiber and its BGS are subject to changes in both parameters. Besides the usage of two independent fibers [26], various approaches have been proposed and demonstrated for achieving this purpose, including approaches to form a sensor matrix, and other approaches based on a combination of using Brillouin scattering and other parameters.

The below sensor matrix (Equation (1)) can be used to describe the sensing system, where the $C_{T1,2}$ and $C_{\varepsilon 1,2}$ represent the temperature and strain coefficients, $\Delta\nu_{B\_peak1,2}$ represent the central frequency shifts of the Brillouin spectra peaks. By solving the inverse matrix problem, distributed measurement of the temperature change ΔT and the strain change $\Delta\varepsilon$ along the fiber can be calculated.

$$\begin{bmatrix} \Delta\nu_{B\_peak1} \\ \Delta\nu_{B\_peak2} \end{bmatrix} = \begin{bmatrix} C_{T1} & C_{\varepsilon 1} \\ C_{T2} & C_{\varepsilon 2} \end{bmatrix} \begin{bmatrix} \Delta T \\ \Delta\varepsilon \end{bmatrix} \quad (1)$$

The sensor matrix can be realized by exploiting the multi-peak feature of the BGS of the dispersion compensating fiber [27], dispersion shifted fiber [28] and specialty fibers, which is due to different acoustic modes or/and different optical modes.

Ideally, by solving the inverse matrix of Equation (1), the temperature change ΔT and the strain change $\Delta\varepsilon$ along the fiber can be calculated. The ability of accurately inverting the matrix determines measurement performance. Some sensor matrices can result in a larger error. For example, when using LEAF, the BFS-temperature/strain coefficients between different Brillouin peaks are very small, and solving the equations produces large errors [14,29]. The condition number (CN) of a matrix A is defined cond(A) = ||A||||A^(−1)||, where ||·|| could be any of the norms of A. In this Section, 2-norm is used to calculate cond(A) [30]. A larger CN results in a larger upper bound of the relative error [30]. Thus, a smaller CN of the sensor matrix is desirable in the sensor system. The calculated CN of the sensor matrices for the different techniques is used to provide an intuitive comparison of the measurement accuracy between the different techniques and are listed in Table 1. As can be seen, two techniques by using different fiber deployment [26], and the PM-PCF in the temperature range from 5 °C to 80 °C [31] have much smaller CNs than other techniques or other temperature ranges of the same configuration [31], indicating better achievable measurement accuracy with a smaller value of upper bound of the relative error.

Table 1. Summary of reported approaches to achieve discriminative sensing using specialty fibers.

Approaches	Specialty Fiber Type	Temperature Coefficient C_T (MHz/°C)	Strain Coefficient C_ε (MHz/$\mu\varepsilon$)	CN of the Sensor Matrix
Multi-peak BGS due to multiple acoustic modes	PCFs with a partially germanium-doped core [32,33]			
	Peak 1	0.96	0.048	
	Peak 2	1.25	0.055	346
	Large effective area fiber (LEAF) [29]			260 (peak1 BFS, peak1 BGS)
	Peak 1 BFS	1.18	0.055	270 (peak2 BFS, peak2 BGS)
	Peak 1 BGS linewidth	−0.111	−0.00059	2580 (peak1 BFS, peak2 BFS)
	Peak 2 BFS	1.2	0.055	260 (peak1 BFS, peak2 BGS)
	Peak 2 BGS linewidth	−0.118	−0.00091	270 (peak1 BGS, peak2 BFS) 836 (peak1 BGS, peak2 BGS)
Multi-peak BGS due to multiple optical modes	Circular core few mode fiber (FMF) [34]			
	LP_{01}	1.0169	0.05924	221
	LP_{11}	0.99099	0.04872	

Table 1. *Cont.*

Approaches	Specialty Fiber Type	Temperature Coefficient C_T (MHz/°C)	Strain Coefficient C_ε (MHz/$\mu\varepsilon$)	CN of the Sensor Matrix
Multi-peak BGS due to multiple acoustic modes and optical modes	Elliptical core FMF [35]			96 (LP_{01}–P_1, LP_{01}–P_2)
	LP_{01}-P_1	1.242	0.0613	107 (LP_{11e}–P_1, LP_{11e}–P_2)
	LP_{01}-P_2	1.278	0.0364	1133 (LP_{01}–P_1, LP_{11e}–P_1)
	LP_{11e}-P_1	1.287	0.0658	119 (LP_{01}–P_1, LP_{11e}–P_2)
	LP_{11e}-P_2	1.501	0.0484	88 (LP_{01}–P_2, LP_{11e}–P_1) 538 (LP_{01}–P_2, LP_{11e}–P_2)
Multiple core/fibers	Different fiber deployment [26]	1.2 ± 0.2	0.054 ± 0.7	44
	Dual core fiber [36]			
	Core 1	0.971	0.0532	95
	Core 2	0.9593	0.0729	
	Multicore fiber (MCF) [37]			380
	Central core	1.08	0.0485	
	Outer core	1.03	0.0517	
	MCF with heterogeneous cores [38]			
	Central core	1.05 ± 0.0095	0.0485	
	Outer core	1.15 ± 0.0343	0.0485	501
BFS and birefringence	Polarization-maintaining photonic crystal fiber (PM-PCF) [31]			
	Birefringence	−40 °C to −15 °C: 27.4 −15 °C to 5 °C: 8 5 °C to 80 °C: 0	−0.154	−40 °C to −15 °C: 495 −15 °C to 5 °C: 115 5 °C to 80 °C: 7
	BFS	1.15	0.049	
BFS and fluorescence	Erbium-doped optical fiber [39]			
	Fluorescence intensity ratio	5.6×10^{-4}/°C	0	28,303
	BFS	0.87	0.0479	
Athermal/atensic optical fibers	Highly Ge dope fibers [40]	0.07	0.0214	Not applicable

The reported approaches to achieve discriminative measurement of temperature and strain or single parameter sensing measurement using specialty fibers are summarized in Table 1, based on the principles of multiple acoustic modes, multiple optical modes, multiple cores/fibers, BFS and other parameter such as birefringence or fluorescence, and athermal optical fibers. The temperature and strain coefficients of reported approaches are listed in Table 1.

Multi-peak BGS due to multiple acoustic modes: Firstly, the non-zero overlapping between multiple acoustic modes and fundamental optical mode field gives rise to multi-peak features on the Brillouin gain (loss) spectrum in PCFs [41]. The strain and temperature coefficients of each peak are different, enabling the discriminative sensing of two physical parameters, based on PCFs with a partially germanium-doped core as shown in Figure 2 (multi peaks are indicated by the letters a, b, c, d, e) [32,33]. Multi-peak BGS was also reported for a highly nonlinear PCF with a Ge-doped core and triangularly arranged by the F-doped buffer. However, this fiber structure does not exhibit distinct temperature and strain dependence for different peaks in the BGS and is not desirable for discriminative sensing [42]. For large effective area fiber (LEAF), such as that which has a triangle refractive index profile, the overlapping of the first three acoustic modes are comparable to the optical effective area, resulting in a multi-peak BGS. Unlike SMFs, the LEAF exhibited an observable BGS linewidth variation with strain due to greater waveguide contribution. Simultaneous detection of temperature and strain were achieved by monitoring the central frequency shift and the linewidth of the first two peaks [29].

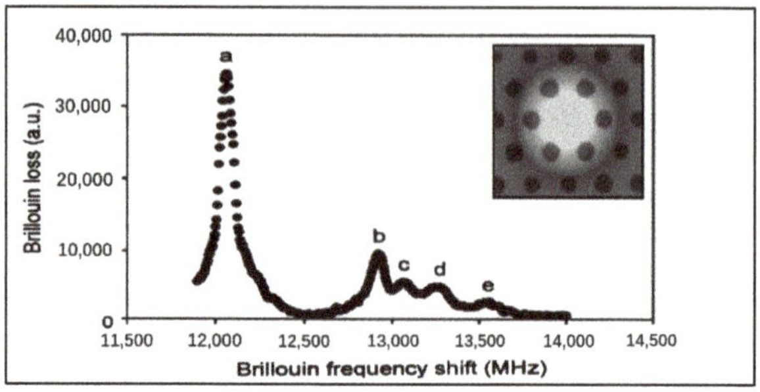

Figure 2. Multi-peak BGS due to multiple acoustic modes (indicated by the letters a, b, c, d, e). (Reprinted/Adapted) with permission from [32] © The Optical Society.

Multi-peak BGS due to multiple optical modes: Secondly, the optical fiber with multiple optical modes to interact with the fundamental acoustic mode, such as few-mode fiber (FMF), is another promising candidate for discriminative sensing based on different optical mode properties including different effective refractive indices, chromatic dispersion and losses. The Brillouin scattering for various linear polarized modes will create different BGSs and the corresponding BFSs will have different temperatures and strain coefficients [15]. A circular core-FMF based BOTDA was used to demonstrate the discriminative distributed measurement of temperature and strain [34]. The free-space mode launcher was used to generate specific spatial modes from the spatial light modulator (SLM) as shown in Figure 3. The BFS for LP_{01} and LP_{11} mode were measured and showed estimated temperature and strain sensitivities that were (1.0169 MHz/°C, 0.05924 MHz/$\mu\varepsilon$) and (0.99099 MHz/°C, 0.04872 MHz/$\mu\varepsilon$), respectively.

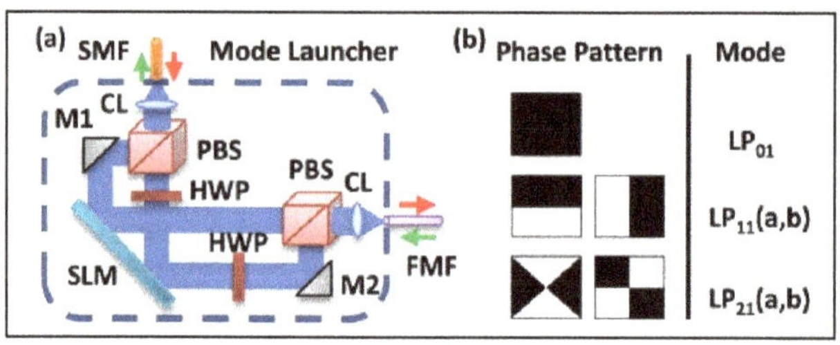

Figure 3. (**a**) Schematic diagram of a free-space mode launcher. (**b**) SLM phase patterns and the corresponding spatial modes. CL: collimating lens, M1/M2: turning mirror, HWP: half-wave plate, PBS: polarization beam splitter. (Reprinted/Adapted) with permission from [34] © The Optical Society.

Multi-peak BGS due to multiple acoustic and optical modes: More recently, an elliptical-core FMF based Brillouin optical time domain reflectometry (BOTDR) was reported to achieve distributed temperature and strain measurement, levering the multiple Brillouin peaks from the interaction between both higher-order optical modes (LP_{01} and LP_{11} mode) and higher-order acoustic modes. Optimum sensing performance can be achieved by choosing proper optical-acoustic mode pairs [35]. This scheme of using selected higher-order optical and modes and higher-order acoustic modes in the elliptical-core FMF offers an alternative approach to achieve temperature and strain discrimination in BOTDA, requiring a configuration of the selective mode launcher.

Multiple core/fibers: Thirdly, using different optical fibers or different optical fiber cores is another approach to produce BFS with different thermal and strain coefficients for simultaneous measurement of temperature and strain. In [26], the optical fibers are deployed with the configuration that the first half of the sensing is tightly bonded to the structure to be monitored, which is sensitive to both strain and temperature, and the second half is loosely deployed and collocated to sense the temperature only. Optical fibers with dual core structure [36,43] have been reported for distributed Brillouin sensing applications. Through appropriate designing of the fiber core profile, the strain and temperature coefficients of the two sensing fiber cores are different, forming the sensing matrix equation (Equation (1)) and enable discriminative sensing of both parameters. By measuring the BGS in two cores (the central core and one of the outer cores) of a seven-core multicore fiber (MCF), the BFS in these two cores show different strain and temperature coefficients, suggesting the feasibility of using MCF based Brillouin sensing for discriminative distributed measurement of temperature and strain [37].

Further improvement can be achieved by optimizing the material and the structure of the MCF to enlarge the difference from the cores [11,37]. MCF with heterogeneous cores as shown in Figure 4 is reported to show ~70 MHz BFS difference between the central core and the outer cores, which are made from different preforms. The experimental results showed the heterogeneous cores have enlarged difference in the temperature sensitivities and indistinguishable strain sensitivities, enabling a simultaneous and discriminative dual parameter measurement [38]. Distributed bending sensing application deduced from strain measurements of the MCF based Brillouin sensors are explored subsequently [44].

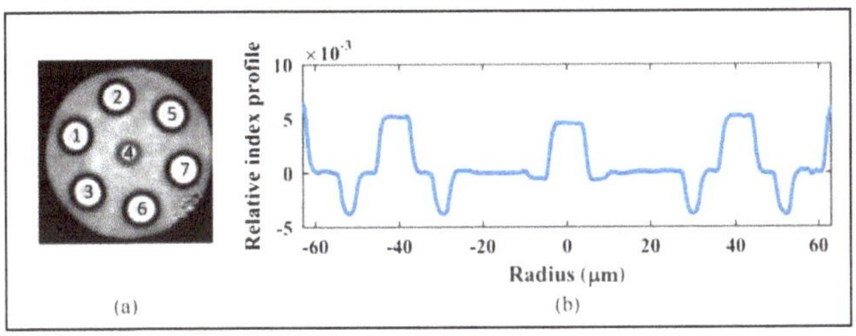

Figure 4. (**a**) Cross-sectional view of the MCF used in the experiment. (**b**) Relative index profile of the MCF. (Reprinted/Adapted) with permission from [38] © The Optical Society.

BFS and birefringence/fluorescence: The fourth approach is through simultaneous monitoring two independent physical parameters of BFS and the fiber birefringence or fluorescence. The birefringence is characterized through the measurement of the diffraction spectrum of the dynamic acoustic grating generated in SBS, or Brillouin dynamic grating (BDG). The combination of BFS and the birefringence was firstly explored using polarization-maintaining fiber (PMF), reporting a discrimination accuracy of 3 $\mu\varepsilon$ and 0.08 °C [45]. Compared with the conventional PMF, polarization-maintaining photonic crystal fiber (PM-PCF) acquires fiber birefringence from the non-circular core in the all-silica holey cladding structure, thus inherently possessing much less temperature sensitivity. Discriminative measurement of strain and temperature using the configuration as depicted in Figure 5, is achieved by monitoring the BFS and BDG of the PM-PCF for the temperature range of −40 °C to 80 °C and the strain range of 0–800 $\mu\varepsilon$. In addition, leveraging the weak temperature sensitivity of the fiber birefringence in the temperature range from 5 °C to 80 °C, temperature-insensitive strain measurement is possible by using the BDG measurement alone [31]. Combination of BFS and fluorescence are used to demonstrate discriminative strain and temperature measurement in erbium-doped optical fiber [39].

Athermal/atensic optical fibers: If complete discrimination of temperature and strain measurement is not required, an optical fiber with immunity to one parameter is desirable for single sensing parameter measurement. It has been shown in Figure 6 that highly Ge doped fibers presented negligible temperature dependence, and the temperature coefficient even decreased to zero when removing the fiber coating, paving the way to develop athermal optical fibers for temperature-free Brillouin strain sensing [40]. Likewise, optical fibers that are insensitive to strain or atensic fibers can be designed and developed for strain-free Brillouin temperature sensing [11]. However, these approaches to develop athermal or atensic fibers through fiber material composition optimization suffer from a significant reduction of the sensitivity to strain or temperature and in the Brillouin gain [11].

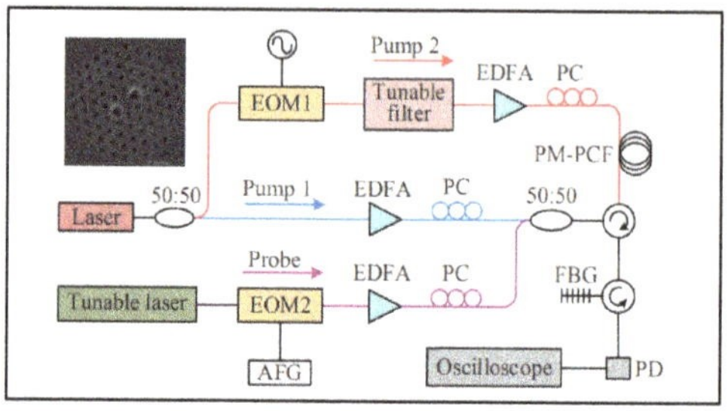

Figure 5. Experimental setup of distributed temperature and strain sensing using BDG and BGS. Reproduced from [31]. Copyright (2017) The Japan Society of Applied Physics.

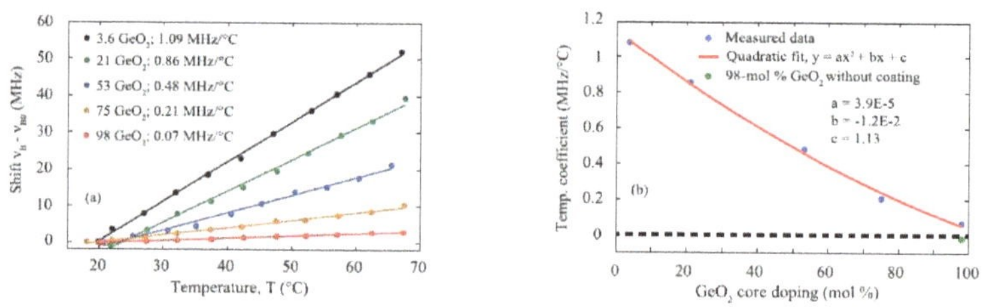

(**a**) (**b**)

Figure 6. (**a**) Brillouin frequency shift versus the temperature increase in five different fibers; (**b**) SBS temperature coefficient versus GeO2-core doping levels. Reproduced with permission [40]. Licensed under a Creative Commons Attribution (CC BY) license.

3.3. Multiplexed Distributed Sensing Systems of BOTDA and Other Distributed Fiber Sensor Technologies

This section is focused on the reported work of specialty fibers in hybrid distributed sensing systems based on Brillouin scattering phenomena, e.g., BOTDA or BOTDR, and other DFS technologies based on Raman or Rayleigh scattering phenomena. For example, utilizing the space-division multiplexed (SDM) system configuration, MCF represent a desirable platform to develop a compact system to multiplex different distributed sensing techniques. A recent review paper introduced the research progress of MCF based hybrid distributed fiber sensor systems [46].

Hybrid Brillouin-Rayleigh sensing: Firstly, hybrid Brillouin-Rayleigh sensing systems have been reported to achieve simultaneous dual parameter measurements of temperature and strain, providing benefits of improved data processing and analytics for distance compensation and accuracy enhancement [47]. Standard optical fibers are used in the reported hybrid sensing system. The BFS and Rayleigh frequency shifts are measured to decouple the strain and temperature based on the following two equations, e.g., Equations (2) and (3):

$$\Delta v_B = C_{11}\Delta\varepsilon + C_{12}\Delta T \tag{2}$$

$$\Delta v_R = C_{21}\Delta\varepsilon + C_{22}\Delta T \tag{3}$$

Δv_B and Δv_R are the shift of the Brillouin and Raman gain spectrum, respectively. The strain and temperature coefficients are denoted by C_{11} and C_{12} in Brillouin sensing system; C_{21} and C_{22} in Rayleigh sensing system, respectively. Subsequently, a more compact system of the hybrid Brillouin-Rayleigh sensing system through SDM configuration based on the MCF is reported for distributed temperature sensing as shown in Figure 7. Two spatially separated sensors of BOTDA and Φ-OTDR are realized in the central core and outer core, respectively. Distributed temperature sensing using the 1.565 km MCF was demonstrated to achieve 2.5 m spatial resolution, large measurement range (10 °C) by BOTDA, and 0.1 °C small temperature variation detection by Φ-OTDR with ~0.001 °C resolution [48].

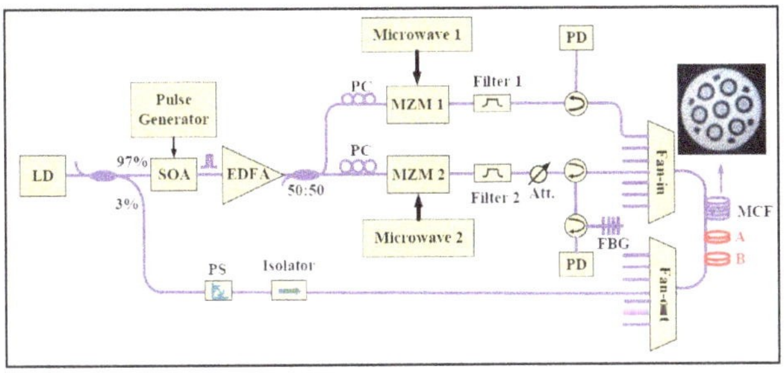

Figure 7. Experimental setup of the MCF based SDM hybrid BOTDA and Φ-OTDR system. (Reprinted/Adapted) with permission from [48] © The Optical Society.

Hybrid Brillouin-Raman sensing: Hybrid Raman/Brillouin optical time domain reflectometry/analysis (RODTR and BOTDR or BOTDA) configurations using standard optical fibers are reported to realize distributed sensing of temperature and strain. The temperature measurement is determined by Raman based sensor without other cross-sensitivity from strain. Based on the temperature measurement, the strain measurement is computed and discriminated from both a strain and temperature dependent Brillouin frequency shift in the BOTDR sensor [49]. Recent advances of the hybrid Raman and Brillouin sensing systems have been reviewed, presenting promising techniques based on pulse coding to address the challenges of the tradeoff between the sensing distance and spatial resolution in the hybrid sensing system [50]. To address the incompatibility of input power requirement for hybrid ROTDR and BOTDR in standard single mode fibers, an MCF based hybrid ROTDR and BOTDR is reported for discriminative and distributed sensing of temperature and strain using a single laser source, shared pump generation devices and separate interrogation fiber channels (Figure 8). The reported system achieved worst temperature and strain resolutions about 2.2 °C and 40 $\mu\varepsilon$, respectively, in a 6 km sensing range with a 3 m spatial resolution [51]. The research progress of MCF based hybrid distributed fiber sensor systems is reviewed in [46].

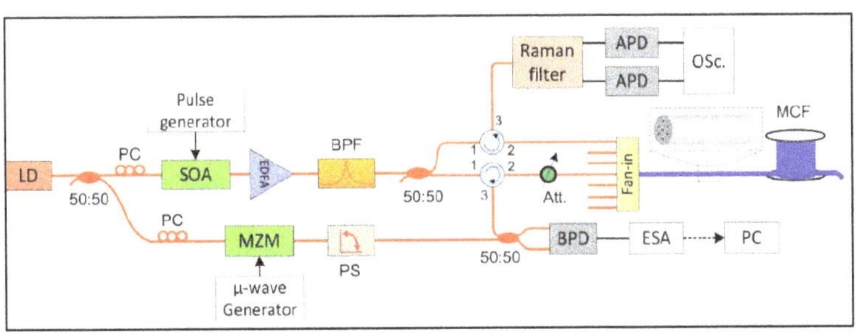

Figure 8. Experimental setup of the MCF based SDM hybrid ROTDR and BOTDR system. (Reprinted/Adapted) with permission from [51] © The Optical Society.

3.4. Distributed Sensing of Other Parameters

Curvature sensing: The dependence of the BFS on longitudinal strain variations (compression/elongation) is exploited to develop distributed curvature sensing using an optical fiber. Since the first sensing demonstration with an SMF (BFS ~ 1.6 MHz for a 1.5 m of

SMF winded around a rod of 25 mm diameter [52]), different optical fibers have been investigated. Among them, a few-mode fiber was used to improve the sensitivity (~5 time for a radius curvature of 0.8 cm) and to extend the range of curvature measurements (radius curvature from 0.7 cm to 3.4 cm) [53]. MCF was successfully used for demonstrating curvature radius and orientation measurements (with a BFS sensitivity of 2.06 MHz/m^{-1}) by correlating the difference of BFS between the central core and outer six cores, paving the way to full 3D shape determination using an optical fiber [44]. The interest of an RCF, with an inner and outer core diameter of 7.38 µm and 15.14 µm, respectively, and a $\Delta n_{\text{core-cladding}}$ ~0.0131, was demonstrated for measuring distributed small curvature radius by exploiting its excellent bending loss resistance (α < 0.01 dB for a radius of 0.5 cm) in comparison to an SMF that suffers from macro-bending losses (α > 10 dB). In addition to a higher sensitivity than SMFs (with BFS ~ 32.9 MHz for a radius of 0.5 cm), the RCF enables distributed curvature sensing without temperature or axial strain sensitivity by measuring its Brillouin gain variation [54], as shown in Figure 9.

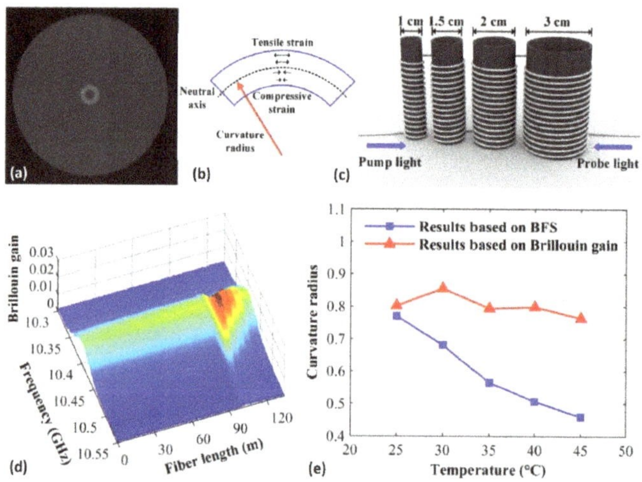

Figure 9. (**a**) Optical microscope image of the cross section of an RCF, (**b**) schematic illustration of strain applied in a bent fiber, (**c**) illustration of distributed curvature measurements by winding the RCF around different rod diameters, (**d**) measured Brillouin gain along the RCF including four sections where the RCF is winded. (**e**) Estimation of the curvature from BFS or Brillouin gain measurements versus temperature variations. Reproduced from [54], with the permission of Chinese Laser Press Publishing.

Pressure sensing: PM-PCF composed of only silica and air, which exhibits a very low thermal sensitivity, was used for hydrostatic pressure sensing by measuring the birefringence changes through exciting and probing the BDG (with a differential pulse-width pair Brillouin optical time domain analysis, DPP-BOTDA) [55]. A linear dependence of the Birefringence BFS difference of about 199 MHz/MPa has been measured for hydrostatic pressure from 0 to 1.1 PM with a measurement error of 0.025 MPa.

Characterization of Fiber-Optic Parametric Amplifier (FOPA) and nonlinear optical fibers: FOPA gain that relies on strict phase-matching conditions imposed by the four-wave mixing process was measured along the length of the highly non-linear fiber by using BOTDA in a novel scheme where the parametric amplification acts on the BOTDA pump [56]. The fluctuation measurements of the zero-dispersion wavelength (ZDW) along the highly nonlinear fiber of a FOPA was also demonstrated (with ZDW fluctuations as low as 0.02 nm along 200 m of the fiber and 2 m longitudinal resolution) [57,58]. These demonstrations offer new possibilities for precisely characterizing the chromatic dispersion along highly nonlinear fibers, i.e., including local fluctuation, which is a key parameter for numerous nonlinear optic applications.

Salinity measurement: A PM-PCF with a polyimide coating was used to demonstrate salinity measurements based on BOTDA and BDG [59]. This coating shrinks in contact with a NaCl solution. A frequency shift is measured with BDG-BOTDA corresponding to a birefringence variation from modification of the hydrostatic pressure applied on the PM-PCF by the coating depending on the salinity concentration. Linear NaCl sensitivity from 30.1 MHz/(mol/L) to 139.6 MHz/(mol/L) have been reported for different coating thickness ranging from 5 μm to 20 μm, respectively, with a maximum salinity accuracy of 0.072 mol/L (considering a measurement uncertainty less than 10 MHz). This demonstration with a spatial resolution of 15 cm enables distributed salinity sensor with both high accuracy and high resolution.

Refractive index measurements of liquid: The parameters previously presented are measured by sensing the dependence of the BFS on strain variations applied on the fiber. The linear dependency of the BFS to the effective index of the propagated light into the fiber was exploited for sensing the refractive index of a liquid surrounding a side-polished optical fiber [60]. The fraction of evanescent field interacting with the liquid along the side-polished section of 3 cm was further increased by depositing a thin film with high refractive index (SU-8 photoresist, $n = 1.56$), leading to a refractive index sensitivity of the BFS about 293 MHz/RIU around $n = 1.40$. The measurements were realized with a high spatial resolution BOFDA apparatus.

Monitoring dihydrogen diffusion: The diffusion of dihydrogen (H_2) within an SMF was demonstrated with BOTDA measurements. The dependence of the BFS to H_2 concentration in the silica was reported with a rate of approximately 0.21 MHz/%H_2 [61]. Simultaneous Rayleigh and Brillouin backscattering measurements during H_2 diffusion have enabled to decorrelate the BFS induced by a refractive index variation and by a linear increase of acoustic velocity of about 5.2 (m/s)/(%mol H_2) with H_2 concentration in silica [62]. Further studies are developed to increase the BFS sensitivity to H_2 concentration by inserting palladium particles (H_2 sensitive material) into specialty optical fibers [63].

4. Prospects and Conclusions

The flexibilities and functionalities of the specialty optical fibers have demonstrated great potential in sensor performance enhancement and application development for BOTDA systems. In this review, several key aspects of the achievements using the specialty fiber in BOTDA systems are summarized, including the Brillouin gain amplification through the integration of highly nonlinear gas medium into the hollow-core fiber, novel polarization properties of specialty fiber based BOTDA systems, sensitivity enhancements by specially coated optical fibers, discriminative sensing of temperature and strain, multiplexed distributed fiber sensing systems of BOTDA and other systems, and novel sensing parameters enabled by specialty fibers. It is believed that specialty fibers are becoming an important design and implementation considerations of the BOTDA system for distributed sensing applications. In addition, the developments of specialty fibers specially designed for a sensing application, by optimizing the topology (core shape, number of modes, multicores) and by associating material properties (silica/air, gas, highly Ge-doped) such as multi-material fibers, will offer new prospects for further enhancing the performance of BOTDA systems in growing applications areas such as the oil and gas industry, power industry, construction industry, transportation industry and security monitoring [9].

Author Contributions: Conceptualization: D.J.J.H., H.D. and H.Z.; writing-original draft preparation, D.J.J.H. and G.H.; writing-review and editing: H.D., H.Z., Q.S. and J.H. All authors have read and agreed to the published version of the manuscript.

Funding: This work presented in this paper is a result of the research effort through Enhancing Offshore System Productivity, Integrity and Survivability in Extreme Environments (ENSURE) programme financed by A*STAR under its RIE 2020 Industry Alignment Fund (Grant No: A19F1a0104).

Institutional Review Board Statement: Not applicable.

Informed Consent Statement: Not applicable.

Conflicts of Interest: The authors declare no conflict of interest.

Nomenclature

The symbols used in the manuscript are listed in the table below.

Symbols	Description
$\Delta\nu_{B\_peak1,2}$	The central frequency shifts of the Brillouin spectra peaks
$C_{T1,2}$	Temperature coefficients of the Brillouin peak 1, 2, respectively
$C_{\varepsilon 1,2}$	Strain coefficients of the Brillouin peak 1, 2, respectively
ΔT	Temperature change
$\Delta\varepsilon$	Strain change
Δv_B	The shift of the Brillouin gain spectrum
Δv_R	The shift of the Raman gain spectrum
C_{11}	Strain coefficient of the Brillouin sensing system
C_{12}	Temperature coefficient of the Brillouin sensing system
C_{21}	Strain coefficient of the Raman sensing system
C_{22}	Temperature coefficient of the Raman sensing system

References

1. Liu, Y.; Li, H.; Wang, Y.; Men, Y.; Xu, Q. Damage detection of tunnel based on the high-density cross-sectional curvature obtained using strain data from BOTDA sensors. *Mech. Syst. Signal Process.* **2021**, *158*, 107728. [CrossRef]
2. Xu, J.; Dong, Y.; Li, H. Research on fatigue damage detection for wind turbine blade based on high-spatial-resolution DPP-BOTDA. In Proceedings of the SPIE 9061, Sensors and Smart Structures Technologies for Civil, Mechanical, and Aerospace Systems, San Diego, CA, USA, 10–13 March 2014. [CrossRef]
3. Zhang, D.; Yang, Y.; Xu, J.; Ni, L.; Li, H. Structural Crack Detection Using DPP-BOTDA and Crack-Induced Features of the Brillouin Gain Spectrum. *Sensors* **2020**, *20*, 6947. [CrossRef]
4. Jamioy, C.A.G.; Lopez-Higuera, J.M. Brillouin Distributed Fiber Sensors: An Overview and Applications. *J. Sens.* **2012**, *2012*, 204121. [CrossRef]
5. Motil, A.; Bergman, A.; Tur, M. State of the art of Brillouin fiber-optic distributed sensing. *Opt. Laser Technol.* **2016**, *78*, 81–103. [CrossRef]
6. Feng, C.; Kadum, J.E.; Schneider, T. *The State-of-the-Art of Brillouin Distributed Fiber Sensing*; IntechOpen: London, UK, 2019. [CrossRef]
7. Dong, Y. High-Performance Distributed Brillouin Optical Fiber Sensing. *Photonic Sens.* **2021**, *11*, 69–90. [CrossRef]
8. Lu, P.; Lalam, N.; Badar, M.; Liu, B.; Chorpening, B.T.; Buric, M.P.; Ohodnicki, P. Distributed optical fiber sensing: Review and perspective. *Appl. Phys. Rev.* **2019**, *6*, 041302. [CrossRef]
9. Bao, X.; Zhou, Z.; Wang, Y. Review: Distributed time-domain sensors based on Brillouin scattering and FWM enhanced SBS for temperature, strain and acoustic wave detection. *PhotoniX* **2021**, *2*, 14. [CrossRef]
10. Bai, Z.; Yuan, H.; Liu, Z.; Xu, P.; Gao, Q.; Williams, R.; Kitzler, O.; Mildren, R.; Wang, Y.; Lu, Z. Stimulated Brillouin scattering materials, experimental design and applications: A review. *Opt. Mater.* **2018**, *75*, 626–645. [CrossRef]
11. Dragic, P.; Ballato, J. A Brief Review of Specialty Optical Fibers for Brillouin-Scattering-Based Distributed Sensors. *Appl. Sci.* **2018**, *8*, 1996. [CrossRef]
12. Bastianini, F.; Di Sante, R.; Falcetelli, F.; Marini, D.; Bolognini, G. Optical Fiber Sensing Cables for Brillouin-Based Distributed Measurements. *Sensors* **2019**, *19*, 5172. [CrossRef]
13. Soto, M.A.; Ramírez, J.A.; Thévenaz, L. Intensifying the response of distributed optical fibre sensors using 2D and 3D image restoration. *Nat. Commun.* **2016**, *7*, 10870. [CrossRef]
14. Wang, B.; Wang, L.; Guo, N.; Zhao, Z.; Yu, C.; Lu, C. Deep neural networks assisted BOTDA for simultaneous temperature and strain measurement with enhanced accuracy. *Opt. Express* **2019**, *27*, 2530–2543. [CrossRef] [PubMed]
15. Bai, Q.; Wang, Q.; Wang, D.; Wang, Y.; Gao, Y.; Zhang, H.; Zhang, M.; Jin, B. Recent Advances in Brillouin Optical Time Domain Reflectometry. *Sensors* **2019**, *19*, 1862. [CrossRef] [PubMed]
16. Kobyakov, A.; Sauer, M.; Chowdhury, D. Stimulated Brillouin scattering in optical fibers. *Adv. Opt. Photonics* **2009**, *2*, 1–59. [CrossRef]
17. Fortier, C.; Fatome, J.; Pitois, S.; Smektala, F.; Millot, G.; Troles, J.; Desevedavy, F.; Houizot, P.; Brilland, L.; Traynor, N. Experimental investigation of Brillouin and Raman scattering in a 2SG sulfide glass microstructured chalcogenide fiber. *Opt. Express* **2008**, *16*, 9398–9404. [CrossRef] [PubMed]
18. Yang, F.; Gyger, F.; Thévenaz, L. Intense Brillouin amplification in gas using hollow-core waveguides. *Nat. Photonics* **2020**, *14*, 700–708. [CrossRef] [PubMed]

19. Yang, F.; Gyger, F.; Thévenaz, L. Giant Brillouin Amplification in Gases. *Opt. Photonics News* **2020**, *31*, 33. [CrossRef]
20. Beugnot, J.-C.; Sylvestre, T.; Alasia, D.; Maillotte, H.; Laude, V.; Monteville, A.; Provino, L.; Traynor, N.; Mafang, S.F.; Thévenaz, L. Complete experimental characterization of stimulated Brillouin scattering in photonic crystal fiber. *Opt. Express* **2007**, *15*, 15517–15522. [CrossRef] [PubMed]
21. Iyer, A.; Xu, W.; Antonio-Lopez, J.E.; Correa, R.A.; Renninger, W.H. Ultra-low Brillouin scattering in anti-resonant hollow-core fibers. *APL Photonics* **2020**, *5*, 096109. [CrossRef]
22. Dong, Y.; Qiu, L.; Lu, Y.; Teng, L.; Wang, B.; Zhu, Z. Sensitivity-Enhanced Distributed Hydrostatic Pressure Sensor Based on BOTDA in Single-Mode Fiber with Double-Layer Polymer Coatings. *J. Lightwave Technol.* **2020**, *38*, 2564–2571. [CrossRef]
23. Gu, H.; Dong, H.; Zhang, G.; He, J.; Pan, H. Effects of Polymer Coatings on Temperature Sensitivity of Brillouin Frequency Shift Within Double-Coated Fibers. *IEEE Sens. J.* **2012**, *13*, 864–869. [CrossRef]
24. Lu, X.; Soto, M.A.; Thevenaz, L. Impact of the Fiber Coating on the Temperature Response of Distributed Optical Fiber Sensors at Cryogenic Ranges. *J. Lightwave Technol.* **2017**, *36*, 961–967. [CrossRef]
25. McElhenny, J.E.; Pattnaik, R.; Toulouse, J. Polarization dependence of stimulated Brillouin scattering in small-core photonic crystal fibers. *J. Opt. Soc. Am. B* **2008**, *25*, 2107–2115. [CrossRef]
26. Bao, X.; Webb, D.J.; Jackson, D.A. Combined distributed temperature and strain sensor based on Brillouin loss in an optical fiber. *Opt. Lett.* **1994**, *19*, 141–143. [CrossRef] [PubMed]
27. Li, Z.; Yan, L.; Zhang, X.; Pan, W. Temperature and Strain Discrimination in BOTDA Fiber Sensor by Utilizing Dispersion Compensating Fiber. *IEEE Sens. J.* **2018**, *18*, 7100–7105. [CrossRef]
28. Lee, C.; Chiang, P.; Chi, S. Utilization of a dispersion-shifted fiber for simultaneous measurement of distributed strain and temperature through Brillouin frequency shift. *IEEE Photonics Technol. Lett.* **2001**, *13*, 1094–1096. [CrossRef]
29. Liu, X.; Bao, X. Brillouin Spectrum in LEAF and Simultaneous Temperature and Strain Measurement. *J. Lightwave Technol.* **2011**, *30*, 1053–1059. [CrossRef]
30. Leon, S.J. *Linear Algebra with Applications*; Pearson Prentice Hall: Upper Saddle River, NJ, USA, 2006.
31. Zhang, H.; Yuan, Z.; Liu, Z.; Gao, W.; Dong, Y. Simultaneous measurement of strain and temperature using a polarization-maintaining photonic crystal fiber with stimulated Brillouin scattering. *Appl. Phys. Express* **2016**, *10*, 012501. [CrossRef]
32. Zou, L.; Bao, X.; Chen, L. Brillouin scattering spectrum in photonic crystal fiber with a partially germanium-doped core. *Opt. Lett.* **2003**, *28*, 2022–2024. [CrossRef]
33. Zou, L.; Bao, X.; Afshar, S.; Chen, L. Dependence of the brillouin frequency shift on strain and temperature in a photonic crystal fiber. *Opt. Lett.* **2004**, *29*, 1485–1487. [CrossRef]
34. Li, A.; Wang, Y.; Fang, J.; Li, M.-J.; Kim, B.Y.; Shieh, W. Few-mode fiber multi-parameter sensor with distributed temperature and strain discrimination. *Opt. Lett.* **2015**, *40*, 1488–1491. [CrossRef]
35. Fang, J.; Milione, G.; Stone, J.; Peng, G.; Li, M.-J.; Ip, E.; Li, Y.; Ji, P.N.; Huang, Y.-K.; Huang, M.-F.; et al. Multi-parameter distributed fiber sensing with higher-order optical and acoustic modes. *Opt. Lett.* **2019**, *44*, 1096–1099. [CrossRef]
36. Zaghloul, M.A.S.; Wang, M.; Milione, G.; Li, M.-J.; Li, S.; Huang, Y.-K.; Wang, T.; Chen, K.P. Discrimination of Temperature and Strain in Brillouin Optical Time Domain Analysis Using a Multicore Optical Fiber. *Sensors* **2018**, *18*, 1176. [CrossRef]
37. Mizuno, Y.; Hayashi, N.; Tanaka, H.; Wada, Y.; Nakamura, K. Brillouin scattering in multi-core optical fibers for sensing applications. *Sci. Rep.* **2015**, *5*, 11388. [CrossRef]
38. Zhao, Z.; Dang, Y.; Tang, M.; Li, B.; Gan, L.; Fu, S.; Wei, H.; Tong, W.; Shum, P.; Liu, D. Spatial-division multiplexed Brillouin distributed sensing based on a heterogeneous multicore fiber. *Opt. Lett.* **2016**, *42*, 171. [CrossRef]
39. Ding, M.; Mizuno, Y.; Nakamura, K. Discriminative strain and temperature measurement using Brillouin scattering and fluorescence in erbium-doped optical fiber. *Opt. Express* **2014**, *22*, 24706–24712. [CrossRef] [PubMed]
40. Deroh, M.; Sylvestre, T.; Chretien, J.; Maillotte, H.; Kibler, B.; Beugnot, J.-C. Towards athermal Brillouin strain sensing based on heavily germania-doped core optical fibers. *APL Photonics* **2019**, *4*, 030801. [CrossRef]
41. McElhenny, J.E.; Pattnaik, R.K.; Toulouse, J.; Saitoh, K.; Koshiba, M. Unique characteristic features of stimulated Brillouin scattering in small-core photonic crystal fibers. *J. Opt. Soc. Am. B* **2008**, *25*, 582–593. [CrossRef]
42. Zou, W.; He, Z.; Hotate, K. Experimental investigation on Brillouin scattering property in highly nonlinear photonic crystal fiber with hybrid core. *Opt. Express* **2012**, *20*, 11083–11090. [CrossRef] [PubMed]
43. Li, M.-J.; Li, S.; Derick, J.A.; Stone, J.S.; Chow, B.C.; Bennett, K.W.; Sutherlin, D.M. Dual core optical fiber for distributed Brillouin fiber sensors. In Proceedings of the Asia Communications and Photonics Conference 2014, Shanghai, China, 11–14 November 2014; p. AW4I.3. [CrossRef]
44. Zhao, Z.; Soto, M.A.; Tang, M.; Thévenaz, L. Distributed shape sensing using Brillouin scattering in multi-core fibers. *Opt. Express* **2016**, *24*, 25211–25223. [CrossRef] [PubMed]
45. Zou, W.; He, Z.; Hotate, K. Complete discrimination of strain and temperature using Brillouin frequency shift and birefringence in a polarization-maintaining fiber. *Opt. Express* **2009**, *17*, 1248–1255. [CrossRef]
46. Zhao, Z.; Tang, M.; Lu, C. Distributed multicore fiber sensors. *Opto-Electron. Adv.* **2020**, *3*, 19002401–19002417. [CrossRef]
47. Kishida, K.; Yamauchi, Y.; Guzik, A. Study of optical fibers strain-temperature sensitivities using hybrid Brillouin-Rayleigh system. *Photonic Sens.* **2013**, *4*, 1–11. [CrossRef]

48. Dang, Y.; Zhao, Z.; Tang, M.; Zhao, C.; Gan, L.; Fu, S.; Liu, T.; Tong, W.; Shum, P.P.; Liu, D. Towards large dynamic range and ultrahigh measurement resolution in distributed fiber measurement resolution in distributed fiber. *Opt. Express* **2017**, *25*, 20183–20193. [CrossRef] [PubMed]
49. Alahbabi, M.N.; Cho, Y.T.; Newson, T.P. Simultaneous temperature and strain measurement with combined spontaneous Raman and Brillouin scattering. *Opt. Lett.* **2005**, *30*, 1276–1278. [CrossRef] [PubMed]
50. Muanenda, Y.; Oton, C.J.; Di Pasquale, F. Application of Raman and Brillouin Scattering Phenomena in Distributed Optical Fiber Sensing. *Front. Phys.* **2019**, *7*, 155. [CrossRef]
51. Zhao, Z.; Dang, Y.; Tang, M.; Duan, L.; Wang, M.; Wu, H.; Fu, S.; Tong, W.; Shum, P.P.; Liu, D. Spatial-division multiplexed hybrid Raman and Brillouin optical time-domain reflectometry based on multi-core fiber. *Opt. Express* **2016**, *24*, 25111–25118. [CrossRef]
52. Minardo, A.; Bernini, R.; Zeni, L. Bend-Induced Brillouin Frequency Shift Variation in a Single-Mode Fiber. *IEEE Photonics Technol. Lett.* **2013**, *25*, 2362–2364. [CrossRef]
53. Wu, H.; Tang, M.; Wang, M.; Zhao, C.; Zhao, Z.; Wang, R.; Liao, R.; Fu, S.; Yang, C.; Tong, W.; et al. Few-mode optical fiber based simultaneously distributed curvature and temperature sensing. *Opt. Express* **2017**, *25*, 12722. [CrossRef]
54. Shen, L.; Wu, H.; Zhao, C.; Zhang, R.; Tong, W.; Fu, S.; Tang, M. Distributed curvature sensing based on a bending loss-resistant ring-core fiber. *Photonics Res.* **2020**, *8*, 165. [CrossRef]
55. Teng, L.; Dong, Y.K.; Zhou, D.; Jiang, T.; Zhou, D.W. Temperature-compensated distributed hydrostatic pressure Brillouin sensor using a thin-diameter and polarization-maintaining photonics crystal fiber. In Proceedings of the Asia-Pacific Optical Sensors Conference 2016, Shanghai, China, 11–14 October 2016; p. W4A.55. [CrossRef]
56. Vedadi, A.; Alasia, D.; Lantz, E.; Maillotte, H.; Thevenaz, L.; Gonzalez-Herraez, M.; Sylvestre, T. Brillouin Optical Time-Domain Analysis of Fiber-Optic Parametric Amplifiers. *IEEE Photonics Technol. Lett.* **2007**, *19*, 179–181. [CrossRef]
57. Alishahi, F.; Vedadi, A.; Denisov, A.; Soto, M.A.; Mehrany, K.; Bres, C.S.; Thevenaz, L. Highly sensitive dispersion map extraction from highly nonlinear fibers using BOTDA probing of parametric amplification. In Proceedings of the 2013 Conference on Lasers & Electro-Optics Europe & International Quantum Electronics Conference CLEO EUROPE/IQEC, Munich, Germany, 12–16 May 2013; p. 1. [CrossRef]
58. Alishahi, F.; Vedadi, A.; Shoaie, A.; Mehrany, K.; Salehi, J.A. Low resolution distributed measurement of fiber optical parametric amplifiers gain. In Proceedings of the Optical Fiber Communication Conference 2012, Los Angeles, CA, USA, 4–8 March 2012; p. JW2A.30.
59. Zhang, H.; Teng, L.; Dong, Y. Distributed Salinity Sensor with a Polymide-Coated Photonic Crystal Fiber Based on Brillouin Dynamic Grating. *J. Lightwave Technol.* **2020**, *38*, 5219–5224. [CrossRef]
60. Bernini, R.; Persichetti, G.; Catalano, E.; Zeni, L.; Minardo, A. Refractive index sensing by Brillouin scattering in side-polished optical fibers. *Opt. Lett.* **2018**, *43*, 2280–2283. [CrossRef] [PubMed]
61. Delepine-Lesoille, S.; Bertrand, J.; Lablonde, L.; Pheron, X. Distributed Hydrogen Sensing with Brillouin Scattering in Optical Fibers. *IEEE Photonics Technol. Lett.* **2012**, *24*, 1475–1477. [CrossRef]
62. Leparmentier, S.; Auguste, J.-L.; Humbert, G.; Pilorget, G.; Lablonde, L.; Delepine-Lesoille, S. Study of the hydrogen influence on the acoustic velocity of single-mode fibers by Rayleigh and Brillouin backscattering measurements. In Proceedings of the 24th International Conference on Optical Fibre Sensors, Curitiba, Brazil, 28 September–2 October 2015; Volume 9634, p. 963432. [CrossRef]
63. Delepine-Lesoille, S.; Girard, S.; Landolt, M.; Bertrand, J.; Planes, I.; Boukenter, A.; Marin, E.; Humbert, G.; Leparmentier, S.; Auguste, J.-L.; et al. France's State of the Art Distributed Optical Fibre Sensors Qualified for the Monitoring of the French Underground Repository for High Level and Intermediate Level Long Lived Radioactive Wastes. *Sensors* **2017**, *17*, 1377. [CrossRef] [PubMed]

hv *photonics*

MDPI

Review

Review of a Specialty Fiber for Distributed Acoustic Sensing Technology

Yixiang Sun, Hao Li, Cunzheng Fan, Baoqiang Yan, Junfeng Chen, Zhijun Yan and Qizhen Sun *

National Engineering Laboratory for Next Generation Internet Access System (NGIA), Wuhan National Laboratory for Optoelectronics (WNLO), School of Optical and Electronic Information, Huazhong University of Science and Technology, Wuhan 430074, China; m202072360@hust.edu.cn (Y.S.); lhbeyond@hust.edu.cn (H.L.); buckt@hust.edu.cn (C.F.); yanbaoqiang@hust.edu.cn (B.Y.); junfengchen@hust.edu.cn (J.C.); yanzhijun@hust.edu.cn (Z.Y.)

* Correspondence: qzsun@mail.hust.edu.cn

Abstract: Specialty fibers have introduced new levels of flexibility and variability in distributed fiber sensing applications. In particular, distributed acoustic sensing (DAS) systems utilized] the unique functions of specialty fibers to achieve performance enhancements in various distributed sensing applications. This paper provides an overview of recent preparations and developments of specialty-fiber-based DAS systems and their sensing applications. The specialty-fiber-based DAS systems are categorized and reviewed based on the differences in scattering enhancement and methods of preparation. The prospects of using specialty fibers for DAS systems are also discussed.

Keywords: specialty fibers; distributed acoustic sensing; DAS; distributed fiber sensor; scattering enhancement fiber

Citation: Sun, Y.; Li, H.; Fan, C.; Yan, B.; Chen, J.; Yan, Z.; Sun, Q. Review of a Specialty Fiber for Distributed Acoustic Sensing Technology. *Photonics* **2022**, *9*, 277. https://doi.org/10.3390/photonics9050277

Received: 29 January 2022
Accepted: 15 March 2022
Published: 20 April 2022

1. Introduction

The identification of the internal properties of a medium is an important way to understand nature and explore the world, which can be obtained using both direct and indirect measurements. Regarding acoustic waves, the transmission form of sound travels in different directions with the help of various media in the environment. Therefore, the internal properties of the medium can be recognized by measuring and tracking the information carried by the acoustic field [1–3]. In recent years, acoustic sensing technology has been widely used in the analysis of material properties at different scales, such as in geological structure detection [4], resource exploration [5], structural health monitoring [6], and perimeter security [7].

Acoustic sensors allow people to analyze media through acoustic information. With the increases in the range and scale of acoustic detection, the demand for distributed high-capacity acoustic sensors is also expanding. Furthermore, distributed acoustic sensing (DAS) has become a research hotspot because of its advantages in terms of anti-electromagnetic interference, high sensitivity [8], and the small volume and light weight of the systems. DAS is a rising acoustic sensing technology that uses passive optical fibers as the transmission and sensing medium at the same time, squeezing information from optical fiber Rayleigh backscatters to achieve acoustic detection in the surrounding environment. Most DAS systems are based on phase-sensitive optical time domain reflectometry (φ-OTDR), since the phase change in Rayleigh backscattered light has a linear relationship with the acoustic wave acting on the optical fiber [9].

Common DAS systems usually use a single-mode fiber (SMF) as the sensing fiber. However, with such fibers the Rayleigh backscattered light is extremely weak and the signal-to-noise ratio (SNR) of the optical sensing signal is low, which leads to a poor SNR of the demodulation phase signal in DAS systems based on φ-OTDR [10]. In addition, since a laser pulse with a narrow linewidth and high coherence is injected into the fiber, the interference between the Rayleigh scattering points in one pulse will produce interference cancellation, resulting in coherent fading of the backscattered light, forming a "dead zone" on the sensing fiber [11]. In recent years, in order to solve the defects of the poor SNR values and the sensitivity of phase demodulation in SMF-DAS systems, many suppression technologies for noise have been studied, such as coherent fading, while at the same time the complexity and cost of the systems has increased, leading to sacrifices in terms of the response frequency band and sensing distance [12–18]. Furthermore, it is also important to enhance the performance of the sensing fiber itself. The development of a specialty fiber for DAS technology will have broad prospects.

This paper intends to provide a comprehensive review of the different sensing features and performance enhancements provided by the specialty fibers that are used in DAS systems. The principle of the DAS technology is introduced in Section 2, followed by a discussion of the specialty-fiber-based DAS system with enhanced sensing capability and performance in Section 3. Firstly, two types of specialty fibers are introduced, a continuous scattering-enhanced (CSE) fiber and discrete scattering-enhanced (DSE) fiber, with the latter one being more widely used because of its ability to completely suppress coherent fading. Secondly, several fabrication techniques for introducing scattering enhancement points (SEP) into the fiber are summarized, including femtosecond writing and UWFBG inscription. Thirdly, a variety of methods to enhance the performance of DAS systems for discrete scattering are reviewed. Fourthly, the applications of the specialty fiber DAS system are reviewed in Section 4. Lastly, the prospects of using specialty fibers for DAS systems are discussed in Section 5.

2. Fundamental Principle

2.1. Principle of DAS Technology

DAS is achieved by measuring the optical phase change caused by the axial strain variation of the optical fiber [19]. When the sound wave acts on the fiber, the fiber will produce axial strain, which changes the phase of the Rayleigh backscattering signal in the fiber. According to the photoelastic effect, there is a linear relationship between the axial strain and the optical phase change:

$$\Delta\varphi = \beta\left[1 - \frac{n^2}{2}(P_{12} + 2P_{11})\right]\Delta L \tag{1}$$

where β is the propagation constant of light, n is the refractive index of the optical fiber, P_{12} and P_{11} are tensor coefficients of the optical fiber, ΔL is the change in fiber length from equation $\Delta L = \varepsilon_s L$, ε_s is the axial strain of the fiber, and L is the length of the fiber. Thus, as shown in Figure 1, as long as the changes in the phase difference between the two points A and B of the probe light can be extracted, the changes in axial strain can be known and the quantitative perception of acoustic waves or vibrations can finally be measured. It is crucial for DAS to achieve distortion-free phase demodulation in all positions of the fiber.

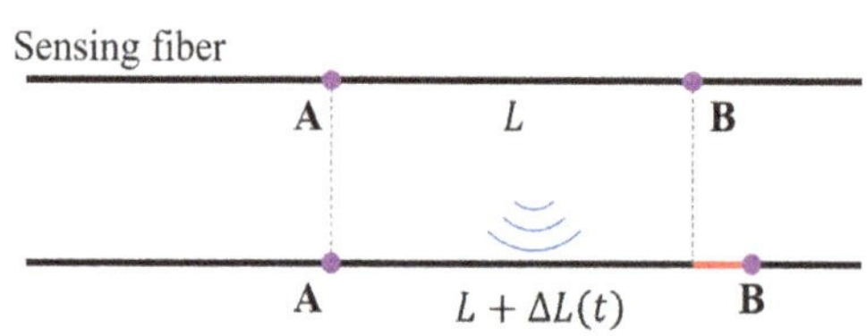

Figure 1. The sensing principle of the fiberoptic DAS.

Since Rayleigh scattering is an elastic scattering method without any nonlinear effect, and as the Rayleigh scattered light at different positions can be distinguished via the time of reflection back to the fiber launch end, the optical fiber phase extraction technique based on phase-sensitive optical time domain reflection (φ-OTDR) is widely used in DAS systems [20,21]. In 2011, using heterodyne coherent detection technology, the phase of the

Rayleigh scattering signal was successfully demodulated for the first time [22]. The scheme of the heterodyne coherent detection is shown in Figure 2. The local oscillator light and the Rayleigh scattering signal at point A can be expressed as:

$$E_{LO} = A_{LO} exp[j2\pi ft + j\varphi_{LO}] \quad (2)$$

$$E_A = A_A exp[j2\pi(f + \Delta f)t + j\varphi_A] \quad (3)$$

where A_{LO} and A_A are the amplitude of the local oscillator light and the Rayleigh scattering signal at point A, f is the frequency of the probe light, Δf is the frequency shift of the pulse modulator, φ_{LO} and φ_A are the optical phases of the local oscillator light and the Rayleigh scattering signal at point A, respectively. After these two signals are coupled by a 3 dB coupler and interfere with the photosensitive surface of the balanced photoelectric detector (BPD), the light intensity detected by the BPD can be expressed as:

$$I \propto 2E_{LO}E_A cos(\Delta ft + \varphi'_A) \quad (4)$$

where $\varphi'_A = \varphi_A - \varphi_{LO}$ is the phase difference between the two signals. Then, φ'_A can be calculated using a digital coherent IQ demodulation algorithm. Similarly, φ'_B can also be calculated at point B by the same algorithm. Through the spatial differential calculation of the phase, the phase difference φ_{AB} between two points A and B can be finally obtained, which linearly shows the axial strain of the optical fiber and finally allows the quantitative perception of sound waves and vibrations.

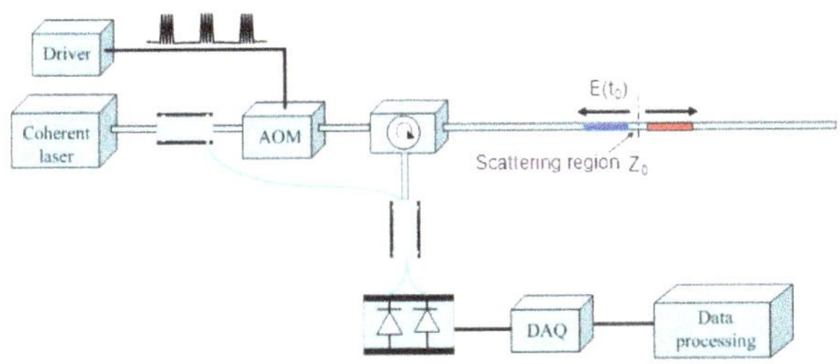

Figure 2. Typical heterodyne coherent detection scheme [22].

In addition to the heterodyne-based coherent detection scheme, researchers have also proposed other phase demodulation schemes, such as the 3 × 3 coupler scheme [23], phase-generated carrier (PGC) [24], and linear frequency sweeping pulse method [25]. Some representative schemes are shown in Figure 3, all of which can achieve distributed optical phase demodulation.

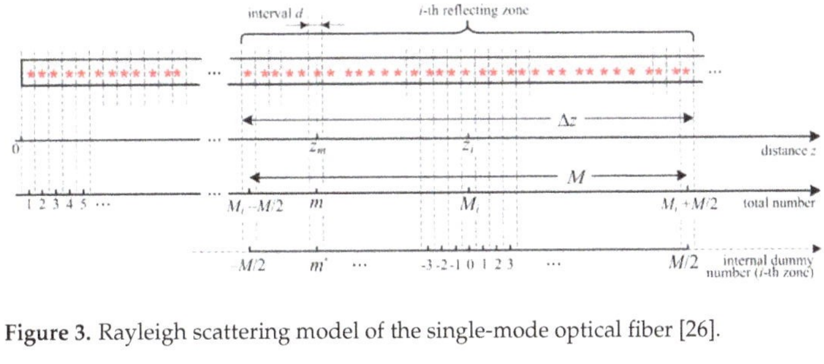

Figure 3. Rayleigh scattering model of the single-mode optical fiber [26].

2.2. Limitations of Single-Mode Fiber (SMF) DAS

Despite SMF DAS technology having been applied in many fields, such as geological monitoring, pipeline monitoring, and oil exploration, and with the advantages of distributed detection, high spatial resolution, and high sensitivity, it still has been limited by interference fading and poor signal consistency. The reasons are analyzed below.

In particular, when the pulse width is narrow enough, the Rayleigh backscattering signals of point A and B as discussed in Section 2.1 can be obtained according to the arrival time of the Rayleigh backscattering light. In fact, the probe pulse has a certain pulse width, and the Rayleigh scattering signals of different scattering points in the pulse width will interfere with each other because of the high coherence of the probe light. The interference result of the Rayleigh scattering can only represent a section fiber near point A and B. Due to the inhomogeneous doping in the preparation of the optical fiber, the interference phenomenon of the Rayleigh scattering is random, which will inevitably affect the phase demodulation performance. In order to describe the Rayleigh scattering phenomenon in SMF accurately, researchers established a SMF scattering model [26].

The Rayleigh scattering model is shown in Figure 4. First, the SMF is discretized and the nonuniform distribution of impurities in the core is regarded as a series of equivalent Rayleigh scattering points (ERSP) of micron or submicron size, which are numbered as $1, 2, 3, \cdots, M$, respectively. Each backscattering coefficient and the position of the scattering point can be presented as (a_m, z_m); that is, the M-th scattering point is on z_m, while the reflectivity is a_m. Since the ERSP can be regarded as the superposition of multiple scattering points with a smaller scale, its position can be expressed as $z_m = md + \Delta d$, where Δd obeys the uniform distribution in $\left[-\frac{d}{2}, \frac{d}{2}\right]$ and a_m obeys the Rayleigh distribution. According to the above analysis, the Rayleigh scattering signal received at the i-th time can be expressed as the superposition of interference of the Rayleigh scattering signal of a section of fiber:

$$E_i = \sum\nolimits_{z_{i1} < z_m < z_{i2}} A a_m \cdot \exp\left(j\frac{2\pi}{\lambda} \cdot 2nz_m\right) \tag{5}$$

where n is the refractive index of the fiber, z_{i1} and z_{i2} are related to the pulse width, which satisfies $z_{i2} - z_{i1} = Wc/2n$, W is the pulse width of the probe pulse, and c is the speed of light in vacuum. Researchers have also verified this model [8]; when the interference signal involves interference cancellation, the signal intensity $|E_i|$ will become weaker, leading to a worse optical SNR. Further, the light intensity declines to the same level as the acquisition noise and the demodulated phase information will be submerged in the noise, forming a blind detection area, which is called interference fading. In addition, due to the randomness of the reflectivity a_m of the equivalent scattering point, the equivalent positions of the signals received at different times are different. The signal received at time i in Equation (5) is taken as an example. Since the scattered signal at this time is the superposition of multiple complex numbers, its intensity Aa_m is the mode of the complex number, and the phase $j\frac{2\pi}{\lambda} \cdot 2nz_m$ related to position z_m is the argument for the complex number, so Equation (5) can be expressed as the superposition of multiple vectors. As depicted in Figure 4, when the Rayleigh scattering intensity of the front-end equivalent scattering point is strong, the signal received at time i can be regarded as Rayleigh scattering near the z_{i1} position, while when the Rayleigh scattering intensity of the back-end equivalent scattering point is strong, the signal received at the same time can be regarded as Rayleigh scattering near the z_{i2} position.

In general, it is the randomness of the intensity and position of Rayleigh scattering of ordinary single-mode fibers that leads to the problems of interference fading and poor signal time consistency in DAS, making it difficult to perform high-fidelity sound wave tracking when analyzing the internal properties of the medium.

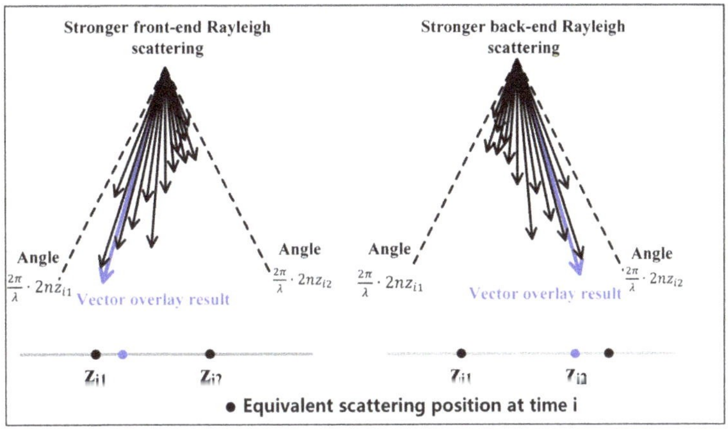

Figure 4. Schematic diagram of equivalent scattering positions in pulse interference.

3. Reviews of Specialty-Fiber-Based DAS Technology

In order to solve the interference fading and poor consistency in SMF DAS, researchers have conducted a series of studies [13,27–30]. Among them, the most effective method is to improve the fibers to enhance the fiber backscattering, such as continuous scattering-enhanced fibers and discrete scattering-enhanced fibers formed by inscribing microstructures into the fibers. In addition, due to the characteristics of microstructure optical fibers, researchers also designed and improved the matching optical scheme to improve the performance of the DAS system. In this section, a comprehensive review of the specialty-fiber-based DAS with enhanced sensing capability and performance is presented.

3.1. Continuous Scattering-Enhanced Fiber-Based DAS

Continuous scattering enhancement (CSE) allows the suppression of coherent fading by enhancing the Rayleigh scattering of the whole fiber, whose essence is to enhance the scattering intensity a_m of the ERSP model described in Section 2.2 by doping or writing continuous grating in the fiber, so as to enhance the light intensity after interference superposition, finally allowing the suppression of interference fading.

In 2017, the OFS laboratory in the United States used the phase mask method to efficiently and continuously inscribe Bragg gratings in multicore fibers through ultraviolet (UV) exposure. The backscattering intensity of the fiber was increased by 14 dB, and the reflection spectrum is shown in Figure 5a. Then, in cooperation with Fotech in the UK, CSE fiber was used in a DAS system, and the SNR of 1 km of CSE fiber was increased by 15 dB [31–34]. Although this method effectively improves the SNR, the bandwidth of the Bragg grating reflection spectrum is narrow, and the reflection wavelength will drift with temperature and stress variation. When the external environment changes, the wavelength of the incident light and grating reflection may not correspond to each other, resulting in the loss of the scattering enhancement effect. Therefore, the application range of this kind of fiber is limited and cannot be used in fields where high temperature and high pressure are required on site. In addition to inscribing continuous gratings, changing the fiber doping is another way to increase the intensity of the Rayleigh scattering. In 2018, Butov et al. used nitrogen-doped fiber as the sensing fiber, which increased the SNR of the acoustic wave detection by 3 dB [35]. In the same year, Feng S et al. used an erbium-doped optical fiber for distributed optical fiber sensing, adopted a phase-generated carrier (PGC) optical scheme, which converts homodyne interference into heterodyne interference by modulating the frequency of the signal, decreased the phase noise by 14 dB, and achieved a high-SNR acoustic measurement on 1.9 km optical fiber [36].

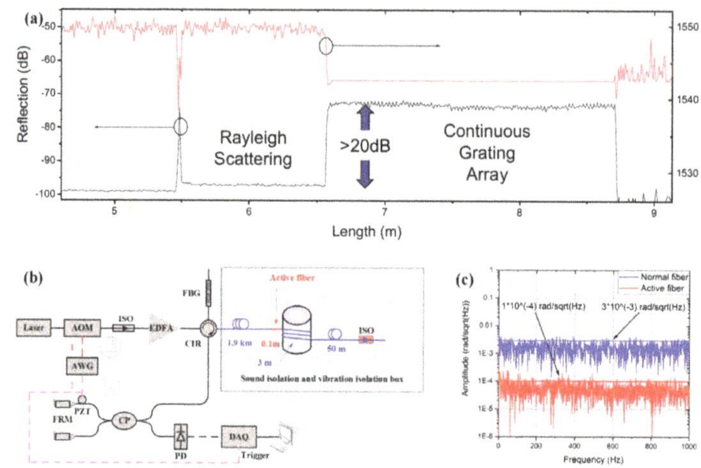

Figure 5. Typical scheme of continuous scattering-enhanced optical fiber [33,36]: (**a**) the backscattering signal distribution of a typical continuous scattering-enhanced fiber; (**b**) scheme of improved Rayleigh scattering achieved by changing the fiber doping; (**c**) scattering spectrum of high Rayleigh scattering fiber.

For the CSE scheme, although it can effectively improve the scattering intensity and suppress the coherent fading noise, the enhanced fiber scattering results in doubling of the light loss, which greatly limits the detection distance. Taking the optical fiber doping scheme as an example, if the front-end SNR is increased by 10 dB, the Rayleigh scattering of the fiber needs to be increased by ten times, which reduces the detectable distance to 1/10. More importantly, the SCE fiber does not fundamentally change the nature of the optical interference in the pulse, and still cannot completely eliminate the coherent fading phenomenon.

3.2. Discrete Scattering-Enhanced Fiber Based DAS

In addition to continuous scattering enhancement, more researchers are focusing on the application of discrete scattering-enhanced (DSE) fiber in DAS. As shown in Figure 6, the essence of this fiber is to enhance the scattering intensity of one single ERSP with a certain interval in the model illustrated in Section 2.2. The intensity of the scattering-enhanced points (SEP) is far greater than that of the ordinary ERSP. In this way, the Rayleigh scattering interference light containing the SEP can be expressed as:

$$E_i = \sum\nolimits_{z_{i1}<z_m<z_{i2}, z_m \neq z_j} a_m \cdot \exp\left(j\frac{2\pi}{\lambda}\cdot 2nz_m\right) + a_j \cdot \exp\left(j\frac{2\pi}{\lambda}\cdot 2nz_j\right) \tag{6}$$

where the j-th ERSP is the scattering enhancement point. Because the scattering intensity of the SEP is much larger than that of the ordinary ERSP, the first term of the above formula can be ignored, which shows that the Rayleigh scattering signal is only determined by the SEP. Through the phase demodulation method in Section 2.1, the phase difference between the two SEP can be accurately obtained, and because of its high scattering intensity and high SNR, the interference fading can be completely eliminated, ensuring the good consistency of the signal.

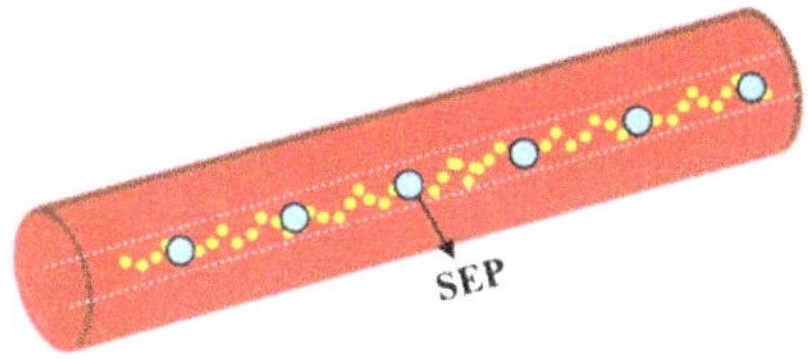

Figure 6. Schematic diagram of discrete scattering-enhanced fiber.

To verify its effectiveness, we have theoretically analyzed the DSE model and the SMF scattering model. According to the parameters shown in Table 1, three SEPs are set at 3, 7, and 11 m on the fiber at intervals of 4 m. Figure 7a,b shows the distribution results of the scattering intensity obtained from the simulation. It can be seen that the DES fiber completely suppresses the random variation of the signal amplitude and interference fading. To further explore the suppression effect of the DSE fiber on the phase noise, we solve the phase differences between 3 and 7 m on the SMF and DSE fibers, respectively, the of which results are shown in Figure 7c. Compared with SMF, the DSE fiber is not affected by the random distribution of ordinary ERSPs, but is determined by the position of the SEPs, so the phase noise is greatly suppressed.

Table 1. Parameters of discrete enhanced scattering model.

Fiber Length	Spacing of SEP	Pulse Width	Spacing of ERSP d	Repeat Times
13 m	4 m	20 ns	1 um	100

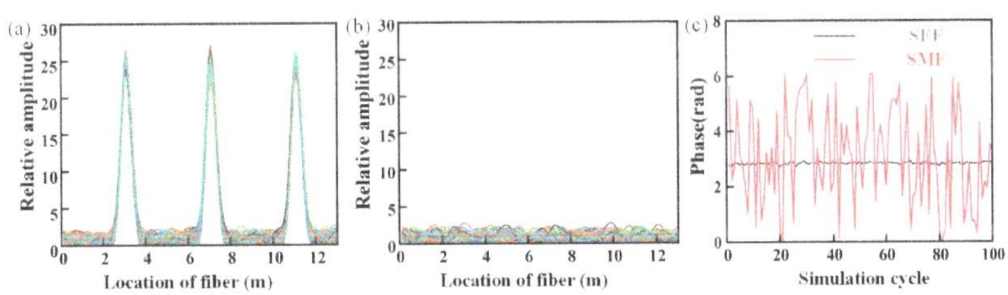

Figure 7. Simulation results of discrete enhanced scattering model: (**a**) reflection distribution of discrete enhanced fiber; (**b**) reflection distribution of single mode fiber; (**c**) phase demodulation result.

3.3. Preparation and Implementation of DSE Fiber

At present, there are two ways to manufacture the discrete scattering-enhanced fiber, one is by inscribing the periodic ultra-weak Bragg grating (UWFBG) array [29,37] into the fiber, and the other is by introducing a local wavelength-independent weak reflection array into the fiber [38–40].

UWFBG usually involves the use of ultraviolet light [41–43] or a femtosecond laser [44,45] to form a permanent periodic change in the refractive index of the optical fiber, so as to allow the backward reflection of a specific light wavelength. The most classic preparation system is shown in Figure 8a [46]. In the process of fiber drawing, through strict dynamic control, the UWFBG was written using the phase mask method, using periodic interference fringes of the $\pm$1st diffraction light to irradiate the photosensitive fiber and periodically change the refractive index of the fiber core. Additionally, an FBG writing platform was mounted on the draw tower near the first coating to weaken vibrations in the fiber. The preparation results are shown in Figure 8b [47], and the scattering rate of the scattering-enhanced array is much higher than that of the single-mode fiber. However, as a Bragg grating method, UWFBG is sensitive to temperature and stress. When the temperature and strain of the environment are changed, the reflection wavelength of the UWFBG will drift. As shown in Figure 8c, the bandwidth of UWFBG is usually less than

2 nm and the central wavelength is determined according to the light source used in the specific experiment. Due to the limited bandwidth of UWFBG, there may be a mismatch between the reflective spectrum of UWFBG and the wavelength of the detection light when the UWFBG array fiber operates in special environments, such as for large temperature vibrations or high pressure, while the fiber will also suffer from twisting or pulling forces, meaning the reflection power from the UWFBG will degenerate into Rayleigh scattering, resulting in a blind sensing area. This characteristic of UWFBG hinders its application in underground, underwater, and other special environments [48,49].

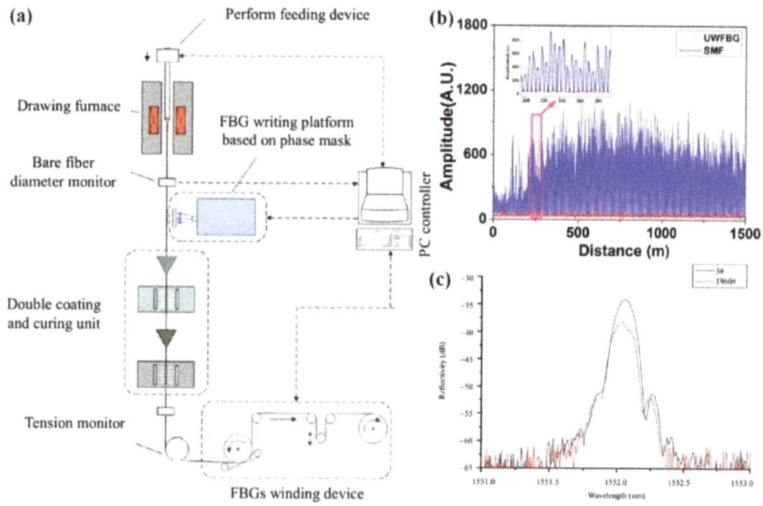

Figure 8. Typical UWFBG preparation method: (**a**) online UWFBG preparation system [46]; (**b**) OTDR measurement results of the UWFBG array [47]; (**c**) reflection spectrum of UWFBG [46].

In recent years, the enhancement scheme of a local achromatic weak reflection array has also been used in DAS [50,51]. In order to prepare a more universal discrete scattering enhancement array, Sun et al. proposed a preparation method for microstructured optical fibers with a colorless reflection point. The mechanism of the microstructure involves the local refractive index change through UV exposure without the help of the phase mask to form a Fresnel reflecting surface. The fabrication system is shown in Figure 9a [52], while the preparation results are shown in Figure 9b–d [53]. A discrete scattering enhancement array with an interval of 5 m and 15 dB enhanced intensity was achieved. When the temperature changes, the scattering intensity of the microstructure is extremely stable. In addition, the University of Southampton and other universities in the UK also used a femtosecond laser to prepare a weak reflection array [40]. The preparation system, scheme, and results are shown in Figure 10. Compared with UWFBG, whose reflection bandwidth is limited, the weak reflection points are directly engraved in the optical fiber, without wavelength selectivity, avoiding the influence of temperature and strain [54].

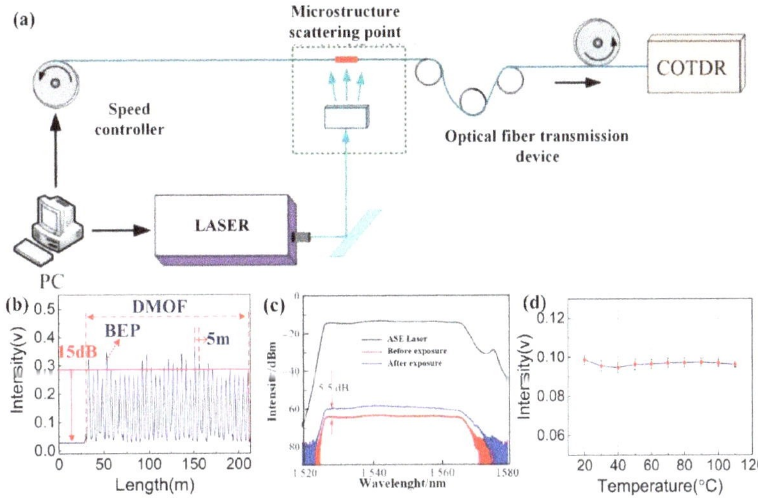

Figure 9. Preparation method for a colorless microstructure array based on UV exposure: (**a**) automatic colorless microstructure preparation system [51]; (**b**) OTDR measurement results of colorless microstructure array [53]; (**c**) spectra of colorless microstructure arrays [51]; (**d**) temperature stability of colorless microstructure arrays [53].

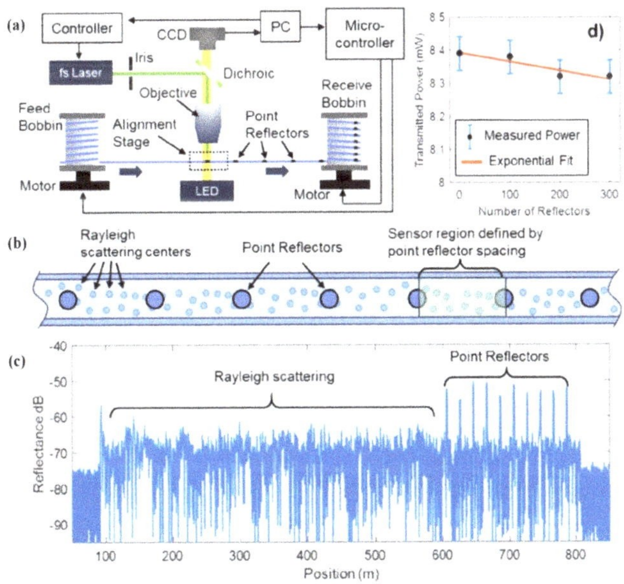

Figure 10. Preparation method for a weak reflection array based on a femtosecond laser [40]: (**a**) weak reflection array preparation system; (**b**) schematic diagram of weak reflection array preparation method; (**c**) OTDR measurement results for the weak reflection array; (**d**) the result of a cut-back measurement on a separate fiber containing 300 reflectors.

3.4. Methods to Improve the Performance of Specialty Fiber DAS

In the early stages, researchers mainly focused on using UWFBG in various DAS schemes. In 2015, Wang et al. used UWFBG in DAS based on a 3 × 3 coupler demodulation for hydroacoustic testing, which can detect a water pressure of 0.112 Pa [55]. Subsequently, the team at Huazhong University of Science and Technology successively used DSE fiber in various DAS schemes, such as double-pulse, coherent detection, and PGC demodulation schemes [29,39,40,50,56–58]. These schemes are shown in Figure 11. In recent years, more attention has been paid to improving the DAS system performance by using the advantages of discrete scattering enhancement fiber, including low-frequency phase shift compensation, polarization fading suppression, pulse width compression, and system sampling rate expansion.

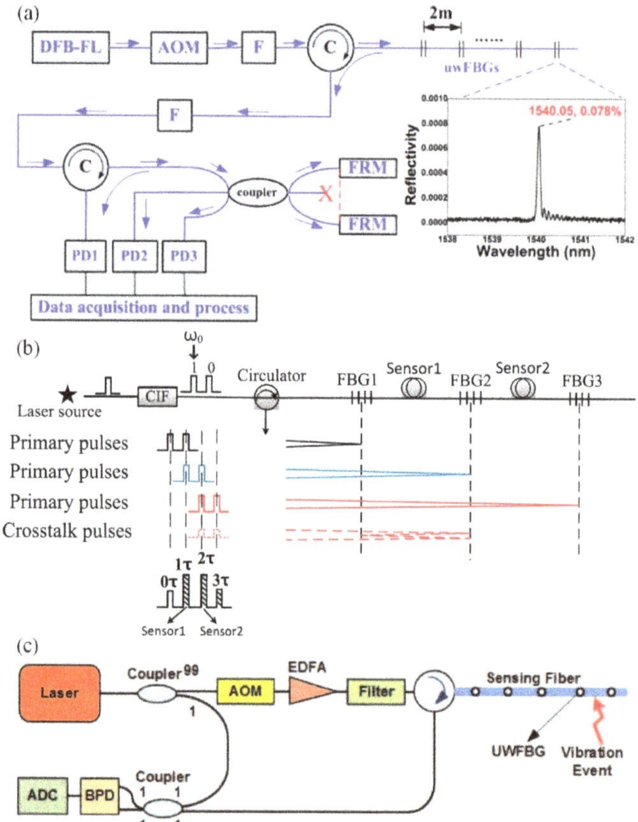

Figure 11. Representative DAS scheme based on UWFBG: (**a**) scheme using a 3 × 3 coupler [54]; (**b**) PGC scheme [55]; (**c**) heterodyne coherent detection scheme [29].

3.4.1. Low-Frequency Phase Drift Compensation Technology

The DSE fiber DAS has the potential for extremely low-frequency detection due to its good signal consistency and lac of interference of random Rayleigh scattering. In the DAS system, the low-frequency drift of the laser phase and the slow change of the external temperature will introduce greater low-frequency noise to the system, limiting the accuracy of the low-frequency detection [59]. The phase drift of the laser can be compensated by the auxiliary interferometer. As shown in Figure 12a, the phase noise of the laser is measured in real time by the auxiliary interferometer, and the measurement results are compensated to suppress the low-frequency noise. In this way, Fan et al. achieved a strain resolution of $3.84\text{p}\varepsilon/\sqrt{\text{Hz}}$ [60,61] at a frequency of 10 Hz. The team from Huazhong University of Science and Technology proposed a reference fiber compensation scheme via the discrete enhancement of UWFBG. The scheme is shown in Figure 12b, which can compensate for laser drift and temperature changes at the same time, achieving a resolution of 0.57n$\varepsilon/\sqrt{\text{Hz}}$ at a frequency of 0.01 Hz and a resolution of 3.4n$\varepsilon/\sqrt{\text{Hz}}$ at a frequency of 10 Hz [62]. Furthermore, the team introduced an LS-SVM operator during demodulation to track and compensate for the hysteresis effect of the temperature, allowing high-precision acoustic measurements in the 0.001 Hz frequency band [63].

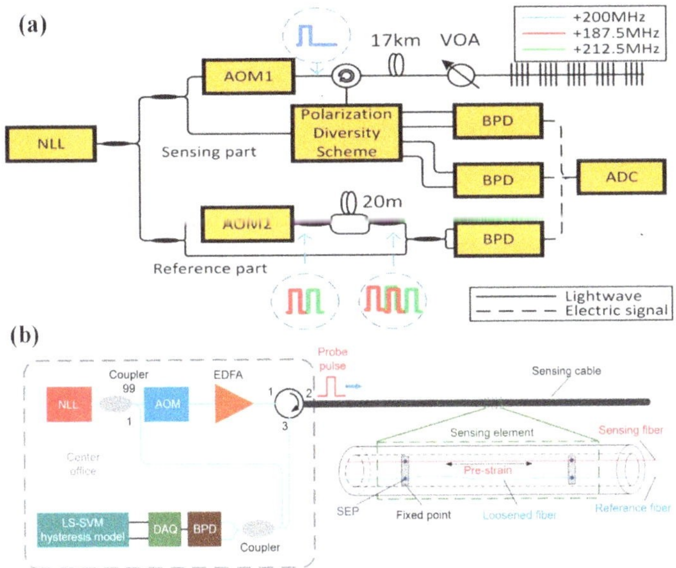

Figure 12. Representative scheme for low-frequency noise suppression: (**a**) auxiliary interferometer compensation [60,61]; (**b**) reference fiber compensation [62].

3.4.2. Polarization Fading Suppression Technology

In the DSE-fiber-based DAS system, the coherent fading is completely suppressed, while the polarization random fading problem must be urgently solved [16]. The random polarization fading is caused by the random change in polarization states of the two interference lights. When the polarization directions of the two interference lights are parallel, the interference light signal is the largest and the phase noise is the smallest. However, when the directions are orthogonal, the interference light intensity is decreased to zero and the phase information is completely overwhelmed by noise. In order to solve this problem, Zhang et al. improved the dual-wavelength DAS scheme. As shown in Figure 13, the polarization composite double pulse was used as the detection pulse to achieve polarization noise suppression [16,64]. For the coherent detection DAS scheme, Sun et al. used a polarization beam splitter to divide the reflected light into two orthogonal polarization states to interfere with the local oscillator light, so that the random polarization noise was reduced by 9.5 dB [65]. However, in this scheme, when the polarization states of the local oscillator light and the polarization beam splitter do not match, fading may still occur. To resolve this problem, the Huazhong University of Science and Technology team proposed a method for emitting dual-wavelength probe pulses and using the random polarization states of different wavelength parameters to greatly reduce the probability of polarization fading and to reduce polarization noise by 19.2 dB [66].

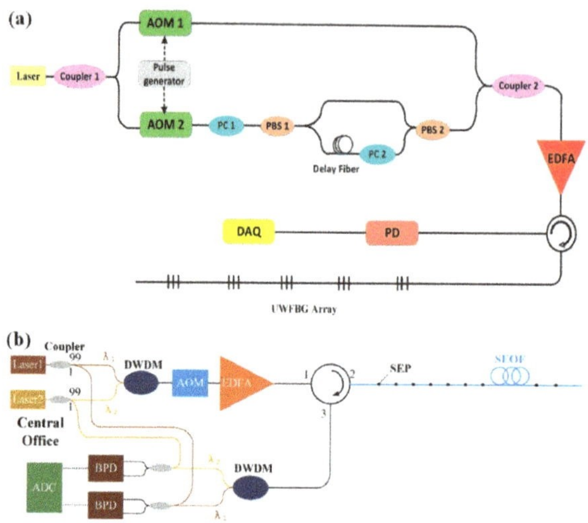

Figure 13. Representative scheme for polarization noise suppression: (**a**) orthogonal polarization scheme [16]; (**b**) multi-parameter scheme [66].

3.4.3. Pulse Width Compression Technology

In the DAS system, the interval between the SEPs determines the resolution of the system and also limits the width of the detection pulse. Assuming that the interval between the SEPs is d, the detection pulse is limited to $\tau \leq 2dn/c$ (where n is the refractive index of the fiber and c is the speed of light in the vacuum). In a radar system, a longer detection pulse is usually transmitted and the pulse is compressed at the receiving end, so as to obtain a signal with a high signal-to-noise ratio under the premise of ensuring the resolution. The common schemes are pulse linear frequency sweeping and pulse coding [67–69]. These techniques have also been used in DSE fiber DAS systems. Fan et al. adopted the sweeping pulse transmission scheme, whereby the receiving end compresses the pulses through matched filtering, achieving a noise level of -93.16dB re rad$/\sqrt{\text{Hz}}$ at 500–2500 Hz as shown in Figure 14a [70,71]. Wang et al. successively used pulse coding technology in UWFBG-based DAS systems and improved the systems' SNR values by 6.9 dB as shown in Figure 14b [72].

3.4.4. Sampling Frequency Expansion Technology

In DAS system, the sampling rate is mainly limited by the length of the sensing fiber. After the system transmits the probe pulse into the optical fiber, it needs to ensure that the Rayleigh backscattering signal transmitted to the farthest end of the pulse returns to the receiving end before transmitting the next pulse, otherwise crosstalk will occur. However, during railway safety detection and power grid partial discharge detection applications, it is necessary for the DAS system to perform long-distance and high-frequency detection simultaneously. The way to solve these problems is to transmit multiple probe lights in parallel and distinguish them through a certain feature to avoid crosstalk. By transmitting pulses with different frequencies into the optical fiber and distinguishing them by using different pulse frequencies, Zhang et al. achieved a sampling rate of 440 kHz on a 330 m optical fiber and expanded the distance bandwidth product by three times [73]. Sun et al. put forward a time slot multiplexing scheme. As shown in Figure 15, multiple pulses are inserted into the time slot between two adjacent SEPs of an optical fiber, with a sampling rate of 300 kHz being achieved on a 1020 m optical fiber, expanding the distance bandwidth product by six times [74].

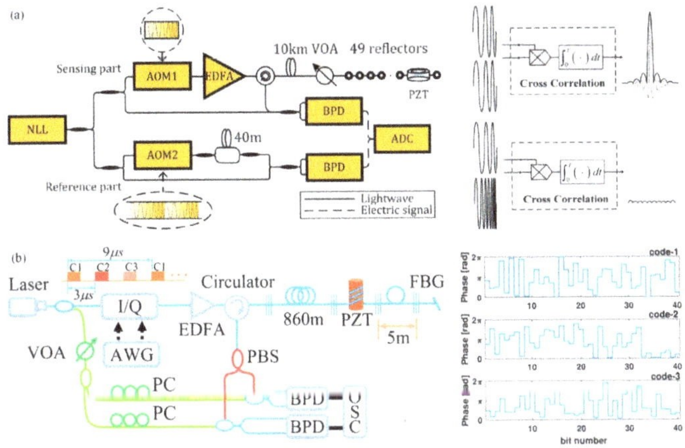

Figure 14. Representative scheme for pulse width compression: (**a**) orthogonal polarization scheme [70,71]; (**b**) multi-parameter scheme [72].

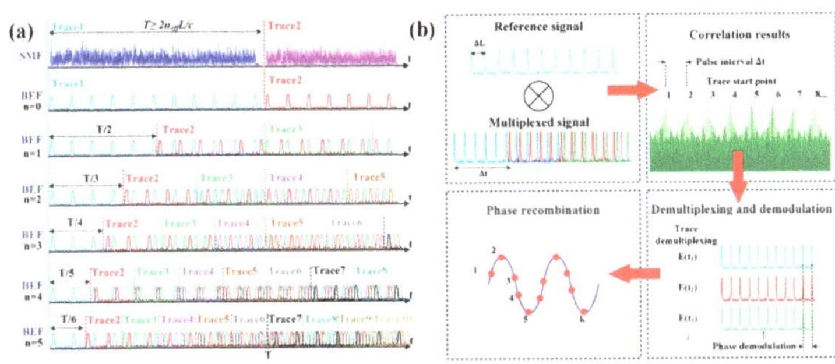

Figure 15. Time slot multiplexing scheme [74]: (**a**) principle; (**b**) demodulation algorithm.

Above, the two specialty fibers, namely CSE fiber and DSE fiber, are analyzed and compared in terms of SNR enhancements, sensing distance, and methods to improve the performance. As shown in the Table 2, it can be seen that the DSE fiber performs significantly better than the CSE fiber in terms of the sensing distance, while the greater number of matching DAS schemes also further improves the performance of DAS system in multiple dimensions.

Table 2. Performance summary for CSE and DSE fibers.

	Fabrication Method	SNR Enhancements	Sensing Distance
CSE fiber	Continuously inscribe Bragg gratings [31–34]	15 dB [31]	1 km [31]
	Highly doped fiber [35,36]	14 dB [36]	1.9 km [36]
DSE fiber	UV exposure [41–43,47,51–53]	5.5–21.1 dB [41–43,47,51]	50 km [51]
	Femtosecond laser inscription [40,44,45,54]	13–15.8 dB [40,54]	9.8 km [54]

4. Significant Application Progress

With the continuous improvement and development of the DAS system based on the specialty scattering-enhanced optical fiber, researchers have applied the system to practical engineering applications in many fields, including geological and resource exploration, structural health monitoring, and hydroacoustic exploration.

4.1. Geological and Resource Exploration

The vertical seismic profile (VSP) is a commonly used parameter for petroleum exploration seismic observations. With the development of DAS technology, fiberoptic logging has become a research hotspot.

Sun et al. conducted a walkaway VSP experiment using a self-made microstructure fiber DAS system [75]. The microstructure fiber is arranged vertically in the test well and the explosive source is artificially introduced on the ground to generate seismic waves, which are transmitted along the stratum to generate direct waves and reflected waves. Figure 16 shows the VSP results recorded at the "zero bias" position and 2.5 km away from the "non-zero bias" position. Figure 16a is the seismic wave transmission diagram recorded by the microstructure fiber DAS system under the condition of "zero bias". The specific details are enlarged, as shown in Figure 16b. It can be observed that clear direct transmission of the p-wave and s-wave is demonstrated, and that the peaks and troughs of multi-channel signals correspond to each other, highlighting the phase difference caused by the transmission time, which proves the high consistency of the microstructure fiber DAS system. In addition, as shown in Figure 16c, for the "non-zero biased" signal away from 2.5 km, the microstructure fiber DAS can also record direct waves and reflected waves with high SNR values. The DAS system based on microstructure fiber has the advantages of a high SNR, high signal consistency, and high reliability, meaning it could become an efficient, fast, and reliable acquisition tool for oil data acquisition, potentially allowing intelligent resource exploration.

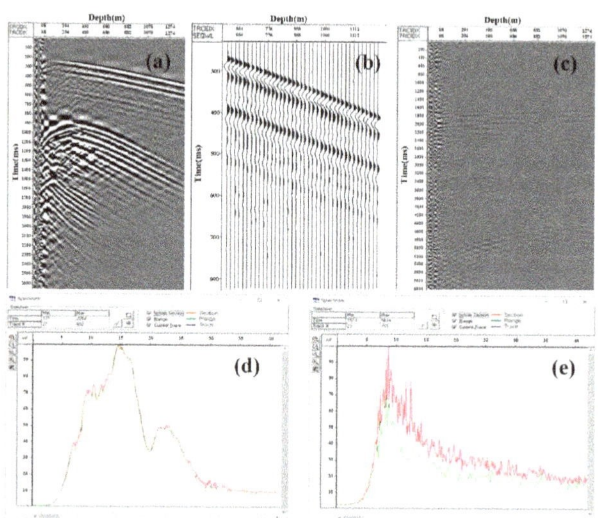

Figure 16. VSP test results for the microstructure fiber DAS system [75]: (**a**) "zero bias" test results; (**b**) schematic diagram of signal consistency; (**c**) "non-zero bias" test results; (**d**) "zero bias" spectrum diagram; (**e**) "non-zero bias" spectrum.

In 2019, Yang built a DAS system based on a UWFBG array [76], using UWFBG to enhance the intensity of the fiber's backscattered signal as shown in Figure 17. In the experiment, the results measured by optical fiber were compared with the waveforms measured by geophone, and the time-frequency results were in good agreement. Furthermore, it was applied to VSP logging, and a clear VSP waveform with a high SNR was obtained without any data processing.

Figure 17. (**a**) The VSP test site. (**b**) The VSP test result [76].

4.2. Structural Health Monitoring

The health status of various infrastructure, such as railways, tunnels, and pipelines, is related to operational safety, the national economy, and social production. The DAS system based on scattering-enhanced fibers was applied to the field of structural health monitoring to ensure high reliability.

4.2.1. Pipeline Monitoring

The transportation of resources such as oil and natural gas is closely related to the national economy and people's lives, but transmission pipelines may fail due to external invasion, corrosion, and other reasons, threatening the safety of personal, profits, and property. The DAS system based on discrete enhanced optical fibers was applied for pipeline safety monitoring due to its high SNR and high stability [77,78].

A team from the University of Pittsburgh [79] used femtosecond writing technology to artificially introduce optical fiber SEPs. As shown in Figure 18, the high-SNR DAS system tracks the propagation characteristics of soundwaves in the pipeline, combined with neural-network-based machine learning algorithms, and the data collected by the DAS system are analyzed to identify external intrusion events and the internal corrosion status of the pipeline. The system's recognition accuracy for different external intrusion events exceeds 85%, the defect recognition accuracy rate exceeds 94% under supervised learning conditions, while the recognition accuracy rate reaches 71% under unsupervised learning conditions. Furthermore, the team also conducted a test of the sand content and the solid–liquid two-phase flow of the pipeline by laying a discrete scattering enhancement fiber on the pipeline wall. By tracking the acoustic emission signals induced by the sand particles in the liquid hitting the pipe wall, the measurement of different sand particle concentrations was achieved.

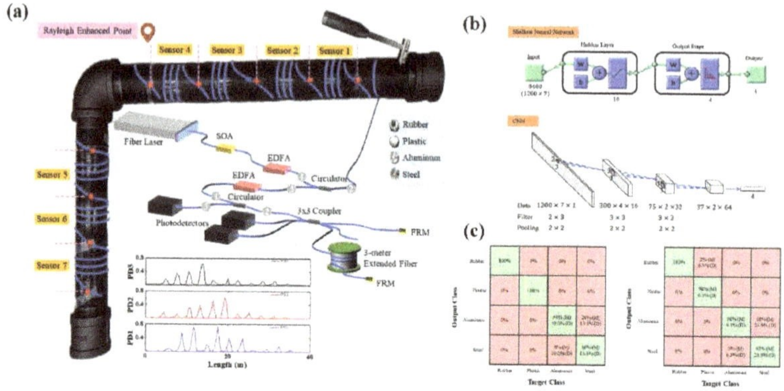

Figure 18. Schematic diagram of a pipeline survey [79]: (**a**) schematic diagram of DAS system for pipeline protection; (**b**) architectures of neural networks in the pipeline protection system; (**c**) the confusion matrix of four acoustic events.

4.2.2. Track Defect Monitoring

With the development of high-speed railway technology, track defect monitoring has become more important. A variety of mature detection methods have been developed, but most of them involve electromagnetic sensors, which are vulnerable to harsh environments [80]. In 2019, Sun et al. used a DAS system based on scattering-enhanced fibers for distributed detection of a railway defect [81]. As shown in Figure 19, the scattering-enhanced fiber is set in the middle of the rail. At the position where the defect occurs, the interaction between the wheel and rail will form an impact sound signal, which is transmitted along the two directions of the track. The DAS system is used to track the transmission signal, and then the accurate positioning of the defect is identified through the positioning algorithm, the results of which show that the maximum error of the defect is only 2.1 m. This method can provide an effective and reliable solution for rail health monitoring.

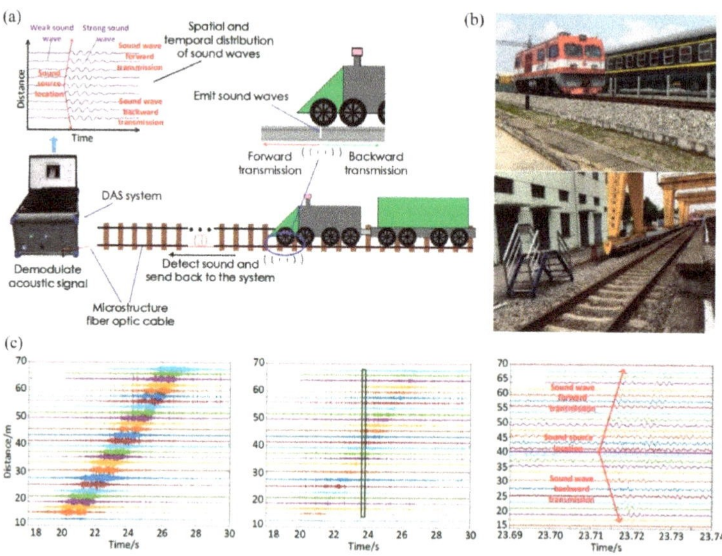

Figure 19. Rail defect detection based on DAS system [80]: (**a**) the process of detecting and analyzing sound waves based on fiber DAS; (**b**) photographs of the field test environment; (**c**) sound distribution measured by DAS.

4.2.3. Tunnel Safety Monitoring

In addition, the health of infrastructure such as tunnels is related to the traffic operation safety. An optical fiber DAS system can also be used for safety detection in tunnels, such as in subways and highways. However, the effectiveness of the reinforcement segment and steel loop in the tunnel structure maybe be a problem. The partial separation of the segment and steel loop will mean the steel loop is unable to effectively support the segment, resulting in potential safety hazards. A resonant cavity will be formed between the segment and the steel loop due to the separation, with different degrees of invalidity corresponding to the different resonant frequencies. In 2021, Sun et al. carried out intelligent monitoring of a tunnel steel loop structure based on a scattering-enhanced optical fiber DAS system [81]. As shown in Figure 20, the sensing fiber is laid on the steel loop structure to obtain the resonant frequency changes induced by the active sound source. Furthermore, the machine learning algorithm based on the BP neural network is used to effectively classify different failure levels, so as to achieve the online detection of the bonding state of the reinforced steel loop. The experimental results show that the system has a recognition rate of up to 97.8%.

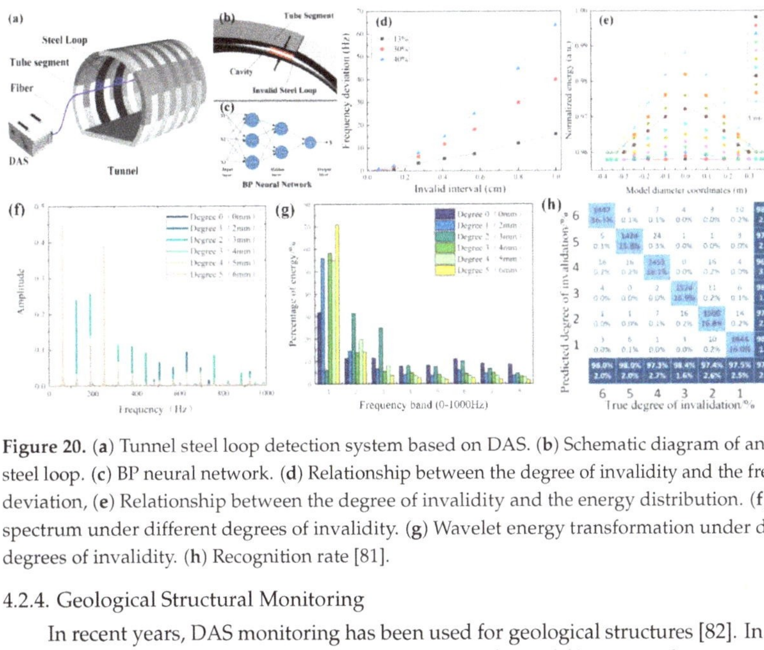

Figure 20. (**a**) Tunnel steel loop detection system based on DAS. (**b**) Schematic diagram of an invalid steel loop. (**c**) BP neural network. (**d**) Relationship between the degree of invalidity and the frequency deviation, (**e**) Relationship between the degree of invalidity and the energy distribution. (**f**) Power spectrum under different degrees of invalidity. (**g**) Wavelet energy transformation under different degrees of invalidity. (**h**) Recognition rate [81].

4.2.4. Geological Structural Monitoring

In recent years, DAS monitoring has been used for geological structures [82]. In 2017, a team from the University of California Berkeley transformed fiberoptic telecommunication cables into sensor arrays enabling meter-scale recording over tens of kilometers of linear fiber length to record the nearly vertically incident arrival of an earthquake from the Geysers Geothermal Field and to estimate its backazimuth and speed [83]. In 2018, Jousset et al. demonstrated the possibility of dynamic strain determination with conventional fiber cables deployed for telecommunication [84]. Then, by using DAS, they recorded seismic signals from natural and man-made sources with 4 m spacing along a 15-km-long fiber cable, identifying new dynamic fault processes with unprecedented resolution, opening a new path for earth hazard assessments and the exploration of structural features. In 2019, Sladen et al. reported measurements on a 41.5-km-long telecom cable that was deployed offshore of Toulon, France, demonstrating the capability to monitor the ocean's solid earth interactions from the coast to the abyssal plain [85]. In the same year, Lindsey et al. reported the use of an optical fiber cable with a DAS operating onshore, creating a ~10,000-component, 20-km-long seismic array. As depicted in Figure 21, they recorded a minor earthquake wavefield, identified multiple submarine fault zones, and tracked sea-state dynamics during a storm cycle in the northern Pacific Ocean [86].

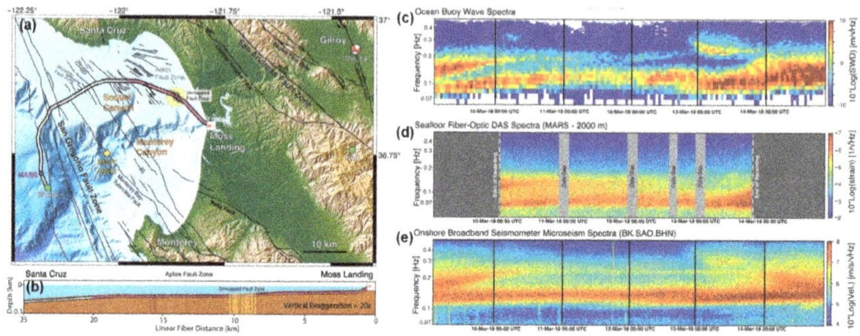

Figure 21. (**a**) Map of Monterey Bay, CA, showing the Monterey Accelerated Research System (MARS) cable (DAS, pink portion), mapped faults, the Gilroy earthquake (red and white beach ball), and major bathymetric features. (**b**) Cross-section illustration of the MARS cable used for DAS. (**c**) The 10 min average wave height and spectral wave density (SWD) measurements outside Monterey Bay, which were measured using a buoy. (**d**) Seafloor DAS strain from a 2 km cable location averaged over a 15 min sliding window. (**e**) North component of ground velocity from an onshore broadband inertial seismometer averaged over a 15 min sliding window [86].

4.3. Hydroacoustic Exploration

Ocean exploration is of great significance to homeland security and resource exploration. The acoustic medium can carry a wealth of information and can transmit signals for long distances underwater, so hydrophones are an important tool for underwater information acquisition. Optical fiber sensors have been preliminarily used in the field of underwater acoustic measurement due to their unique advantages. Single-point sensors and hydrophone arrays based on FBG and interferometers were common methods used for early optical fiber hydrophones. However, due to the limited multiplexing capacity, it is difficult to meet the requirements for large-scale exploration in the ocean and other application environments. In recent years, the DAS system has been well applied in the field of terrestrial acoustic detection. Because of its distributed, long-distance, and high-sensitivity characteristics, the DAS system has been introduced into the field of hydrophones. In the preliminary stage, researchers measured the underwater acoustic sensitivity and other parameters of an optical fiber in the laboratory. Usually, a distributed underwater acoustic sensing system is built based on a UWFBG array and unbalanced interferometer demodulation structure [54,87]. The underwater acoustic signal is obtained by demodulating the change in phase difference between two adjacent UWFBGs, in which the sensitivity obtained usually reaches about $-160\sim-150$ dB re rad/μPa.

In 2021, Sun et al. designed a lightweight and fully distributed hydroacoustic optical sensing cable [88]. An acoustic sensitizing layer is superimposed on the inner core layer, the microstructure scattering-enhanced fiber is wound on the sensitizing layer, and a further protective layer is added to the outermost layer. As depicted in Figure 22, the coherent detection system has been utilized to conduct experiments, with the results showing that the hydroacoustic detection sensitivity is as high as -127 dB re rad/μPa with a flat frequency response range of 100–2000 Hz. Furthermore, the team conducted an on-site lake test, the results of which are depicted in Figure 23.

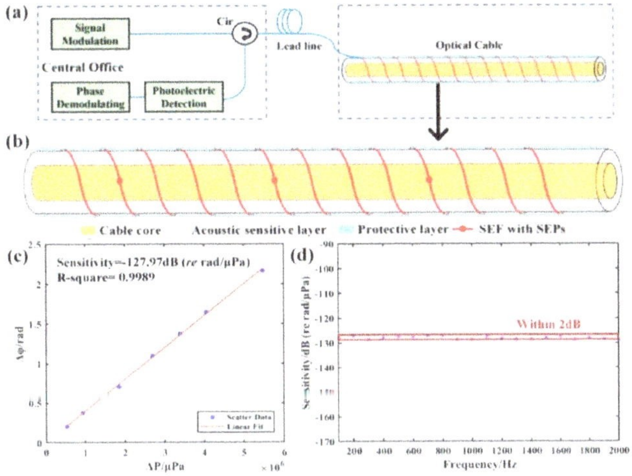

Figure 22. (**a**) Fully distributed underwater acoustic sensor system. (**b**) Lightweight fully distributed underwater acoustic fiberoptic cable based on scattering-enhanced fiber. (**c**) The relationship between the phase demodulation change and acoustic pressure. (**d**) The frequency response curve within the frequency range of 100–2000 Hz [88].

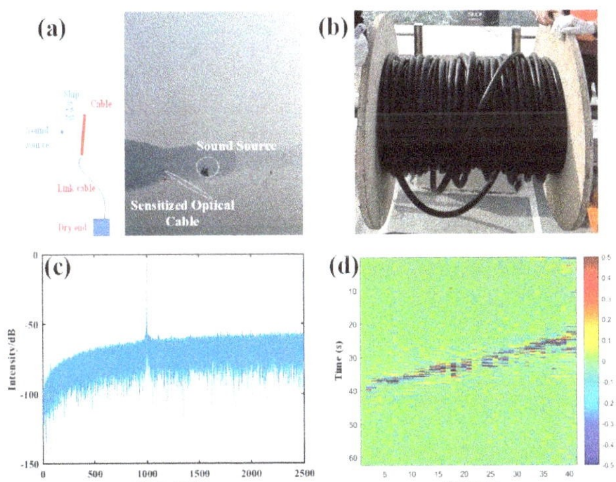

Figure 23. The underwater acoustic test: (**a**) the setup for the field test; (**b**) picture of the sensitized optical cable; (**c**) the static test results for the sound source; (**d**) motion trajectory tracking of the sound source.

5. Conclusions and Prospects

The variability and functionality of specialty optical fibers have demonstrated their major potential for sensing performance enhancement and application development for DAS systems. In this review, the scattering characteristics and noise suppression principles of specialty optical fibers are analyzed, and the fabrication technologies used for DSE are introduced. Based on the discrete scattering-enhanced fiber, the performance enhancement technologies of the DAS system are summarized, including the low-frequency phase noise compensation, polarization fading suppression, pulse width compression, and system sampling rate expansion. Further, due to the unique advantages of the scattering-enhanced fiber, the DAS system has been widely applied in various fields, such as in resource exploration, structural health monitoring, and for distributed hydrophones. Looking to the future, the research on the specialty-fiber based DAS will develop rapidly for many aspects, such as the doping of new materials for high sensitization, the development of fabrication techniques for higher efficiency, the introduction of AI techniques for higher measurement accuracy, and the exploration of wider application fields. Owing to the distinct advantages, it is believed that the specialty-fiber-based DAS system will have bright market prospects in geological, resource, and hydroacoustic exploration, as well as structural health monitoring.

Author Contributions: Conceptualization, Q.S. and Z.Y.; methodology, H.L. and C.F.; formal analysis, Y.S. and J.C.; investigation, B.Y.; writing—original draft preparation, Y.S.; writing—review and editing, Y.S.; funding acquisition, Q.S. All authors have read and agreed to the published version of the manuscript.

Funding: This research was supported by National Natural Science Foundation for Distinguished Young Scholars through Grant No. 61922033; the Marine economic development project of Guangdong Province through Grant No. GDNRC 2020(045); the National Natural Science Foundation of China through Grant No. 61775072; and the Innovation Fund of WNLO.

Institutional Review Board Statement: Not applicable.

Informed Consent Statement: Not applicable.

Data Availability Statement: Not applicable.

Conflicts of Interest: The authors declare no conflict of interest.

References

1. Yatman, G.; Üzumcü, S.; Pahsa, A.; Mert, A.A. Intrusion detection sensors used by electronic security systems for critical facilities and infrastructures: A review. *Safe* **2015**, *151*, 131–141. [CrossRef]
2. Shelef, E.; Oskin, M. Deformation processes adjacent to active faults: Examples from eastern California. *J. Geophys. Res. Earth Surf.* **2010**, *115*. [CrossRef]
3. Mu, D.; Lee, E.J.; Chen, P. *Seismic Imaging, Fault Damage and Heal*; De Gruyter: Berlin, Germany, 2014; ISBN 9787040390988.
4. Thun, J.; Lokmer, I.; Bean, C.; Eibl, E.; Bergsson, B.H.; Braiden, A. Micrometre-scale deformation observations reveal fundamental controls on geological rifting. *Sci. Rep.* **2016**, *6*, 36676. [CrossRef] [PubMed]
5. Hartog, A.; Frignet, B.; Mackie, D.; Clark, M. Vertical seismic optical profiling on wireline logging cable. *Geophys. Prospect.* **2014**, *62*, 693–701. [CrossRef]
6. Tejedor, J.; Ahlen, C.H.; Gonzalez-Herraez, M.; Macias-Guarasa, J.; Martins, H.F.; Pastor-Graells, J.; Martin-Lopez, S.; Guillen, P.C.; De Pauw, G.; De Smet, F.; et al. Real Field Deployment of a Smart Fiber-Optic Surveillance System for Pipeline Integrity Threat Detection: Architectural Issues and Blind Field Test Results. *J. Light. Technol.* **2017**, *36*, 1052–1062. [CrossRef]
7. Lyu, C.; Huo, Z.; Cheng, X.; Jiang, J.; Alimasi, A.; Liu, H. Distributed Optical Fiber Sensing Intrusion Pattern Recognition Based on GAF and CNN. *J. Light. Technol.* **2020**, *38*, 4174–4182. [CrossRef]
8. Masoudi, A.; Belal, M.; Newson, T.P. A distributed optical fibre dynamic strain sensor based on phase-OTDR. *Meas. Sci. Technol.* **2013**, *24*, 085204. [CrossRef]
9. Zhou, L.; Wang, F.; Wang, X.; Pan, Y.; Sun, Z.; Hua, J.; Zhang, X. Distributed Strain and Vibration Sensing System Based on Phase-Sensitive OTDR. *IEEE Photon. Technol. Lett.* **2015**, *27*, 1884–1887. [CrossRef]
10. Gabai, H.; Eyal, A. On the sensitivity of distributed acoustic sensing. *Opt. Lett.* **2016**, *41*, 5648–5651. [CrossRef]
11. Healey, P. Fading in heterodyne OTDR. *Electron. Lett.* **1984**, *20*, 30–32. [CrossRef]
12. Zabihi, M.; Chen, Y.; Zhou, T.; Liu, J.; Shan, Y.; Meng, Z.; Wang, F.; Zhang, Y.; Zhang, X.; Chen, M. Continuous Fading Suppression Method for Φ-OTDR Systems Using Optimum Tracking Over Multiple Probe Frequencies. *J. Light. Technol.* **2019**, *37*, 3602–3610. [CrossRef]
13. Zhou, J.; Pan, Z.; Ye, Q.; Cai, H.; Qu, R.; Fang, Z. Characteristics and Explanations of Interference Fading of a Φ-OTDR With a Multi-Frequency Source. *J. Light. Technol.* **2013**, *31*, 2947–2954. [CrossRef]
14. Ren, M.; Lu, P.; Chen, L.; Bao, X. Theoretical and Experimental Analysis of Ο-OTDR Based on Polarization Diversity Detection. *IEEE Photon. Technol. Lett.* **2015**, *28*, 697–700. [CrossRef]
15. Mompo, J.J.; Shiloh, L.; Arbel, N.; Levanon, N.; Loayssa, A.; Eyal, A. Distributed Dynamic Strain Sensing via Perfect Periodic Coherent Codes and a Polarization Diversity Receiver. *J. Light. Technol.* **2019**, *37*, 4597–4602. [CrossRef]
16. Wang, F.; Liu, Y.; Wei, T.; Zhang, Y.; Ji, W.; Zong, M.; Zhang, X. Polarization fading elimination for ultra-weak FBG array-based Φ-OTDR using a composite double probe pulse approach. *Opt. Express* **2019**, *27*, 20468–20478. [CrossRef]
17. Qin, Z.; Zhu, T.; Chen, L.; Bao, X. High Sensitivity Distributed Vibration Sensor Based on Polarization-Maintaining Configurations of Phase-OTDR. *IEEE Photon. Technol. Lett.* **2011**, *23*, 1091–1093. [CrossRef]

18. He, X.; Zhang, M.; Gu, L.; Xie, S.; Liu, F.; Lu, H. Performance Improvement of Dual-Pulse Heterodyne Distributed Acoustic Sensor for Sound Detection. *Sensors* **2020**, *20*, 999. [CrossRef]
19. Juarez, J.C.; Maier, E.W.; Choi, K.N.; Taylor, H.F. Distributed fiber-optic intrusion sensor system. *J. Lightwave Technol.* **2005**, *23*, 2081–2087. [CrossRef]
20. Rao, Y.-J.; Luo, J.; Ran, Z.-L.; Yue, J.-F.; Luo, X.-D.; Zhou, Z. Long-distance fiber-optic Φ-OTDR intrusion sensing system. In Proceedings of the Spie the International Society for Optical Engineering, San Diego, CA, USA, 3–4 August 2009; Volume 7503, p. 75031O. [CrossRef]
21. Juarez, J.C.; Taylor, H.F. Polarization discrimination in a phase-sensitive optical time-domain reflectometer intrusion-sensor system. *Opt. Lett.* **2005**, *30*, 3284–3286. [CrossRef]
22. Pan, Z.; Liang, K.; Ye, Q.; Cai, H.; Qu, R.; Fa Ng, Z. Phase-sensitive OTDR system based on digital coherent detection. In Proceedings of the SPIE Proceedings (SPIE SPIE/OSA/IEEE Asia Communications and Photonics-Shanghai, China), Shanghai China, 13–16 November 2011; Volume 8311, p. 83110S-2. [CrossRef]
23. Zhang, X.; Sun, Z.; Shan, Y.; Li, Y.; Wang, F.; Zeng, J.; Zhang, Y. A high performance distributed optical fiber sensor based on Φ-OTDR for dynamic strain measurement. *IEEE Photon. J.* **2017**, *9*, 1–12. [CrossRef]
24. Fang, G.; Xu, T.; Feng, S.; Li, F. Phase-Sensitive Optical Time Domain Reflectometer Based on Phase-Generated Carrier Algorithm. *J. Light. Technol.* **2015**, *33*, 2011–2016. [CrossRef]
25. Chen, D.; Liu, Q.; Fan, X.; He, Z. Distributed Fiber-Optic Acoustic Sensor with Enhanced Response Bandwidth and High Signal-to-Noise Ratio. *J. Light. Technol.* **2017**, *35*, 2037–2043. [CrossRef]
26. Liokumovich, L.; Ushakov, N.; Kotov, O.; Bisyarin, M.; Hartog, A.H. Fundamentals of Optical Fiber Sensing Schemes Based on Coherent Optical Time Domain Reflectometry: Signal Model Under Static Fiber Conditions. *J. Light. Technol.* **2015**, *33*, 3660–3671. [CrossRef]
27. Zhang, J.; Wu, H.; Zheng, H.; Huang, J.; Yin, G.; Zhu, T.; Qiu, F.; Huang, X.; Qu, D.; Bai, Y.; et al. 80 km Fading Free Phase-Sensitive Reflectometry Based on Multi-Carrier NLFM Pulse Without Distributed Amplification. *J. Light. Technol.* **2019**, *37*, 4748–4754. [CrossRef]
28. Li, H.; Liu, Y.; He, T.; Fan, C.; Liu, T.; Yan, Z.; Liu, D.; Sun, Q. Dual-pulse Complex Superposition Based Noise Suppression for Distributed Acoustic Sensing. In Proceedings of the CLEO: Applications and Technology, Washington, DC, USA, 10–15 May 2020; p. JW2E.28. [CrossRef]
29. Ai, F.; Sun, Q.; Zhang, W.; Liu, T.; Yan, Z.; Liu, D. Wideband Fully-Distributed Vibration Sensing by Using UWFBG Based Coherent OTDR. In Proceedings of the Optical Fiber Communications Conference & Exhibition, Los Angeles, CA, USA, 19–23 March 2017; p. W2A.19. [CrossRef]
30. Lin, S.T.; Wang, Z.N.; Xiong, J.; Wu, Y.; Rao, Y.J. Progresses of Anti-Interference-Fading Technologies for Ray-leigh-Scattering-Based Optical Fiber Sensing. *Laser Optoelectron. Prog.* **2021**, *58*, 1306008. [CrossRef]
31. Westbrook, P.S.; Feder, K.S.; Ortiz, R.M.; Kremp, T.; Monberg, E.M.; Wu, H.; Simoff, D.A.; Shenk, S. Kilometer length low loss enhanced back scattering fiber for distributed sensing. In Proceedings of the 25th International Conference on Optical Fiber Sensors, Jeju, Korea, 24–28 April 2017; pp. 1–5. [CrossRef]
32. Handerek, V.A.; Karimi, M.; Nkansah, A.; Yau, A.; Westbrook, P.S.; Feder, K.S.; Ortiz, R.M.; Kremp, T.; Monberg, E.M.; Wu, H.; et al. Improved Optical Power Budget in Distributed Acoustic Sensing Using Enhanced Scattering Optical Fibre. In Proceedings of the 26th International Conference on Optical Fiber Sensors, Vaud, Switzerland, 24–28 September 2018; p. TuC5. [CrossRef]
33. Westbrook, P.S.; Kremp, T.; Feder, K.S.; Ko, W.; Monberg, E.M.; Wu, H.; Simoff, D.A.; Ortiz, R.M. Improving distributed sensing with continuous gratings in single and multi-core fibers. In Proceedings of the Optical Fiber Communication Conference, San Diego, CA, USA, 11–15 March 2018; p. W1K.1. [CrossRef]
34. Lalam, N.; Westbrook, P.S.; Li, J.; Lu, P.; Buric, M.P. Phase-Sensitive Optical Time Domain Reflectometry with Rayleigh Enhanced Optical Fiber. *IEEE Access* **2021**, *9*, 114428–114434. [CrossRef]
35. Butov, O.V.; Chamorovskii, Y.K.; Golant, K.M.; Fotiadi, A.A.; Jason, J.; Popov, S.M.; Wuilpart, M. Sensitivity of high Rayleigh scattering fiber in acoustic/vibration sensing using phase-OTDR. In Proceedings of the Optical Sensing and Detection V, Strasbourg, France, 9 May 2018. [CrossRef]
36. Feng, S.; Xu, T.; Huang, J.; Yang, Y.; Li, F.; Zhou, J.; Yu, H. Enhanced SNR phase-sensitive OTDR system with active fiber. In Proceedings of the Fiber Optic Sensing and Optical Communication, Beijing, China, 12 December 2018. [CrossRef]
37. Huang, J.B.; Ding, P.; Tang, J.S. Progress in Fabrication, Demodulation and Application of Weak-Reflection Fiber Bragg Grating Array. *Laser Optoelectron. Prog.* **2021**, *58*, 1700005. [CrossRef]
38. Wu, M.; Fan, X.; Zhang, X.; Yan, L.; He, Z. Frequency Response Enhancement of Phase-Sensitive OTDR for Interrogating Weak Reflector Array by Using OFDM and Vernier Effect. *J. Light. Technol.* **2020**, *38*, 4874–4882. [CrossRef]
39. Hicke, K.; Eisermann, R.; Chruscicki, S. Enhanced Distributed Fiber Optic Vibration Sensing and Simultaneous Temperature Gradient Sensing Using Traditional C-OTDR and Structured Fiber with Scattering Dots. *Sensors* **2019**, *19*, 4114. [CrossRef]
40. Redding, B.; Murray, M.J.; Donko, A.; Beresna, M.; Masoudi, A.; Brambilla, G. Low-noise distributed acoustic sensing using enhanced backscattering fiber with ultra-low-loss point reflectors. *Opt. Express* **2020**, *28*, 14638–14647. [CrossRef]
41. Hill, K.O.; Fujii, Y.; Johnson, D.C.; Kawasaki, B.S. Photosensitivity in optical fiber waveguides: Application to reflection filter fabrication. *Appl. Phys. Lett.* **1978**, *32*, 647–649. [CrossRef]

42. Meltz, G.; Morey, W.W.; Glenn, W.H. Formation of Bragg gratings in optical fibers by a transverse holographic method. *Opt. Lett.* **1989**, *14*, 823–825. [CrossRef] [PubMed]
43. Zheng, Y.; Yu, H.; Guo, H.; Li, X.; Jiang, D. Analysis of the Spectrum Distortions of Weak Fiber Bragg Gratings Fabricated In-Line on a Draw Tower by the Phase Mask Technique. *J. Light. Technol.* **2014**, *33*, 2670–2673. [CrossRef]
44. Martinez, A.; Dubov, M.; Khrushchev, I.; Bennion, I. Direct writing of fibre Bragg gratings by femtosecond laser. *Electron. Lett.* **2004**, *40*, 1170–1172. [CrossRef]
45. Chen, Z.; He, J.; Xu, X.; He, J.; Xu, B.; Du, B.; Liao, C.; Wang, Y. High-Temperature Sensor Array Based on Fiber Bragg Gratings Fabricated by Femtosecond Laser Point-by-Point Method. *Acta Opt. Sin.* **2021**, *41*, 1306002. [CrossRef]
46. Yang, M.; Bai, W.; Guo, H.; Wen, H.; Yu, H.; Jiang, D. Huge capacity fiber-optic sensing network based on ultra-weak draw tower gratings. *Photon. Sens.* **2016**, *6*, 26–41. [CrossRef]
47. Tang, J.; Cai, L.; Li, C.; Yang, M.; Guo, H.; Gan, W. Distributed acoustic sensors with wide frequency response based on UWFBG array utilizing dual-pulse detection. *Opt. Fiber Technol.* **2021**, *61*, 102452. [CrossRef]
48. Ai, F.; Li, H.; He, T.; Yan, Z.; Liu, D.; Sun, Q. Simultaneous Distributed Temperature and Vibration Measurement with UWFBG based Coherent OTDR. In Proceedings of the Optical Fiber Communication Conference, San Diego, CA, USA, 11–15 March 2018; p. W2A.12. [CrossRef]
49. Yang, M.; Li, C.; Mei, Z.; Tang, J.; Guo, H.; Jiang, D. Thousand of fiber grating sensor array based on draw tower: A new platform for fiber-optic sensing. In Proceedings of the Optical Fiber Sensors, OFS 2018, Lausanne, Switzerland, 24–28 September 2018; p. FB6. [CrossRef]
50. Peng, Z.; Jian, J.; Wen, H.; Wang, M.; Liu, H.; Jiang, D.; Mao, Z.; Chen, K.P. Fiber-optical distributed acoustic sensing signal enhancements using ultrafast laser and artificial intelligence for human movement detection and pipeline monitoring. *Opt. Data Sci. II* **2019**, *10937*, 109370J. [CrossRef]
51. Liu, D.M.; He, T.; Xu, Z.J.; Sun, Q. New type of microstructure-fiber distributed acoustic sensing technology and its applications. *J. Appl. Sci.* **2020**, *38*, 296–309. [CrossRef]
52. Ai, F. Investigation on Discrete Enhanced Fiber Based Distributed Sensing Technologies and Their Applications. Ph.D. Thesis, Huazhong University of Science and Technology, Wuhan, China, 2019.
53. Li, H.; Sun, Q.; Liu, T.; Fan, C.; He, T.; Yan, Z.; Liu, D.; Shum, P.P. Ultra-High Sensitive Quasi-Distributed Acoustic Sensor Based on Coherent OTDR and Cylindrical Transducer. *J. Light. Technol.* **2019**, *38*, 929–938. [CrossRef]
54. Wu, M.; Li, C.; Fan, X.; Liao, C.; He, Z. Large-scale multiplexed weak reflector array fabricated with a femtosecond laser for a fiber-optic quasi-distributed acoustic sensing system. *Opt. Lett.* **2020**, *45*, 3685. [CrossRef] [PubMed]
55. Wang, C.; Shang, Y.; Liu, X.-H.; Wang, C.; Yu, H.-H.; Jiang, D.-S.; Peng, G.-D. Distributed OTDR-interferometric sensing network with identical ultra-weak fiber Bragg gratings. *Opt. Express* **2015**, *23*, 29038–29046. [CrossRef] [PubMed]
56. Jiang, P.; Ma, L.; Wang, W.; Hu, Z.; Hu, Y. Crosstalk Reduction and Demodulation Stability Promotion in Inline Fiber Fabry–Pérot Sensor Array Using Phase Generated Carrier Scheme. *J. Light. Technol.* **2015**, *34*, 1006–1014. [CrossRef]
57. Peng, Z.; Wen, H.; Jian, J.; Gribok, A.; Wang, M.; Huang, S.; Liu, H.; Mao, Z.-H.; Chen, K.P. Identifications and classifications of human locomotion using Rayleigh-enhanced distributed fiber acoustic sensors with deep neural networks. *Sci. Rep.* **2020**, *10*, 21014. [CrossRef] [PubMed]
58. Tang, J.; Liu, Y.; Li, C.; Guo, H.; Yang, M. Distributed Vibration Sensing System with High Signal-to-Noise Ratio Based on Ultra-Weak Fiber Bragg Grating. *Acta Opt. Sin.* **2021**, *41*, 1306014. [CrossRef]
59. Fan, X.; Yang, G.; Wang, S.; Liu, Q.; He, Z. Distributed Fiber-Optic Vibration Sensing Based on Phase Extraction from Optical Reflectometry. *J. Light. Technol.* **2016**, *35*, 3281–3288. [CrossRef]
60. Wu, M.; Fan, X.; Liu, Q.; He, Z. Highly sensitive quasi-distributed fiber-optic acoustic sensing system by interrogating a weak reflector array. *Opt. Lett.* **2018**, *43*, 3594–3597. [CrossRef] [PubMed]
61. Wu, M.; Fan, X.; Liu, Q.; He, Z. Quasi-distributed Fiber-optic Acoustic Sensor using Ultra-weak Reflecting Point Array. In Proceedings of the 26th International Conference on Optical Fiber Sensors, Vaud, Switzerland, 24–28 September 2018; p. WF19. [CrossRef]
62. Liu, T.; Li, H.; Ai, F.; Wang, J.; Fan, C.; Luo, Y.; Yan, Z.; Liu, D.; Sun, Q. Ultra-high Resolution Distributed Strain Sensing based on Phase-OTDR. In Proceedings of the Optical Fiber Communication Conference (OFC), San Diego, CA, USA, 3–7 March 2019; p. Th2A.16. [CrossRef]
63. He, T.; Liu, Y.; Zhang, S.; Yan, Z.; Liu, D.; Sun, Q. High Accuracy Intrusion Pattern Recognition using a Dual-Stage-Recognition Network for Fiber Optic Distributed Sensing System. In Proceedings of the Conference on Lasers and Electro-Optics, Munich, Germany, 21–25 June 2021; p. JW1A.119. [CrossRef]
64. Liu, Y.; Wang, F.; Zhang, X.; Zhang, Y.; Xu, W.; Zhang, L. High performance interrogation by a composite-double-probe-pulse for ultra-weak FBG array. In Proceedings of the 17th International Conference on Optical Communications and Networks (ICOCN2018), International Society for Optics and Photonics. Bellingham, WA, USA, 14 February 2019. [CrossRef]
65. Liu, Y.; Li, H.; Liu, T.; Fan, C.; Yan, Z.; Liu, D.; Sun, Q. Polarization dependent noise suppression for fiber distributed acoustic sensor with birefringence estimation. In Proceedings of the Conference on Lasers and Electro-Optics: Applications and Technology, Washington, DC, USA, 10–15 May 2020; p. JW2E.18. [CrossRef]

66. Liu, T.; Li, H.; He, T.; Fan, C.; Yan, Z.; Liu, D.; Sun, Q. Fading Noise Free Distributed Acoustic Sensor Assisted with Double Wavelength lasers. In Proceedings of the Conference on Lasers and Electro-Optics: Applications and Technology, Anaheim, CA, USA, 23–25 October 2020; p. JW2E.10. [CrossRef]
67. Liu, Z.; Chen, D.W.; Xia, X.X.; Wang, G.Y.; Wang, X.S. A countering technique for LFM PC radar against shift-frequency jamming. *Mod. Radar* **2006**, *28*, 84–86. [CrossRef]
68. Zhou, Y.; Wang, C.; Wei, T.; Shangguan, M.; Xia, H. Simulation Research of Coherent Lidar Based on Golay Coding Technology. *Chin. J. Lasers* **2018**, *45*, 0810004. [CrossRef]
69. Jiang, J.; Xiong, J.; Wang, Z.; Qiu, Z.; Liu, C.; Deng, Z.; Rao, Y.-J. Quasi-Distributed Fiber-Optic Acoustic Sensing with MIMO Technology. *IEEE Internet Things J.* **2021**, *8*, 15284–15291. [CrossRef]
70. Wu, M.; Fan, X.; He, Z. Phase Noise Compensation for Ultra-highly Sensitive Fiber-optic Quasi-distributed Acoustic Sensing System. In Proceedings of the Conference on Lasers and Electro-Optics: Science and Innovations, San Jose, CA, USA, 5–10 May 2019. [CrossRef]
71. Wu, M.; Fan, X.; Liu, Q.; He, Z. Quasi-distributed fiber-optic acoustic sensing system based on pulse compression technique and phase-noise compensation. *Opt. Lett.* **2019**, *44*, 5969–5972. [CrossRef] [PubMed]
72. Wang, Z.T.; Jiang, J.L.; Xiong, J.; Wang, Z.N. Fiber-optic Quasi-distributed Acoustic Sensing System at Doubled Repetition Rate. In Proceedings of the Asia Communications and Photonics Conference (ACP), Chengdu, China, 2–5 November 2019, OSA Technical Digest (Optica Publishing Group). 2019; p. M4A.183.
73. Zhang, Y.X.; Fu, S.Y.; Chen, Y.S.; Ding, Z.W.; Shan, Y.Y.; Wang, F.; Chen, M.M.; Zhang, X.P.; Meng, Z. A visibility enhanced broadband phase-sensitive OTDR based on the UWFBG array and frequency-division-multiplexing. *Opt. Fiber Technol.* **2019**, *53*, 101995. [CrossRef]
74. Li, H.; Fan, C.; Liu, T.; Liu, Y.; Yan, Z.; Shum, P.; Sun, Q. Time-slot multiplexing based bandwidth enhancement for fiber distributed acoustic sensing. *Sci. China Inf. Sci.* **2021**, *65*, 119303. [CrossRef]
75. Li, Y.P.; Karrenbach, M.; Ajo, J. *Distributed Acoustic Sensing in Geophysics: Methods and Applications*; American Geophysical Union: Washington, DC, USA, 2022.
76. Li, C.; Mei, Z.; Cheng, C.; Tang, J.; Jiang, Y.; Yang, M. Draw Tower Grating-based Distributed Acoustic Sensing System and its Applications. In Proceedings of the Asia Communications and Photonics Conference, Chengdu, China, 2–5 November 2019; p. T2A.5. [CrossRef]
77. Li, T.; Fan, C.; Li, H.; He, T.; Qiao, W.; Shi, Z.; Yan, Z.; Liu, C.; Liu, D.; Sun, Q. Nonintrusive Distributed Flow Rate Sensing System Based on Flow-Induced Vibrations Detection. *IEEE Trans. Instrum. Meas.* **2021**, *70*, 7001808. [CrossRef]
78. Jiang, J.; Liu, F.; Wang, H.; Li, S.; Gan, W.; Jiang, R. Lateral positioning of vibration source for underground pipeline monitoring based on ultra-weak fiber Bragg grating sensing array. *Measurement* **2020**, *172*, 108892. [CrossRef]
79. Peng, Z.; Jian, J.; Wen, H.; Gribok, A.; Wang, M.; Liu, H.; Huang, S.; Mao, Z.-H.; Chen, K. Distributed Fiber Sensor and Machine Learning Data Analytics for Pipeline Protection against Extrinsic Intrusions and Intrinsic Corrosions. *Opt. Express* **2020**, *28*, 27277–27292. [CrossRef]
80. Fan, C.; Ai, F.; Liu, Y.; Xu, Z.; Wu, G.; Zhang, W.; Liu, C.; Yan, Z.; Liu, D.; Sun, Q. Rail Crack Detection by Analyzing the Acoustic Transmission Process Based on Fiber Distributed Acoustic Sensor. In Proceedings of the Optical Fiber Communication Conference, San Diego, CA, USA, 3–7 March 2019; p. Th2A.17. [CrossRef]
81. Hu, D.; Tian, B.; Li, H.; Fan, C.; Liu, T.; He, T.; Liu, Y.; Yan, Z.; Sun, Q. Intelligent Structure Monitoring for Tunnel Steel Loop Based on Distributed Acoustic Sensing. In Proceedings of the Conference on Lasers and Electro-Optics, Munich, Germany, 21–25 June 2021; p. ATh1S.4. [CrossRef]
82. Min, R.; Liu, Z.; Pereira, L.; Yang, C.; Sui, Q.; Marques, C. Optical fiber sensing for marine environment and marine structural health monitoring: A review. *Opt. Laser Technol.* **2021**, *140*, 107082. [CrossRef]
83. Lindsey, N.J.; Martin, E.R.; Dreger, D.S.; Freifeld, B.; Cole, S.; James, S.R.; Biondi, B.L.; Ajo-Franklin, J.B. Fiber-Optic Network Observations of Earthquake Wavefields. *Geophys. Res. Lett.* **2017**, *44*, 11792–11799. [CrossRef]
84. Jousset, P.; Reinsch, T.; Ryberg, T.; Blanck, H.; Clarke, A.; Aghayev, R.; Hersir, G.P.; Henninges, J.; Weber, M.; Krawczyk, C.M. Dynamic strain determination using fibre-optic cables allows imaging of seismological and structural features. *Nat. Commun.* **2018**, *9*, 2509. [CrossRef]
85. Sladen, A.; Rivet, D.; Ampuero, J.P.; De Barros, L.; Hello, Y.; Calbris, G.; Lamare, P. Distributed sensing of earthquakes and ocean-solid Earth interactions on seafloor telecom cables. *Nat. Commun.* **2019**, *10*, 5777. [CrossRef]
86. Lindsey, N.J.; Dawe, T.C.; Ajo-Franklin, J.B. Illuminating seafloor faults and ocean dynamics with dark fiber distributed acoustic sensing. *Science* **2019**, *366*, 1103–1107. [CrossRef] [PubMed]
87. Zhou, C.; Pang, Y.; Qian, L.; Chen, X.; Xu, Q.; Zhao, C.; Zhang, H.; Tu, Z.; Huang, J.; Gu, H.; et al. Demodulation of a Hydroacoustic Sensor Array of Fiber Interferometers Based on Ultra-Weak Fiber Bragg Grating Reflectors Using a Self-Referencing Signal. *J. Light. Technol.* **2018**, *37*, 2568–2576. [CrossRef]
88. Chen, J.F.; Li, H.; Liu, T.; Fan, C.Z.; Yan, Z.J.; Sun, Q.Z. Fully distributed hydroacoustic sensing based on lightweight optical cable assisted with scattering enhanced fiber. In Proceedings of the Optical Fiber Communication Conference (OFC), Washington, DC, USA, 6–11 June 2021; p. W7C.3. [CrossRef]

photonics

MDPI

Review

Transmission and Generation of Orbital ANGULAR Momentum Modes in Optical Fibers

Zhuo Wang [1,2], Jiajing Tu [3], Shecheng Gao [3], Zhaohui Li [4], Changyuan Yu [1,2,*] and Chao Lu [1,2]

1 Photonics Research Center, Department of Electronic and Information Engineering, The Hong Kong Polytechnic University, Hong Kong 999077, China; z-eie.wang@connect.polyu.hk (Z.W.); chao.lu@polyu.edu.hk (C.L.)
2 Shenzhen Research Institute, The Hong Kong Polytechnic University, Shenzhen 518057, China
3 Department of Electronic Engineering, School of Information Science and Technology, Jinan University, Guangzhou 510632, China; tujiajing@jnu.edu.cn (J.T.); gaosc825@jnu.edu.cn (S.G.)
4 State Key Laboratory Optoelectronic Materials and Technologies, School of Electronics and Information Technology, Sun Yat-sen University, Guangzhou 510275, China; lzhh88@mail.sysu.edu.cn
* Correspondence: changyuan.yu@polyu.edu.hk

Abstract: The orbital angular momentum (OAM) of light provides a new degree of freedom for carrying information. The stable propagation and generation of OAM modes are necessary for the fields of OAM-based optical communications and microscopies. In this review, we focus on discussing the novel fibers that are suitable for stable OAM mode transmission and conversion. The fundamental theory of fiber modes is introduced first. Then, recent progress on a multitude of fiber designs that can stably guide or generate OAM modes is reviewed. Currently, the mode crosstalk is regarded as the main issue that damages OAM mode stability. Therefore, the coupled-mode theory and coupled-power power theory are introduced to analyze OAM modes crosstalk. Finally, the challenges and prospects of the applications of OAM fibers are discussed.

Keywords: orbital angular momentum; optical fiber; fiber mode crosstalk; vortex mode generator

Citation: Wang, Z.; Tu, J.; Gao, S.; Li, Z.; Yu, C.; Lu, C. Transmission and Generation of Orbital ANGULAR Momentum Modes in Optical Fibers. *Photonics* **2021**, *8*, 246. https://doi.org/10.3390/photonics8070246

Received: 24 May 2021
Accepted: 22 June 2021
Published: 30 June 2021

Publisher's Note: MDPI stays neutral with regard to jurisdictional claims in published maps and institutional affiliations.

1. Introduction

A photon can carry spin angular momentum (SAM) and orbital angular momentum (OAM). The SAM of light is associated with the spin of a photon, and it can be characterized as the circular polarization of a light beam [1]. The values of SAM are $\pm\hbar$ per photon, which correspond to left- and right-handed circularly polarized states, respectively. Different from the SAM, the OAM of light originates from the azimuthal phase distribution at the wave front [2]. In 1992, Allen et al. first recognized that a light beam having a helical phase $\exp(il\phi)$ can carry OAM with the value $l\hbar$ per photon [3]. After this groundbreaking work, detailed studies on optical OAM were constantly carried out [4] and have given rise to developments in many application areas including trapping and guiding atoms with optical OAM [5–7], OAM states' entanglement [8], high-dimensional cryptography [9], vortex beam assisted microscopy [10,11]. and optical communications [12–15].

At present, the high growth rate of the demand for data transmission capacity leads to a huge challenge for the optical communication community. Space-division multiplexing (SDM) is considered as an effective method to solve the problem of capacity crunch in the f long term [16]. It is obvious that, by using multiple space channels, the nominal data transmission capacity can be increased greatly. Fortunately, the optical OAM provides a new degree of physical dimension for SDM [13,14]. OAM modes can be generated by using spatial light modulators (SLMs) [12], spiral phase plates [17], log-polar mapping [18,19], angular gratings [20], fiber couplers [21,22], and fiber gratings [23–27]. These methods provide various choices for loading information onto OAM beams. In addition, Dammann vortex gratings give a feasible way for the wide-range detection of OAM modes [28] and for the multiplexing and demultiplexing of massive OAM channels [29]. In addition, the technique of multi-plane light conversion provides another effective way to multiplex and demultiplex OAM modes [30,31]. However, the stable long-distance transmission of massive OAM modes is still an open issue. In the case of free-space-optics (FSO) data transmission, the atmospheric turbulence can cause the fluctuation of the refractive index of air, which damages the intensity profile and the transverse phase structure of OAM modes [32]. In addition, the diffraction of OAM beams limits the FSO data transmission in long-distance situations. Another way for OAM mode propagation is using optical fiber. The OAM modes form a set of orthogonal bases, which is equivalent to the basis of vector modes in optical fibers. Because of the orthogonality, each OAM mode can carry information independently under ideal conditions. However, the fiber defects introduced in the fabrication and the irregular bending in the application will break the orthogonal condition of fiber modes and cause inter-mode crosstalk [33]. Therefore, the novel designs to reduce the crosstalk of OAM modes in optical fiber and the signal processing algorithm to compensate the overall crosstalk at the receiving end attract extensive attention.

In 2009, Ramachandran et al. first demonstrated a kind of ring-core fiber (RCF) in which the first-order vector modes can be transmitted with outstanding stability [34]. Later, a fiber with ring-core structure was used to transmit OAM modes, achieving terabit-scale SDM [14]. To further increase the stability of OAM modes, a kind of air-hole ring-core fiber (AH-RCF) was proposed and tested [35]. It shows good ability to conserve OAM modes even in the presence of strong bend perturbations. Using this kind of fiber, a 13.4 km OAM state transmission was demonstrated [36]. At the same time, a kind of ring-core photonic crystal fiber (RC-PCF) was designed for OAM mode transmission with ultra-low confinement loss and wide bandwidth [37]. In addition, Wong et al. found that helically twisted PCF can preserve the chirality of OAM modes [38]. These works provide good references for the long-distance transmission of OAM modes.

Our group also attaches great importance to the study of OAM-related optical fibers and fiber devices. A kind of microstructure RCF employing negative curvature cladding structures was proposed and fabricated [39,40]. Under experimental test, it showed low transmission loss of the first-order OAM modes with the value of 0.095 dB/m. In addition, all-fiber OAM generators based on asymmetric long-period fiber gratings (LPFGs) were fabricated. Experiments demonstrated that the generation efficiencies from the fundamental mode to second- and third-order OAM modes were 99.7% [24] and 99.8% [25], respectively. Furthermore, we are currently doing the simulation and fabrication of low crosstalk muli-ring-core fibers (MRCFs) for dense SDM.

This paper introduces the fiber-guided OAM modes from the principle of fiber mode theory and highlight the advances in design, fabrication, and application of OAM fibers including conventional solid optical fibers and microstructure optical fibers. The crosstalk of OAM modes under fiber deformations is discussed using coupled-mode theory [33] and coupled-power theory [41]. Additionally, we also review the progress of all-fiber OAM generation techniques. The technical challenges and application prospects of OAM fibers are discussed at the end.

2. OAM Modes in Regular Solid Optical Fibers

Optical fiber is a kind of circularly symmetric waveguide; thus, the electric and magnetic fields of eigenmodes can be expressed in the form of separating variables as

$$\mathbf{E}(r,\phi,z) = [\hat{r}e_r(r)\cos(v\phi+\phi_0) - \hat{\phi}e_\phi(r)\sin(v\phi+\phi_0) + \hat{z}e_z(r)\cos(v\phi+\phi_0)]\exp(i\beta z), \tag{1a}$$

$$\mathbf{H}(r,\phi,z) = [\hat{r}h_r(r)\sin(v\phi+\phi_0) + \hat{\phi}h_\phi(r)\cos(v\phi+\phi_0) + \hat{z}h_z(r)\sin(v\phi+\phi_0)]\exp(i\beta z), \tag{1b}$$

where $\hat{r}$, $\hat{\phi}$, and $\hat{z}$ are unit vectors in the radial, angular, and longitudinal directions of an optical fiber. β is the propagation constant. Shown by the vector mode category in Figure 1, the fiber eigenmodes can be classified into $HE_{l,v}$, $EH_{l,v}$, $TE_{0,v}$, and $TM_{0,v}$ modes [42], where l and v (= 1,2,3$\cdots\cdots$) represent azimuthal order and radial order of modes, respectively. Commonly, the constant phase ϕ_0 in Equation (1) is chosen as 0 or $\pi/2$ to represent even or odd polarization state. The superposition of even and odd states of each HE and EH mode with a $\pm\pi/2$ phase difference can form OAM modes as [43]

$$\mathrm{OAM}^{\pm}_{\pm(l-1),v} = \mathrm{HE}^{\mathrm{e}}_{l,v} \pm i\mathrm{HE}^{\mathrm{o}}_{l,v}, \tag{2a}$$

$$\mathrm{OAM}^{\mp}_{\pm(l+1),v} = \mathrm{EH}^{\mathrm{e}}_{l,v} \pm i\mathrm{EH}^{\mathrm{o}}_{l,v}, \tag{2b}$$

where "e" and "o" represent even and odd polarization states and the superscripts "+" and "−" correspond to left- and right-handed circular polarization, respectively. The elements $l-1$ and $l+1$ are the absolute values of OAM topological charge. When $l = 1$, the HE modes can only form a circularly polarized state without carrying OAM. In addition, because $\mathrm{TE}_{0,v}$ and $\mathrm{TM}_{0,v}$ modes are intrinsically non-degenerate [43], their relative phase difference will vary periodically with propagation; consequently, they can not form stable OAM modes. Therefore, the synthetic modes constructed by $\mathrm{TE}_{0,v}$ and $\mathrm{TM}_{0,v}$ are excluded from the category of OAM modes in Figure 1. When a fiber is in the weakly guiding condition, the effective index difference (Δn_{eff}) between the near-degenerate modes are too small to be distinguished. In this case, mode coupling can easily happen, and the superposition of near-degenerate vector modes can form linearly polarized (LP) states, which is shown by the category in the left side of Figure 1. Mathematically, the LP modes are the solutions of scalar Helmholtz equations under the weakly guiding approximation. Since the polarization effects at the refractive index step are ignored, all the polarization states in the same LP mode group have an identical effective index (n_{eff}).

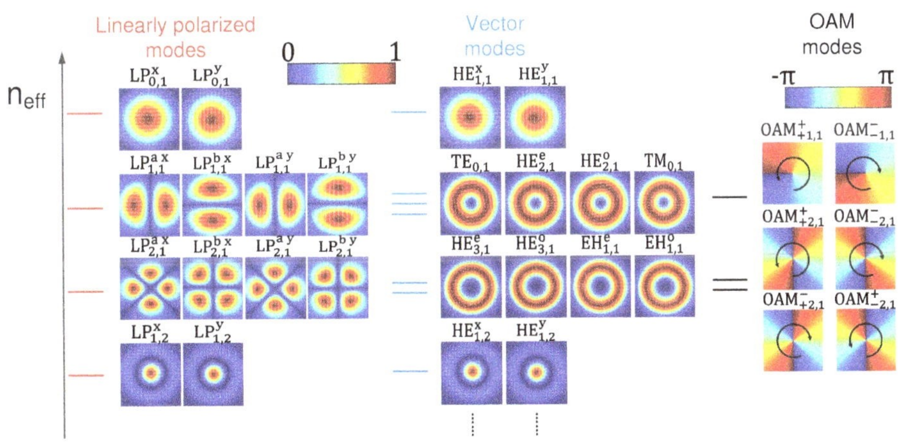

Figure 1. The schematic diagram of mode classification categories in optical fiber.

2.1. OAM Modes in Circular-Core Fibers

The refractive index profile of a conventional step-index circular-core fiber (SI-CCF) is shown in Figure 2a. The fiber guiding condition is determined by the values of core refractive index (n_{co}) and core radius (r_{co}). Vector modes in the SI-CCF can be obtained by solving Helmholtz equations with the tangential continuous boundary conditions. Using Equation (2), the OAM modes having a helical wave-front phase can be formed by the superposition of vector modes. As an example, Figure 1 shows the wave-front phase and circularly polarized chirality of $\mathrm{OAM}^{\pm}_{\pm1,1}$, $\mathrm{OAM}^{\pm}_{\pm2,1}$ and $\mathrm{OAM}^{\mp}_{\pm2,1}$; these modes are formed by $\mathrm{HE}^{e/o}_{2,1}$, $\mathrm{HE}^{e/o}_{3,1}$, and $\mathrm{EH}^{e/o}_{1,1}$, respectively. The variation of n_{eff} for vector modes versus fiber core radius is shown in Figure 2b. We can see that radial higher-order modes (HOMs) such as $\mathrm{HE}_{1,2}$, $\mathrm{TE}_{0,2}$, and $\mathrm{TM}_{0,2}$ will by accompanied by the azimuthal HOMs. These undesired radial HOMs will increase the complexity of mode coupling and decrease the stability of OAM modes in transmission.

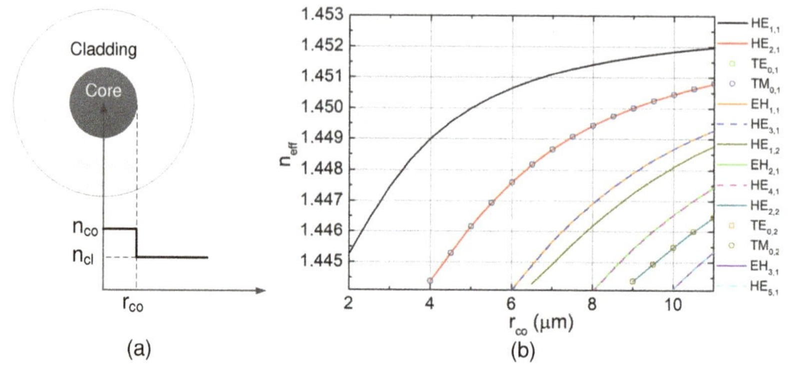

Figure 2. (**a**) The refractive index profile of SI-CCF. (**b**) The variation of n_{eff} versus r_{co} when $(n_{\mathrm{co}} - n_{\mathrm{cl}})/n_{\mathrm{co}} = 0.6\%$ at the wavelength of 1550 nm.

In addition, near-degenerate vector modes such as $\mathrm{TE}_{0,1}$, $\mathrm{HE}_{2,1}$, and $\mathrm{TM}_{0,1}$ have almost the same n_{eff}. The Δn_{eff} around 1×10^{-5} is too small to avoid the accidental phase matching between the near-degenerate modes. Referring to the experiment of polarization maintaining fibers, the Δn_{eff} should be greater than 1×10^{-4} to maintain a polarization state of the $\mathrm{HE}_{1,1}$ mode in length scales exceeding 100 m [43]. Thus, in practice, the fiber index profile deformation and fiber bending can easily cause mode coupling between the near-degenerate vector modes or OAM modes. The superposition of near-degenerate vector modes can form LP modes. As shown in Figure 3, because of strong mode coupling, the intensity patterns of LP modes can be observed at the fiber output end, even though the inputs are true vector modes.

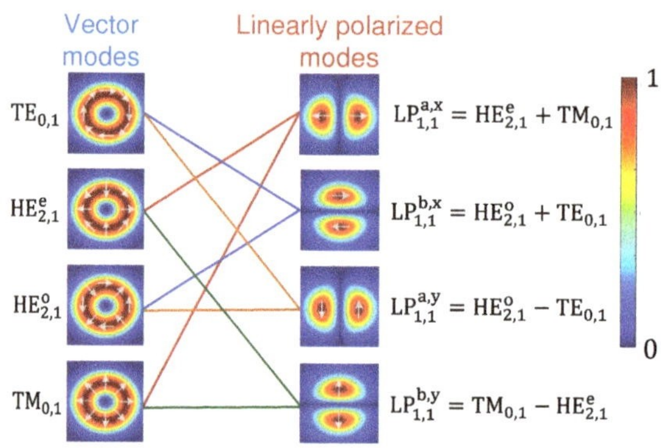

Figure 3. Intensity patterns of the first higher-order mode group in a SI-CCF. Arrows show the polarization of the modal electric field. The left column shows the input exact vector modes, while the right column shows the resultant modes at the output end. Specific linear combinations of mode pairs in the left column, resulting in the variety of $\mathrm{LP}_{1,1}$ modes, are shown by colored lines.

Graded-index circular-core fibers (GI-CCFs) were also studied for OAM mode transmission [44–46]. Because the gradient of the graded-index profile is much slower than the step-index profile, the OAM modes have a higher mode purity in GI-CCFs. In [44], the authors presented a specially designed GI-CCF that can support OAM modes having a purity higher than 99.9%. This kind of design can effectively decrease the intrinsic crosstalk between OAM modes. The possibility of OAM mode transmission in commercially available graded-index multimode fiber was discussed by Chen [45]. Recently, all-fiber OAM (de)multiplexing devices based on graded-index multimode fiber were demonstrated with high excitation purity of over 90% [46].

Although GI-CCF shows better potential for OAM mode transmission than SI-CCF, no experiment demonstrates the long-distance transmission of OAM modes in circular-core fibers so far. OAM modes may easily transform into LP modes under the external disturbance in practice. Such instability is the main barrier for long-distance OAM modes propagation in CCFs. Therefore, new types of optical fiber are required to enhance the OAM mode stability.

2.2. OAM Modes in Ring-Core Fibers

The azimuthal higher-order vector modes or OAM modes in circularly symmetrical optical fiber have a donut-shaped intensity pattern. One may be inspired to think that the waveguide index profile mimicking the donut shape may provide a more stable environment for the transmission of OAM modes. The pioneering works on the stable transmission of vector modes were done by Ramachandran et al., who adopted a first-order perturbation analysis to provide physical insight into the degeneracy of vector modes. Taking the first-order vector modes as an example, the propagation constant of each mode can be expressed by adding a vector correction to the scalar propagation constant $\widetilde{\beta}$, that is [34]

$$\beta_{\mathrm{TE01}} = \widetilde{\beta}, \quad \beta_{\mathrm{TM01}} = \widetilde{\beta} - 2(\delta_1 + \delta_2), \quad \beta_{\mathrm{HE21}} = \widetilde{\beta} + (\delta_2 - \delta_1), \tag{3}$$

where

$$\delta_1 = \frac{\Delta n_{\mathrm{max}}}{2a^2 n_{\mathrm{co}}\beta} \int_0^\infty rE(r)\frac{\partial E(r)}{\partial r}\frac{\partial(\Delta n(r)/\Delta n_{\mathrm{max}})}{\partial r}dr \tag{4a}$$

$$\delta_2 = \frac{\Delta n_{\mathrm{max}}}{2a^2 n_{\mathrm{co}}\beta} \int_0^\infty E^2(r)\frac{\partial(\Delta n(r)/\Delta n_{\mathrm{max}})}{\partial r}dr, \tag{4b}$$

where r is the radial coordinate, $E(r)$ is the electric field for the scalar mode, a is the size of the fiber core, $\Delta n(r)$ is fiber refractive index profile relative to the cladding index, and Δn_{max} is the highest refractive index difference. The expressions of δ_1 and δ_2 indicate that, when the electric field has a high changing rate and large amplitude in the region of refractive index discontinuities, the polarization orientation of modes will cause a large additional propagation constant. Since a ring core has inner and outer discontinuous boundaries of refractive index, the Δn_{eff} between the near-degenerate vector modes can be effectively enlarged. Figure 4a shows the index profile of a fabricated novel RCF and the simulated intensity distribution of the first higher-order mode group (HOMG). Although the fiber has a high refractive index core at the center, it does not damage the guidance of the first HOM in the ring core. Most of the mode energy is guided in the ring core, which is the case expected to help with mode stability. The n_{eff} of $\mathrm{TE}_{0,1}$, $\mathrm{HE}_{2,1}$, and $\mathrm{TM}_{0,1}$ measured by recording grating resonance wavelength are shown in Figure 4b. We can see that the n_{eff} for each pair of modes is separated well. The minimum Δn_{eff} between $\mathrm{TM}_{0,1}$ and the other modes is no less than 1.8×10^{-4}; it is much higher than the value in conventional CCFs, which is of the order of 10^{-5}. The stability of the first-order vector modes in this RCF was measured. Under the perturbations of multiple twists and bends with bending radius down to 1 cm, the mode purity can remain 99.8% after transmitting through the 20 m long fiber.

Later, the stable transmission and manipulation of the first-order OAM modes were achieved in RCF [47]. Theoretically, using the same design philosophy, more stable OAM modes can be accommodated into a RCF by increasing the core size and refractive index contrast. However, the radial HOMs may access fiber accompanying the azimuthal HOMs. To eliminate the mode coupling caused by undesired radial HOMs, an optimized design can be applied to RCF to suppress radial HOMs without damaging azimuthal HOMs [48]. Since a radial HOM has a multi-layer ring intensity pattern, narrowing down the ring core thickness can effectively suppress the existence of radial HOMs. Figure 5a shows the refractive index profile of a RCF; the ring core thickness is determined by r_1 and r_2. The cutoff curves of OAM mode groups are shown in Figure 5b. Actually, the cutoff condition of a mode group $OAM_{l,v}$ consists of cutoff curves of a set of near-degenerate modes $OAM^{\pm}_{\pm l,v}$ and $OAM^{\mp}_{\pm l,v}$; however, they are too close to be clearly distinguished when the relative index difference between the core and cladding is small ($(n_{co} - n_{cl})/n_{co} < 2\%$). Thus, only one cutoff curve is displayed for each OAM mode group. These cutoff conditions define boundaries of parameter regions where a specific number of modes is supported in the fiber. The curve of $OAM_{0,2}$ forms a design boundary for eliminating all the radial HOMs. In the region above this cutoff curve (unshaded region), the radial order of all modes is one. In the gray shaded region below the cutoff curve, the modes with $v > 1$ become guided in the fiber. Therefore, by tuning the values of r_2 and r_1/r_2, a fiber can support a targeted number of $OAM_{l,1}$ without introducing radial HOMs.

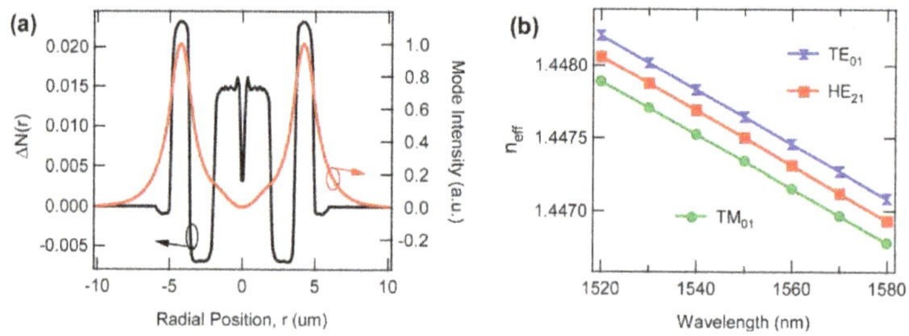

Figure 4. (**a**) Measured index profile of the fabricated fiber and simulated intensity distribution of the LP_{11} mode. (**b**) Measured n_{eff} of $TE_{0,1}$, $HE_{2,1}$, and $TM_{0,1}$ of the fiber shown in (**a**). Reprinted with permission from Ref. [34] © The Optical Society.

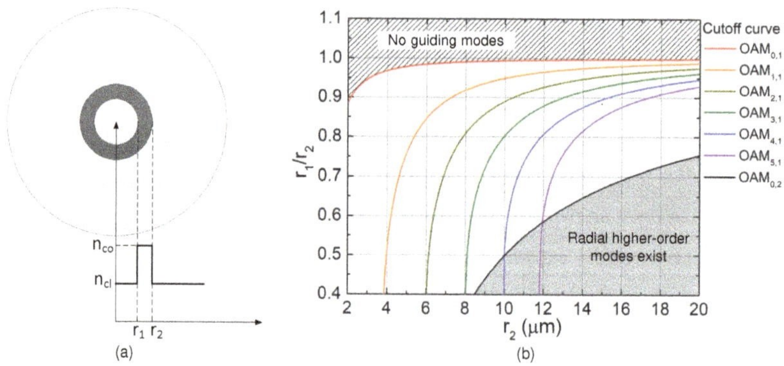

Figure 5. (**a**) The refractive index profile of RCF and (**b**) the cutoff conditions of different order OAM mode groups when $(n_{co} - n_{cl})/n_{co} = 0.6\%$ at the wavelength of 1550 nm.

The stable propagation of multiple azimuthal higher-order OAM modes in a RCF was first demonstrated in 2013 [49]. The authors designed and fabricated a RCF with a central air-hole which maximizes the refractive index gradient at the inner boundary of the ring core. This large refractive index gradient coincides with modal field gradient yields large Δn_{eff} of OAM modes. The best propagation stability was measured on the eighth-order OAM mode, which had ~20 dB mode purity at the output end of a 1 km long fiber. After that, many research works were done to optimize the design and fabrication of AH-RCFs [35,50–52]. An interesting phenomenon found in AH-RCFs was that the OAM longevity increases with mode order [35], which is contrary to the common intuition that fundamental states of systems are the most stable. To test modal stability, circularly polarized OAM modes were generated by a SLM and a quarter-wave plate and then were coupled into a 10 m long AH-RCF. The output OAM modes purities were measured under the condition of arbitrary fiber bending and twisting, as schematically illustrated in Figure 6a. The air-hole radius of the under test AH-RCF is 3 μm, and the outer radius of the ring core is 8.25 μm shown in Figure 6b. The central air-hole forces the modes to distribute more intensity at the outer boundary of the ring core, and modes will "feel" more of the large refractive index difference between the ring core and cladding. Hence, the n_{eff} of modes with OAM and SAM aligned (having the same handedness) will be separated from those with OAM and SAM anti-aligned (Figure 6c). Furthermore, the separation level of n_{eff} increases with the OAM order. A Δn_{eff} around 10^{-4} is considered reasonable for OAM mode propagation [43]; in this fiber, the OAM modes with $|L| = 5, 6$ and 7 can meet the requirement, as indicated in Figure 6d. In addition, the perturbation has weaker effects on modes with the increase of mode order [35]. The perturbation of fiber bends and geometrical deformations can be expanded into Fourier series, where the azimuthal factor of each element can be expressed as $f_p(r)\exp(ip\phi)$. Since the coupling coefficient is determined by the inner product between the first OAM mode $E_1(r)\exp(iL_1\phi)$, the perturbation factor and the second OAM mode $E_1(r)\exp(iL_2\phi)$, mode coupling mainly happens when $p = L_1 - L_2$. For the near-degenerate OAM modes with $|L|$, because of the requirement of spin alignment, the perturbation can cause mode coupling when $p = 2|L|$. However, the $f_p(r)$ becomes increasingly negligible for large p. Therefore, the higher-order OMA modes gain better stability.

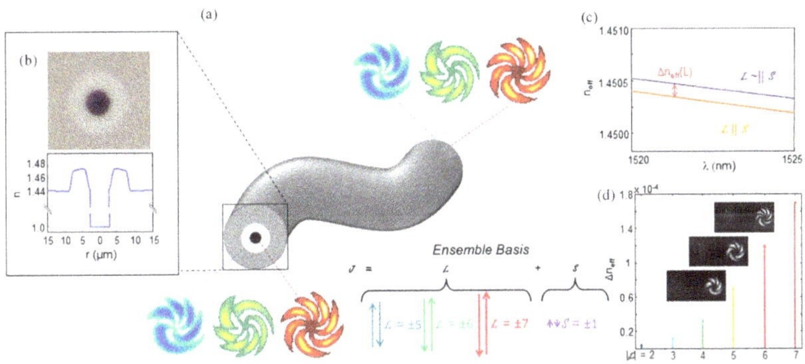

Figure 6. (**a**) Schematic picture of OAM mode transmission through a randomly bended and twisted AH-RCF conserving their original states. Each state has a total angular momentum J that consists of orbital component L and spin part s. (**b**) Cross section image (top) and measured refractive index profile (bottom) for the AH-RCF. (**c**) Example of n_{eff} separation between near-degenerate OAM modes. (**d**) The splitting of n_{eff} for near-degenerate OAM modes among different OAM mode groups and the modal interference images when $|L| = 5$, 6 and 7. Reprinted with permission from Ref. [35] © The Optical Society.

The inter-mode crosstalks of the AH-RCF were measured on a multitude of fiber lengths [36]. Figure 7a illustrates the crosstalks versus fiber length for the $|L| = 5$ and $|L| = 7$ OAM modes. The insets show the modal images at the output end of different fiber lengths. The mode intensities gradually evolve from a donut shape into periodical bright spots in the azimuthal direction with the increase of integrated crosstalk. Longer fiber length leads to heavier crosstalk; the growth rate is 1.5 % per kilometer for the $|L| = 7$ modes and 15 % per kilometer for the $|L| = 5$ modes. The $|L| = 7$ modes maintain better mode purities which have crosstalks lower than -10 dB when the fiber length is up to 5.5 km. The measured results are coincident with the prediction of a $\tanh(hz)$ (where h is the mode coupling coefficient) dependent crosstalk based on coupled-power theory [53]. Similarly, in Figure 6b, the crosstalks of $|L| = 6$ modes measured in a 1 km long fiber fit the crosstalk curve with an h value of $\sim 4.1 \times 10^{-2}$ km^{-1}.

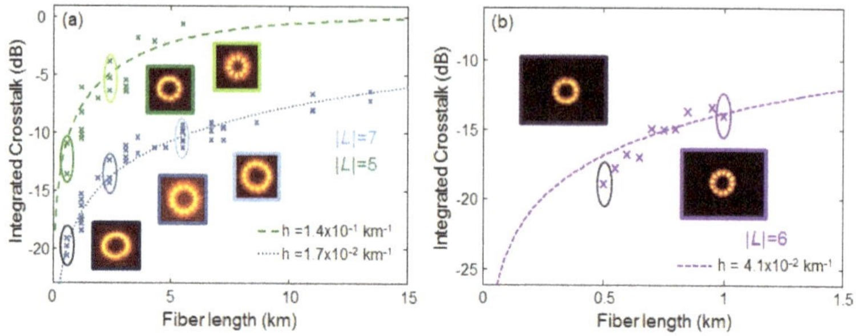

Figure 7. (**a**) Measured crosstalks for $|L| = 5$ and $|L| = 7$ modes in an AH-RCF (stars), and theoretical crosstalks which have the minimum mean-squared error from the measured data (dashed lines). (**b**) Measured and theoretical crosstalks for $|L| = 6$ modes. Insets are modal images at the output of fiber with different length. Reprinted with permission from Ref. [36] © The Optical Society.

Another novel inverse-parabolic graded-index fiber (IP-GIF) was proposed in 2014 [54]. Similar to the RCFs, the IP-GIF has a refractive index valley at the center of cross section; therefore, the fiber modes can feel large refractive gradient in the fiber. Numerical simulation showed that the specially designed IP-GIF enables a large Δn_{eff} over 2.1×10^{-4} between the vector modes $TE_{0,1}$, $HE_{1,1}$ and $TM_{0,1}$. Then, the authors experimentally demonstrated the stable transmission of $OAM^{\pm}_{\pm 1,1}$ modes in the 1.1 km long IP-GIF.

The IP-GIFs and RCFs, especially the one assisted by a central air-hole, show good ability to separate the n_{eff} of near-degenerate OAM modes. It greatly enhances the stability of OAM modes which is hardly achieved in conventional CCFs. These designs are beneficial to increase fiber transmission capacity via adding spatial modes without resorting to digital signal processing or adaptive optics.

3. Microstructure Optical Fibers Supporting OAM Modes

Photonic crystal fibers (PCFs) which have periodic transverse air-hole arrangements were first proposed and fabricated in 1996 [55]. With the improvement of manufacturing technique, the accuracy of microstructure can reach 10 nm on the scale of 1 μm [56]. The design of PCFs is more flexible than conventional solid optical fibers, which allows effective control of key modal properties such as dispersion, birefringence, nonlinearity and cutoff conditions. Inspired by the success of PCFs, other kinds of microstructure fibers such as photonic band gap fibers (FBGFs) [57], Kagome fibers [58,59], and negative-curvature fibers [60,61] were proposed to guide optical modes in an air core. These novel guiding mechanisms of microstructure fibers can be adopted to design special OAM fibers.

3.1. Photonic Crystal Fibers Supporting OAM Modes

The common structure of a PCF is to arrange a regular hexagonal lattice of air-holes symmetrically around a central pure silica core. The equivalent refractive index of the cladding is lower than that of the core due to the air-holes. Thus, optical modes can be well confined in the core. When a PCF is periodically twisted in the longitudinal direction, the cladding light will be guided in the helical lattice of air-holes [62]. The spiral path regulated by air-holes (Figure 8a) will transfer a fraction of the axial momentum into angular momentum, causing the formation of cladding orbital angular momentum states. Furthermore, the wavelengths of the cladding OAM modes vary linearly with the twist rate. However, in this fiber, no stable OAM modes can be transmitted in the core region.

Later, a phenomenon of OAM preservation was detected in a helically twisted PCF with a three-bladed Y-shaped core [38]. The schematic pattern and scanning electron micrograph of fiber cross section are shown in Figure 8b,c, respectively. Based on a helicoidal extension of Bloch wave theory, the authors developed a scalar analysis for the modes in the twisted PCF. Figure 8d shows the calculated axial energy flow of a ring mode when the fiber twist rate is 1.26 rad/mm. In the same fiber, the calculated axial

Poynting vector of the fundamental mode is shown in Figure 8e. An experiment showed that the chirality of the +1 order OAM mode was preserved rather than coupled into the −1 order OAM mode, and vice versa in the twisted PCF, which is in agreement with the theoretical prediction.

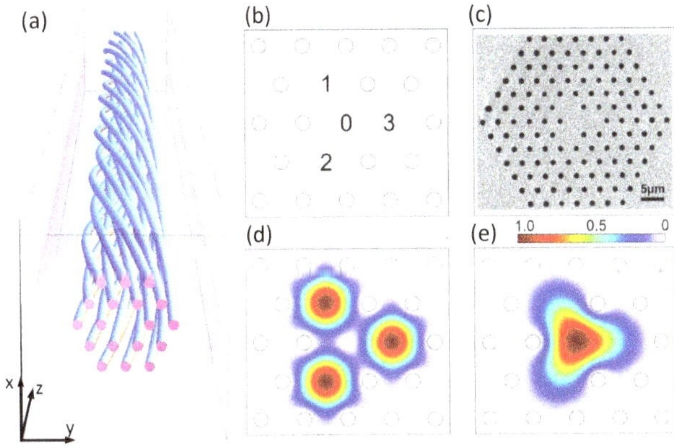

Figure 8. (**a**) The perspective of a helical PCF in a twisting period (blues rods are the air-holes passing through the silica fiber). Adapted with permission from Ref. [62] © AAAS. (**b**) Schematic structure of a Y-shaped core PCF. (**c**) Scanning electron micrograph of the Y-shaped core PCF. (**d**) Calculated axial Poynting vector of a ring mode at the wavelength of 800 nm in the fiber with a twist rate of 1.26 rad/mm. (**e**) Calculated axial Poynting vector of the fundamental mode which carries no OAM modes. Adapted with permission from Ref. [38] © The Optical Society.

In addition to the twisted PCFs, a family of RC-PCFs was proposed for OAM mode transmission [37,63–66]. As shown in Figure 9, the RC-PCFs all adopt circularly symmetric air-hole arrangement, so that the even and odd polarization states of a vector mode will keep well degenerate hence making sure the stability of the combined OAM modes. In [64], three layers of air-holes were used to form the cladding region (Figure 9a). The authors proved that the Δn_{eff} between the even and odd polarization states are around 10^{-9} when the number of air elements in each layer is no less than 24. In this case, the fiber structure-caused circular asymmetry is negligible compared with the deformations that arise from external perturbation such as bending and twisting. In the same year, RC-PCFs supporting 14 [37] and 26 [65] OAM modes were designed and analyzed, successively (Figure 9b,c). In these works, identical air-holes were used to construct four circular cladding layers. This design contributes to keeping a similar swelling force in each air-hole, which helps to maintain ideal cross section geometry during the fabrication. The high air-hole filling ratio of the cladding region leads to an equivalent refractive index much lower than that of pure silica; therefore, both inner and outer boundaries of the ring core have very large index gradients. The electric field of modes with a relatively lower-order concentrate and vary rapidly in the vicinity of the inner refractive index step, thus the Δn_{eff} of near-degenerate modes such as ($HE_{2,1}$, $TE_{0,1}$, $TM_{0,1}$) is very large. However, with the increase of mode order, the Δn_{eff} first reduces, and then rebounds to grow after a certain order. Because the modal fields move to the outer region of the ring core with the increase of mode order, the great index step at the outer core boundary becomes dominant to cause large Δn_{eff} between near-degenerate modes in the HOMGs.

In 2019, a PCF with cladding consists of a circular layer of air-hole and three hexagonal layers of air-hole were designed and fabricated [66]. Figure 9d shows the designed fiber cross section. The circular arrangement of the first layer air-holes can help to minimize the Δn_{eff} between the even and odd polarization states of the same vector mode. Furthermore, three outer layers of air-holes contribute to increasing the light confinement in the ring core. The hexagonal arrangement of the outer air-holes makes the fiber easier to be manufactured by the stack-and-draw method. The scanning electron micrograph of the fabricated RC-PCF is shown in Figure 9e. The ring core maintains a good circular symmetry. However, the fabricated fiber has a relatively large transmission loss of 50 dB/km, and only the transmission of the first- and second-order OAM modes was demonstrated in the experiment.

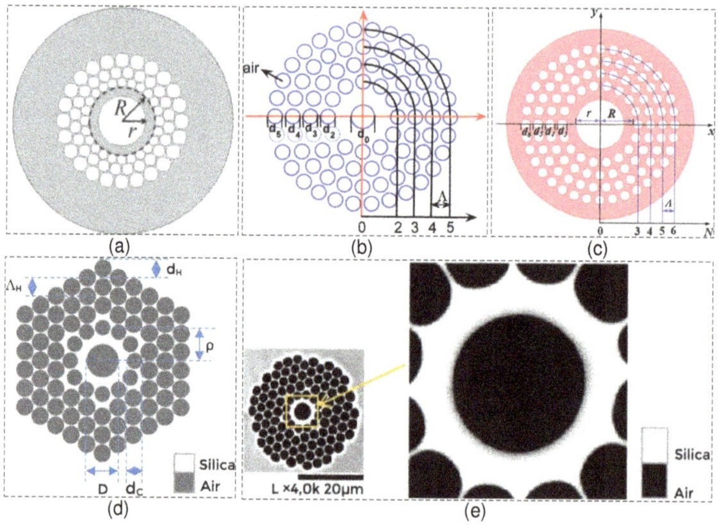

Figure 9. The cross sections of PCFs supporting OAM modes. (**a**) Heterogeneous air-hole cladding design. Adapted with permission from Ref. [64] © IEEE. RC-PCFs with cladding constructed by identical air-holes supporting (**b**) 14 and (**c**) 26 OAM modes. Adapted with permission from Ref. [37] © IEEE and Ref. [65] © Elsevier. (**d**) A design of RC-PCF with cladding that consists of a circular layer of air-holes and three hexagonal layers of air-holes. (**e**) The fabricated RC-PCF based on the design in (**d**). Adapted with permission from Ref. [66] © The Optical Society.

3.2. Negative-Curvature Ring-Core Fibers for OAM Modes Transmission

In 2011, Wang et al. proved that a hypocycloid-shaped core structure can enhance light confinement in Kagome fibers [59]. Furthermore, by simplified the Kagome cladding into a layer of silica tubes, a kind of negative-curvature fiber was proposed to guide light in a quasi-air environment with ultra-wide bandwidth, low scattering loss, and weak nonlinearity [60,61,67–69]. These designs provide valuable references to improve the cladding structure of RC-PCFs.

Enlightened by the negative-curvature fibers, our group studied a new kind of negative-curvature ring-core fiber (NC-RCF). The loss comparison between the multi-layered hollow-core fiber and negative-curvature hollow-core fiber is shown in Figure 10a [39]. When the number of air–silica interfaces is the same (four layers), the confinement loss of negative-curvature hollow-core fiber is much lower. Therefore, to increase the number of supported OAM modes with achieving ultra-low loss for the HOMs, an NC-RCF (Figure 10c) was proposed that originate from RC-PCF (Figure 10b). The thickness of the cladding tubes should meet the antiresonant condition for the central wavelength of the transmission band that [69–71]

$$t = \frac{\lambda(M-0.5)}{2\sqrt{n_g^2-1}}, \tag{5}$$

where λ is the free space wavelength, n_g is the refractive index of silica glass, and M represents the order of antiresonance ($= 1, 2, 3 \ldots$).

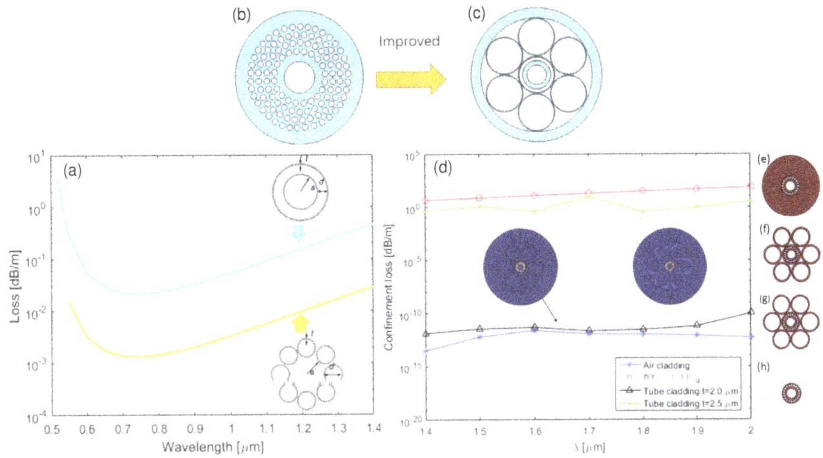

Figure 10. (**a**) Loss comparison between multi-layered annular fiber and negative-curvature fiber. (**b**) RC-PCF. (**c**) Improved NC-RCF. (**d**) Dependence of the confinement loss on the wavelength for (**e**) silica cladding design, (**f**) negative curvature tube cladding with tube thickness of $t = 2$ μm, (**g**) negative curvature tube cladding with tube thickness of $t = 2.5$ μm, and (**h**) air cladding model. Adapted with permission from Ref. [39] © The Optical Society.

Another interesting property of NC-RCF is that the modes formed in the thin membranes of the cladding tubes can couple with certain core modes. Then, the undesired core modes can be effectively filtered out by the coupling caused the large loss. In Figure 10e–h, red regions represent silica and white regions represent air. The simulated confinement losses of the $HE_{3,1}$ mode in the four designs are shown in Figure 10d. It is easy to understand that the silica cladding and air cladding designs have the highest and lowest confinement loss, respectively. The green and black lines in Figure 10d show the confinement losses of NC-RCF when $t = 2.5$ μm and $t = 2$ μm, respectively. When t is 2 μm, the confinement ability of negative curvature tubes is very close to that of air cladding. However, when t increases to 2.5 μm, antiresonant and inhibited coupling conditions are not satisfied. Hence, a lot of energy was coupled into the tube membrane modes and caused a very high loss, which can be understood by observing the electric field distribution of $HE_{3,1}$ mode shown on the right side of Figure 10d.

The detailed structure of NC-RCF and the descriptions of the parameters are illustrated by Figure 11a,b. We took $r = 7$ μm, $\rho = 0.65$, $t = 1.2$ μm, and $d/\Lambda = 0.8$ as an example to discuss the fiber properties. To prevent the HE- and EH-synthesis OAM modes coupling into LP modes during the propagation, Δn_{eff} should be larger than 10^{-4}, which is the minimum modal birefringence of the polarization-maintaining fibers. Figure 11c shows the simulated spectra of Δn_{eff} between the adjacent vector eigenmodes in an OAM mode group. As the $TE_{0,1}$ and $TM_{0,1}$ are filtered out, only $HE_{2,1}$ mode remains in OAM Mode Group #1. Therefore, the case of Group #1 is excluded in Figure 11c. It can be seen that the requirement of $>1 \times 10^{-4}$ is met in the wavelength band from 1.4 to 1.8 μm for OAM Mode Groups #2, #3, #4, #5, #6, and #8. In this fiber, the Δn_{eff} within a mode group decreases with the order of mode group (except for the Mode Group #8). Such variation trend of Δn_{eff} is opposite to that in the AH-RCF illustrated in Section 2.2. It is because the equivalent refractive index of cladding of our designed fiber is close to 1, thus the large core/cladding refractive index difference makes the lower-order mode group distribute approaching the inner air–silica boundary, and hence generates a higher Δn_{eff} between vector eigenmodes in the mode group. This property will be good for the use of lower-order OAM modes. Figure 11d shows the minimum Δn_{eff} between the adjacent OAM mode groups over the wavelength from 1.2 to 2.0 μm. The eigenmodes used for calculating the minimum inter-mode-group Δn_{eff} are described in Figure 11(b2). The values increase with the order of

OAM mode group and wavelength. When the wavelength is larger than 1.2 μm, all the inter-mode-group Δn_{eff} exceed 1×10^{-3}. Consequently, in theory, the designed NC-RCF can stably support 26 OAM states in the telecommunication wavelength band, including OAM Mode Groups #1 (two states), #2 (four states), #3 (four states), #4 (four states), #5 (four states), #6 (four states), and #8 (four states).

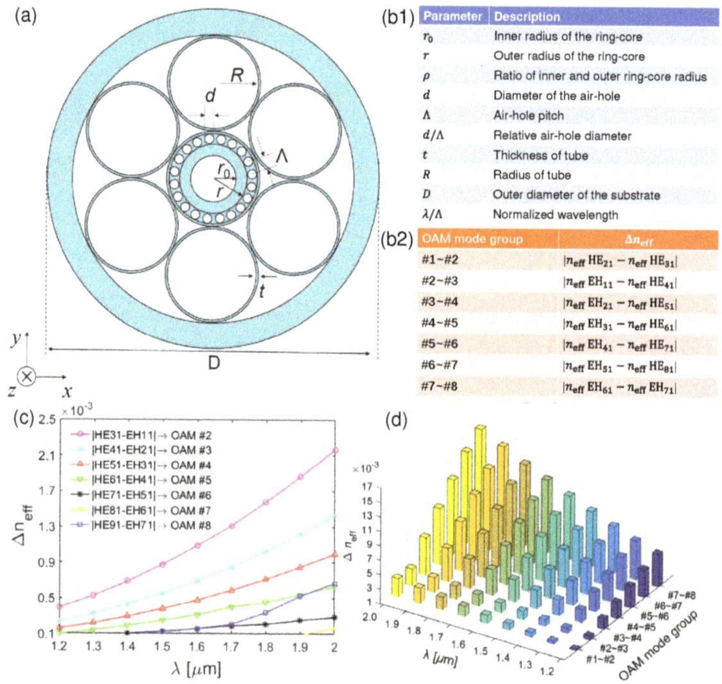

(b1)

Parameter	Description
r_0	Inner radius of the ring-core
r	Outer radius of the ring-core
ρ	Ratio of inner and outer ring-core radius
d	Diameter of the air-hole
Λ	Air-hole pitch
d/Λ	Relative air-hole diameter
t	Thickness of tube
R	Radius of tube
D	Outer diameter of the substrate
λ/Λ	Normalized wavelength

(b2)

OAM mode group	Δn_{eff}
#1~#2	$\lvert n_{\text{eff}}\,HE_{21} - n_{\text{eff}}\,HE_{31}\rvert$
#2~#3	$\lvert n_{\text{eff}}\,EH_{11} - n_{\text{eff}}\,HE_{41}\rvert$
#3~#4	$\lvert n_{\text{eff}}\,EH_{21} - n_{\text{eff}}\,HE_{51}\rvert$
#4~#5	$\lvert n_{\text{eff}}\,EH_{31} - n_{\text{eff}}\,HE_{61}\rvert$
#5~#6	$\lvert n_{\text{eff}}\,EH_{41} - n_{\text{eff}}\,HE_{71}\rvert$
#6~#7	$\lvert n_{\text{eff}}\,EH_{51} - n_{\text{eff}}\,HE_{81}\rvert$
#7~#8	$\lvert n_{\text{eff}}\,EH_{61} - n_{\text{eff}}\,EH_{71}\rvert$

Figure 11. (**a**) The cross section of the proposed NC-RCF. (**b1**) The description of each structure parameter. (**b2**) The modes used to calculate the minimum Δn_{eff} in (**d**). (**c**) The Δn_{eff} between the adjacent vector eigenmodes in each OAM mode group when $r = 7$ μm, $\rho = 0.65$, $t = 1.2$ μm, and $d/\Lambda = 0.8$. (**d**) The minimum Δn_{eff} between OAM mode groups. Adapted with permission from Ref. [39] © The Optical Society.

The fabrication procedure of the NC-RCF is illustrated in Figure 12a. It includes four steps: (1) stack the center preform; (2) draw the center cane; (3) stack the fiber preform; and (4) draw the fiber. In the second and fourth steps, the cane and tubes should be sealed at one end to keep positive air pressure in the hollow holes for maintaining the fiber structure during drawing. Furthermore, in Step 4, a vacuum pump should be used to help reserve the hollow structure during the fiber drawing. Figure 12e,f shows the fibers fabricated under the vacuum level of ~60 kPa at the temperature around 1880 °C and ~40 kPa at the temperature around 1895 °C, respectively. We can see that the intervals between the cladding tubes and the central casing tube are filled in by silica; therefore, the fabricated fibers have much higher confinement loss than that predicted in simulation. The minimum transmission losses of $OAM_{1,1}$ in Fibers (e) and (f) are 0.3 and 0.095 dB/m, respectively. The higher loss in Fiber (e) is caused by the relatively thicker silica layer surrounding the core, which leads to the leakage of energy from the ring core to the cladding silica through mode coupling.

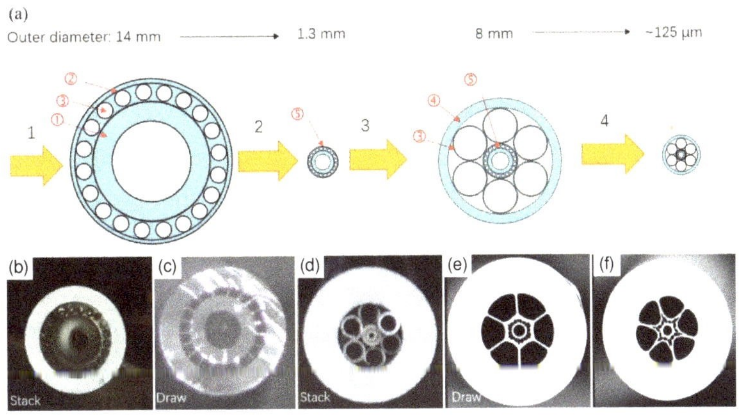

Figure 12. Fabrication of the proposed NC-RCF. (**a**) Schematic figure of the fabrication process using stack-and-draw method. (**b**) Stacked preform of the central cane. (**c**) Cross section of the intermedia cane. (**d**) Stacked preform of the NC-RCF. Scanning electron microscopic photo of the fabricated NC-RCF with (**e**) the thickness of ring core around 2.45 μm, Adapted with permission from Ref. [39] © The Optical Society, and (**f**) the thickness of ring core around 2.65 μm. Adapted with permission from Ref. [40] © IEEE.

Finally, we compare the different kinds of fibers supporting OAM mode in Table 1. The recorded number of OAM mode transmissions is 36, achieved by a AH-RCF in 2014. In addition, the maximum transmission length of OAM modes is 13.4 km using a loop of AH-RCF in 2016. Although, theoretically, RC-PCFs and NC-PCFs provide a larger Δn_{eff} between the near-degenerate modes, which help to enhance the mode stability, no experiment demonstrates the long-distance stable transmission of OAM modes in these two kinds of fibers. The main reason is that RC-PCFs and NC-PCFs may suffer relatively large geometrical deformation during the fabrication. Then, the refractive index fluctuation will cause severe mode coupling. We believe that the performance of RC-PCFs and NC-PCFs will be improved with the progress of the fabrication technique.

Table 1. Comparison among the reported fibers supporting OAM modes.

Year	Fiber Type	Simulation		Experiment			Min. Δn_{eff}	Wavelength (nm)
		Mode Number	OAM Modes	Mode Number	OAM Modes	Max. Transmission Length		
2012	RCF [47]	-	-	2	$\text{OAM}_{\pm1,1}$	20 m	-	1527
2013	AH-RCF [49]	-	-	12	$\text{OAM}_{\pm7,1}-\text{OAM}_{\pm9,1}$	1 km	$\sim 1\times10^{-4}$	1530, 1555
2015	AH-RCF [35]	-	-	12	$\text{OAM}_{\pm5,1}-\text{OAM}_{\pm7,1}$	10 m	7×10^{-5}	1530–1565
2016	AH-RCF [36]	-	-	12	$\text{OAM}_{\pm1,1}-\text{OAM}_{\pm2,1}$	13.4 km	7×10^{-5}	1550
2014	AH-RCF [50]	36	$\text{OAM}_{0,1}-\text{OAM}_{\pm9,1}$	36	$\text{OAM}_{0,1}-\text{OAM}_{\pm9,1}$	0.85 m	1.1×10^{-4}	1550
2014	IP-GIF [54]	6	$\text{OAM}_{\pm1,1}-\text{OAM}_{\pm2,1}$	2	$\text{OAM}_{\pm1,1}$	1.1 km	2.1×10^{-4}	1530–1565
2015	GI-CCF [44]	10	$\text{OAM}_{\pm1,1}-\text{OAM}_{\pm3,1}$	-	-	-	1.16×10^{-4}	1520–1580
2016	RC-PCF [65]	26	$\text{OAM}_{\pm1,1}-\text{OAM}_{\pm7,1}$	-	-	-	$>1\times10^{-4}$	1250–1900
2016	RC-PCF [37]	14	$\text{OAM}_{\pm1,1}-\text{OAM}_{\pm4,1}$	-	-	-	$>1\times10^{-4}$	1250–1810
2019	RC-PCF [66]	14	$\text{OAM}_{\pm1,1}-\text{OAM}_{\pm4,1}$	6	$\text{OAM}_{\pm1,1}-\text{OAM}_{\pm2,1}$	1.2 m	2.13×10^{-3}	1550
2019	NC-RCF [39]	26	$\text{OAM}_{\pm1,1}-\text{OAM}_{\pm7,1}$	14	$\text{OAM}_{\pm1,1}-\text{OAM}_{\pm4,1}$	40 m	2×10^{-3}	1550

4. Crosstalk Analysis of OAM Modes in Optical Fibers

Multi-core fiber (MCF) is regarded as a good solution for increasing data transmission capacity and saving fiber consumption for the next generation of fiber-optic networks.

When the mode coupling between fiber cores is sufficiently small, each core can be regarded as an independent signal channel. Many studies have been done for reducing the inter-core crosstalk, including the studies on heterogeneous and trench-assisted MCFs [72–76]. By optimizing the core structure and arrangement, ultra-low inter-core crosstalk and high core density can be achieved simultaneously [75].

Those experiences in designing MCFs provide good references for the design of MRCFs. Figure 13a shows the refractive index profile of a step-index trench-assisted MRCF. The design region of the size of ring core for supporting a certain number of OAM mode groups can be tuned by changing the central refractive index (n_{ce}) and the core refractive index (n_{co}). Figure 13b illustrates the design region for supporting mode groups from $OAM_{0,1}$ to $OAM_{4,1}$ and cutting off all the radial HOMs in the C+L band (from 1.53 μm to 1.625 μm). The boundaries of each design region are formed by the cutoff curves of $OAM_{0,2}$ and $OAM_{5,1}$ at the wavelength of 1.53 μm and $OAM_{4,1}$ at the wavelength of 1.625 μm, respectively. When the relative index difference of the core ($\Delta_{co} = (n_{co} - n_{cl})/n_{co}$) increases, the available values of the outer radius of ring core (r_2) decreases. The smaller core size is beneficial to the compact design of MRCFs. In addition, the negative relative index difference of the central area ($\Delta_{ce} = (n_{ce} - n_{cl})/n_{ce}$) can force the n_{eff} of $OAM_{0,2}$ fall to the cladding refractive index more quickly. As shown in Figure 13b, the cutoff curve of $OAM_{0,2}$ occurs at a smaller value of r_1/r_2 when Δ_{ce} decreases, which indicates that the available width of the ring core can be extended. In the experiment, a relatively thicker ring core can help to increase the mode stimulation efficiency at the fiber input end and improve modal stability during transmission.

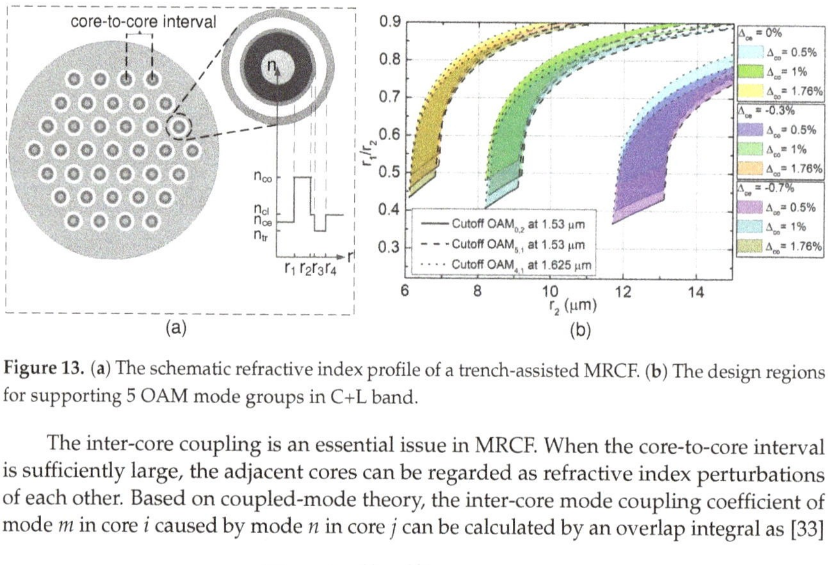

Figure 13. (**a**) The schematic refractive index profile of a trench-assisted MRCF. (**b**) The design regions for supporting 5 OAM mode groups in C+L band.

The inter-core coupling is an essential issue in MRCF. When the core-to-core interval is sufficiently large, the adjacent cores can be regarded as refractive index perturbations of each other. Based on coupled-mode theory, the inter-core mode coupling coefficient of mode m in core i caused by mode n in core j can be calculated by an overlap integral as [33]

$$c_{mn}^{(i)(j)} = \frac{\omega\epsilon_0 \iint_{A\infty}(n^{(j)2} - n_{cl}^2)\mathbf{e}_m^{(i)} \cdot \mathbf{e}_n^{(j)*}\mathrm{d}A}{\sqrt{\iint_{A\infty}\hat{z}\cdot(\mathbf{e}_m^{(i)} \times \mathbf{h}_m^{(i)*} + \mathbf{e}_m^{(i)*} \times \mathbf{h}_m^{(i)})\mathrm{d}A \iint_{A\infty}\hat{z}\cdot(\mathbf{e}_n^{(j)} \times \mathbf{h}_n^{(j)*} + \mathbf{e}_n^{(j)*} \times \mathbf{h}_n^{(j)})\mathrm{d}A}}, \tag{6}$$

where $n^{(j)}$ is the refractive index profile of core j, n_{cl} is the cladding refractive index, and $\mathbf{e}$ and $\mathbf{h}$ are the electric and magnetic components of modes. However, in the case that the cores are not identical, the conventional mode coupling coefficient are not symmetric. Therefore, when only considering the forward propagation modes, the total power may not be conserved. To satisfy the law of power conservation, redefined mode coupling coefficients $C_{mn}^{(i)(j)}$ and $C_{nm}^{(j)(i)}$ were used; their definitions are the average of the usual mode coupling coefficients [77]

$$C_{mn}^{(i)(j)} = \frac{c_{mn}^{(i)(j)} + c_{nm}^{(j)(i)}}{2} = C_{nm}^{(j)(i)}. \tag{7}$$

Because real optical fiber is not perfect, there is a random phase difference between fiber modes, which is a function of propagation length $\delta f(z)$. To consider the effect of the random phase difference, the fiber is divided into finite segments in the longitudinal direction. Furthermore, uniform random phase offsets are applied to all cores in each segment. The overall mode coupling can be obtained by calculating coupled-mode equations in each segment in order.

On the other hand, the inter-core crosstalk can be calculated by coupled-power theory as well [41,77]. Based on the assumption that the random phase fluctuation along the fiber is a stationary random process, it has an autocorrelation function: $R(\Delta z) = \langle \delta f(z + \Delta z) \delta f^*(z) \rangle$. It has been demonstrated that the calculation based on exponential autocorrelation can obtain crosstalk that agrees well with the experiment results in a single-mode MCF. Furthermore, the power coupling coefficient can be expressed by the Fourier transformation of exponential autocorrelation function as [77]

$$h_{mn}^{(i)(j)}(z) = \frac{(C_{mn}^{(i)(j)})^2 L_c}{1 + [\Delta\beta_{mn}^{(i)(j)}(z) L_c]^2}, \tag{8}$$

where $\Delta\beta_{mn}^{(i)(j)}(z)$ represents local propagation constant difference of mode m in core i and mode n in core j and L_c is the correlation length of $R(\Delta z)$. To take into account the effects of fiber bending and twisting without using numerical solutions, a closed-form average power coupling coefficient was proposed for estimating the inter-core crosstalk in MCFs [78]. It represents the average power coupling coefficient over a twist pitch as

$$\bar{h}_{mn}^{(i)(j)} = \frac{\gamma}{2\pi} \int_0^{2\pi/\gamma} h_{mn}^{(i)(j)}(z) \mathrm{d}z, \tag{9}$$

where γ is the rate of fiber twisting. By solving power coupling equations, the crosstalk between the two modes can be expressed as

$$\mathrm{XT}_{mn}^{(i)(j)} = 10 \log_{10}[\tanh(\bar{h}_{mn}^{(i)(j)} z)]. \tag{10}$$

The crosstalk values calculated by using the average power coupling coefficient were demonstrated to agree well with experimental results [78]. It provides a fast and accurate method for estimating inter-core crosstalk in bent and twisted MCFs.

Using the average power coupling coefficient, we can analyze the variation trend of inter-core crosstalk in MRCFs. Figure 14a illustrates the inter-core crosstalk between two identical fiber cores. Each fiber core can support up to the fourth-order OAM mode group. We can find that the relatively heavy crosstalk exists between the modes in the same mode group, which is because modes having similar Δn_{eff} easily meet the phase-matching condition. In addition, the crosstalk increases with the mode order. As the higher-order OAM mode has a larger fraction of energy distribution in the cladding region, it can "feel" more of refractive index perturbation from the adjacent cores. Therefore, the overlapping integration is greater between the higher-order OAM modes. The maximum crosstalk exist between $\mathrm{OAM}_{+4,1}^{+}$ and $\mathrm{OAM}_{-4,1}^{+}$ with the value of -47 dB/m. The variation of the maximum inter-core crosstalk versus the size and position of the trench in the 5-mode-group MRCF is shown in Figure 14b. It is shown that increasing the width of the trench can rapidly reduce the crosstalk, and the gap between core and trench also helps to decrease the crosstalk.

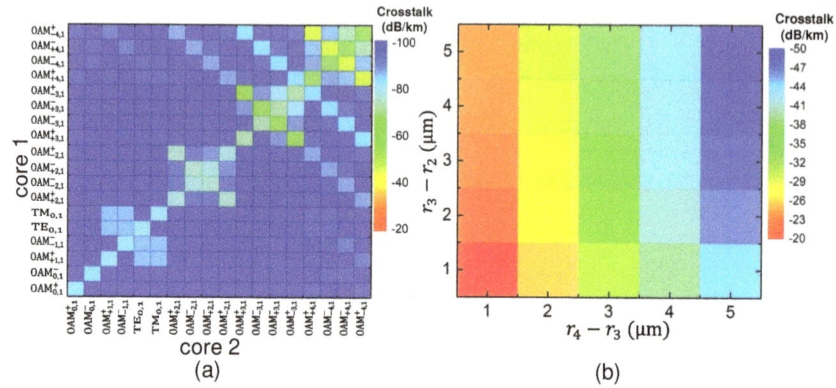

Figure 14. Inter-core crosstalk analysis at the wavelength of 1550 nm when the core-to-core interval is 40 μm. (**a**) The crosstalk between modes in two ring cores with the parameters $r_1 = 4.95$ μm, $r_2 = 9$ μm, $r_3 = 11$ μm, $r_4 = 16$ μm, $\Delta_{co} = 1\%$, $\Delta_{ce} = 0\%$, and $\Delta_{tr} = -0.7\%$. (**b**) The variation of crosstalk between $OAM^{+}_{+4,1}$ and $OAM^{+}_{-4,1}$ in two identical adjacent cores versus the core-trench gap ($r_3 - r_2$) and the trench width ($r_4 - r_3$).

In addition, the inter-mode crosstalk between different OAM modes in the same fiber core was analyzed in strongly guiding ring-core fibers recently [79]. The AH-RCF is a representative of strongly guiding fibers. The large air–silica refractive index contrast can effectively separate the n_{eff} of OAM modes in the same mode group, as discussed in Section 2. However, with the increase of refractive index difference between fiber core and neighboring materials, the purity of OAM modes will decrease. Essentially, an OAM mode in optical fiber has two components contributing opposite circularly polarized OAM states of orders l and $l + 2$. According to Equation (2), for HE-synthesis OAM modes, the spin-orbital aligned component with the OAM order l is dominant, while the spin-orbital anti-aligned component with the OAM order $l + 2$ is relatively weaker. However, for EH-synthesis OAM modes, the weight of the compositions is just opposite to that of HE-synthesis OAM modes. With the increase of the mode guiding level in optical fiber, the intensity of the weak modal component will grow, approaching the dominant modal intensity. Therefore, it is necessary to analyze OAM mode crosstalk in a strongly guiding condition.

The inter-mode crosstalk of OAM modes can be calculated by either coupled-mode theory or coupled-power theory. The only difference from the inter-core crosstalk is that the refractive index perturbation does not originate from adjacent cores (represented by $n^{(j)}$ in Equation (6)) but is caused by stress birefringence, core ellipticity, diameter fluctuation, core axis deviation, and so on. In [79], the OAM modes crosstalk are calculated under the influence of stress birefringence and core ellipticity. The normalized mode coupling coefficients are shown in Figure 15. Stress birefringence and core ellipticity induce the same coupling pattern among OAM modes, while the intensities are different. In the case of stress birefringence, the dominate-dominate coupling mainly occurs between cross-polarized OAM modes having the same OAM topological charge. When the perturbation is core ellipticity, the dominate-dominate coupling only happens between OAM modes having a OAM topological charge difference of two. Moreover, the two mode coupling coefficients caused by perturbations have opposite signs for higher-order OAM modes in the region close to the main diagonal. However, the ellipticity-caused coupling close to the main diagonal is weaker because it is induced by dominant-weak interaction. The two distributions indicate that the effects of birefringence and ellipticity tend to partially compensate one another.

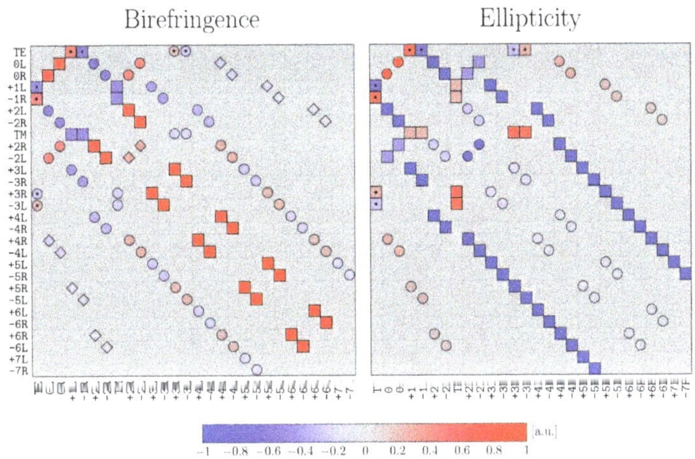

Figure 15. The calculated mode coupling coefficient under the influence of stress birefringence and core ellipticity. "□" represents the coupling between dominant components, "○" is the interaction between dominant and weak components, and "◇" indicates the coupling between weak components. Reprinted with permission from Ref. [79] © The Optical Society.

5. All-Fiber OAM Generator

All-fiber OAM mode generators have the advantages of compatibility with fiber systems, low insertion loss, easy integration, and low cost. Therefore, although OAM beams can be efficiently generated by commercial SLMs, all-fiber OAM mode generators still receive increasing attention in recent years.

Using LPFGs is an effective way to convert the fundamental mode to OAM modes in optical fibers. Our group proposed to adopt asymmetric LPFGs to directly generate higher-order OAM modes from the fundamental. We first consider the generation of second-order OAM modes. In a four-mode fiber, the grating pitch (Λ) is determined by the phase-matching equation $\Lambda = \lambda/(n_{0,1} - n_{2,1})$, where $n_{0,1}$ and $n_{2,1}$ are the effective refractive index of the $LP_{0,1}$ and $LP_{2,1}$ modes, respectively. As shown in Figure 16a, the calculated Λ in the vicinity of $\lambda = 1530$ nm is ~520 μm. The fabrication of the asymmetric LPFG was using a focused CO_2 laser with high power periodically irradiating a 4-mode fiber from one side [24]. Figure 16(a2) shows the image of asymmetric LPFG; the strong geometrical deformation can cause a large modulation on n_{eff} with the value about -1.1×10^{-2}. In addition, since the deformation is asymmetric, the modulation is not uniform but azimuthal varying. As a result, the power of $LP_{0,1}$ mode can be effectively coupled into $LP_{2,1}$ mode. In the same manner, an asymmetric LPFG can be fabricated on a 6-mode fiber to convert $LP_{0,1}$ to $LP_{3,1}$. Since these two modes have a larger Δn_{eff}, the required Λ becomes smaller. As shown in Figure 16b, the calculated Λ is ~200 μm in the vicinity of $\lambda = 1543$ nm. To make large geometrical deformations, the six-mode fiber was scanned six times by a CO_2 laser with increasing power step by step [25]. Figure 16(b2) shows the image of asymmetric LPFG with $\Lambda = 200$ μm. The large geometrical deformation breaks the orthogonality between $LP_{0,1}$ and $LP_{3,1}$, which enhances the coupling between these two modes. The fabricated asymmetric LPFGs tested have mode conversion efficiencies of 99.7% and 99.8% for $LP_{2,1}$ and $LP_{3,1}$, respectively.

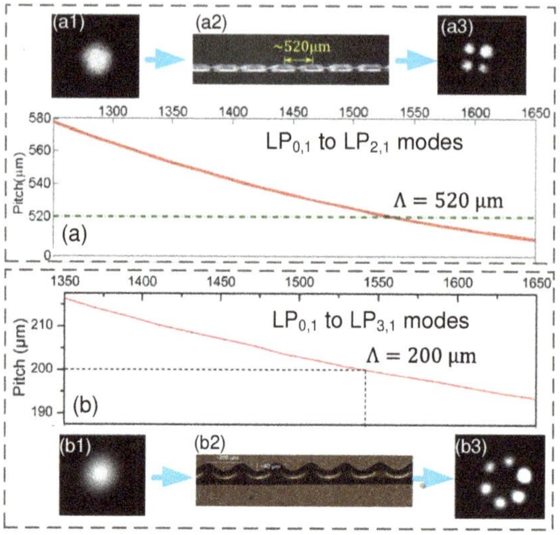

Figure 16. (**a**) Calculated grating pitch for the mode coupling from $LP_{0,1}$ to $LP_{2,1}$ modes in a four-mode fiber. (**a1**,**a3**) The patterns of $LP_{0,1}$ and $LP_{2,1}$ modes. (**a2**) The image of the fabricated LPFG with $\Lambda = 520$ μm. Adapted with permission from Ref. [24] © The Optical Society. (**b**) Calculated grating pitch for the mode coupling from $LP_{0,1}$ to $LP_{3,1}$ modes in a six-mode fiber. (**b1**,**b3**) The patterns of $LP_{0,1}$ and $LP_{3,1}$ modes. (**b2**) The image of the fabricated LPFG with $\Lambda = 200$ μm. Adapted with permission from Ref. [25] © The Optical Society.

The experimental setups for the generation and detection of OAM modes using the fabricated asymmetric LPFGs are shown in Figure 17a. A tunable laser was used as an optical source. Furthermore, the input light was divided into a generation branch (upper one) and a reference branch (lower one) using a 90:10 optical coupler. The 90% energy branch was used for the OAM generation. A single-mode fiber (SMF) was coaxially spliced with the under test few-mode fiber (FMF) to stimulate the fundamental mode. Polarization Controller 1 (PC1) was used to adjust the polarization states of the incident fundamental mode to the asymmetric LPFG. It was used to determine the power ratio of the even and odd polarization states of the generated higher-order LP mode. One end of the LPFG was clamped by a fiber holder, while the other end was mounted at the center of a rotator. The relative phase difference between the even and odd polarization states of the target LP mode can be adjusted by twisting the LPFG with the rotator. The generated mode was collimated by a microscope and then transmitted into the charge-coupled device (CCD) camera. The measured mode intensity patterns are shown in Figure 17(b1–e1). To verify the number of the OAM topological charge, a reference Gaussian beam in the lower branch was adjusted into the same power level as the generated OAM beam and was combined with the OAM beam using a beam splitter (BS). Then, we can see the OAM order according to the interference patterns. Figure 17(b2,c2) indicates that the generated modes were $OAM_{+2,1}$ and $OAM_{-2,1}$. In the same way, the interference patterns in Figure 17(d2,e2) show that the generated modes were $OAM_{+3,1}$ and $OAM_{-3,1}$. These works demonstrated a feasible way to produce low-cost compact all-fiber OAM generators.

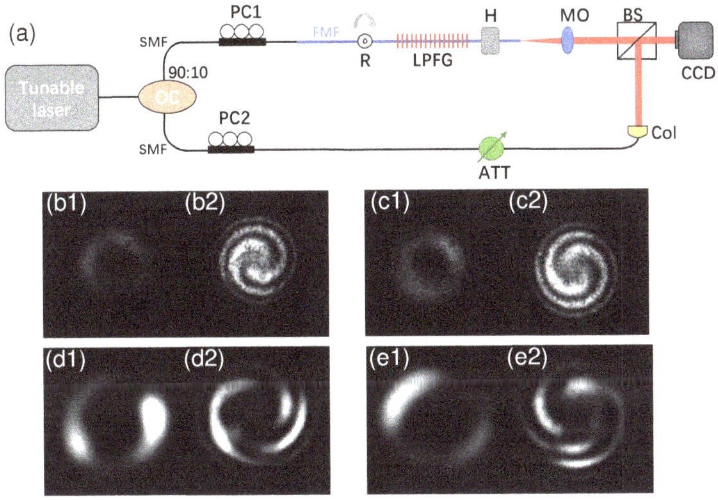

Figure 17. (**a**) Schematic diagram of the experimental setups for the generation and detection of OAM modes using asymmetric LPFGs. OC, optical coupler; H, holder; R, rotator; MO, microscope objective; ATT, attenuator; Col, collimator. (**b1–e1**) The intensity profiles of the generated modes. (**b2–e2**) The interference patterns of the generated OAM modes with a reference Gaussian beam. Adapted with permission from Refs. [24,25] © The Optical Society.

In addition, many valuable works on all-fiber OAM generators were reported by other groups. In 2015, Li et al. first proposed a controllable all-fiber OAM mode converter [23]. As shown in Figure 18a, at the input end, a SMF is spliced coaxially with a two-mode fiber to efficiently stimulate the fundamental mode in the two-mode fiber. Then, a mechanical long-period grating is applied on the two-mode fiber to convert the fundamental mode to the $LP_{1,1}$ mode. Then, a rotator is used to adjust the intensity orientation of the $LP_{1,1}$ mode to the pressure applying slabs. When the orientation angle is 45°, the energy of the vertical and transverse polarization components are the same. By applying pressure to modify the circular core into an elliptical core, a Δn_{eff} is generated between the vertical and transverse polarization components.

Then, the first-order OAM mode can be generated when the two components have a $\pm\pi/2$ phase difference at the output end. Moreover, the grating pitch applied to the fiber can be tuned by placing the fiber at different angles to the mechanical long-period grating. It achieves a scalable scheme to generate OAM modes at different wavelengths.

However, the mechanical long-period grating is relatively bulky, and a more compact OAM generator by using an all-fiber mode selective coupler was proposed in 2018 [22]. In Figure 18b, the coupler is fabricated by tapering a SMF and a graded-index FMF together with a proper diameter ratio. The FMF used here is in the strong guiding condition, which can well separate the n_{eff} of $HE_{2,1}$ from those of $TE_{0,1}$ and $TM_{0,1}$. Hence, the coupling efficiency into $HE_{2,1}$ is increased. In the experiment, a polarization controller was used to adjust the purity of the generated first-order OAM mode. Finally, a purity of ~95% was obtained.

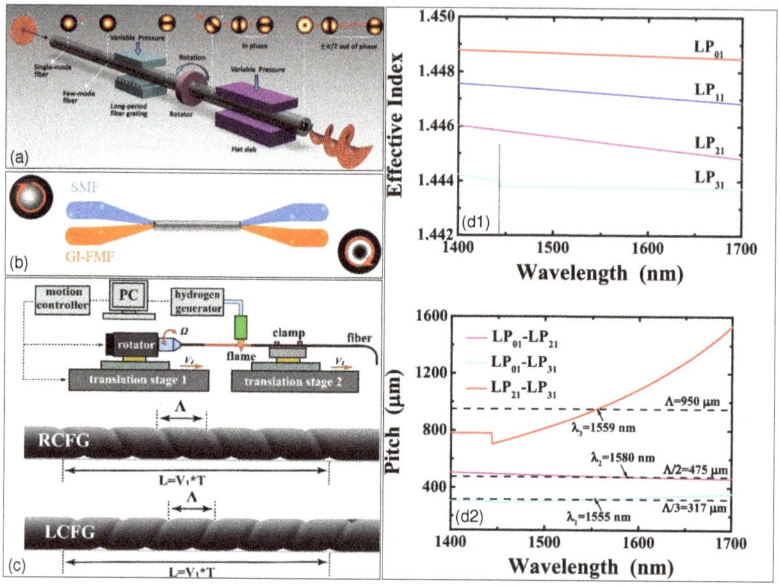

Figure 18. All-fiber OAM mode generators. (**a**) Principle of the controllable all-fiber OAM beam generator based on mechanical long-period grating. Adapted with permission from Ref. [23] © The Optical Society. (**b**) Schematic of the all-fiber mode selective coupler. Adapted with permission from Ref. [22] © The Optical Society. (**c**) Schematic of the helical fiber grating fabrication setup using a hydrogen-oxygen flame, and the periodic helical structures of right-handed chiral and left-handed chiral gratings. Adapted with permission from Ref. [26] © The Optical Society. (**d1**) n_{eff} of LP modes in a four-mode fiber. (**d2**) The resonant pitches of a helical fiber grating versus the wavelength. Adapted with permission from Ref. [27] © The Optical Society.

The above-mentioned all-fiber OAM generators are sensitive to the polarization states of fiber modes, which may have instability in applications. In 2019, a polarization-independent OAM generator based on a helical fiber grating was proposed [26]. As illustrated in Figure 18c, the helical fiber grating was produced by twisting a FMF during hydrogen-oxygen flame heating. Through carefully setting the rotation speed of the rotator and the moving speeds of the two translation stages, the pitch of the helical grating can be well kept. In their works, the resonant wavelengths of the right-handed helical grating and left-handed helical grating are 1550.9 and 1554.5 nm, respectively. The two samples were tested in experiments. The results show that the first-order OAM modes can be generated with a mode purity of around 93%. Furthermore, the polarization state of the output OAM modes is consistent with the polarization state of the input light. Since no coupling was observed between the +1-order and −1-order OAM modes, it is indicated that the OAM chirality was determined only by the twisting direction of the helical fiber grating.

Recently, a further work on helical fiber grating based all-fiber OAM generator was reported. By taking advantage of the second and third diffraction orders of a helical fiber grating, the second- and third-order OAM modes can be simultaneously generated in a single fiber [27]. Figure 18(d1,d2) shows the n_{eff} of modes and grating resonant conditions in a four-mode fiber. In Figure 18(d1), we can see that the fiber can only support the $LP_{3,1}$ mode when the wavelength is smaller than 1440 nm, while the other three LP modes can be well guided in the fiber in the wavelength band from 1400 to 1700 nm. The grating resonant pitches and Δn_{eff} of modes are shown in Figure 18(d2). The resonant coupling from $LP_{0,1}$ to $LP_{2,1}$ occurs at the wavelength of 1580 nm, and that from $LP_{0,1}$ to the cladding mode (noted as $LP_{3,1}$ to be consistent with the notation in Figure 18(d2) occurs at the wavelength of 1555 nm. Therefore, the fundamental fiber mode can be converted directly into $OAM_{2,1}$ and

$OAM_{3,1}$ at the two specific wavelengths, respectively. From the experiment, the maximum purities for the $OAM_{2,1}$ and $OAM_{3,1}$ were 95% and 98%. In addition to OAM modes generators, chiral fiber gratings can also be applied as torsion sensors, band-rejection filters, wave plates and linear- and circular-light polarizers. The mode-coupling theories, fabrication techniques and applications of helical fiber gratings can be found in a detailed and comprehensive review contributed by Zhao Hua and Li Hongpu [80].

6. Conclusions and Prospects

The stability of OAM mode transmission in optical fibers has obtained great improvements in the last decade. Using an AH-RCF, the best record for stable OAM mode transmission reached 13.4 km [36]. In addition, many simulation works have been done on microstructure fibers for OAM mode transmission. Some of these designs were manufactured and tested. Although the produced microstructure fibers have geometrical deformations from the ideal design, they performed the transmission ability for OAM modes in experiments. Therefore, these exploratory works provide flexible designs for OAM fiber and expand the application of OAM modes in the aspects of enhancing data transmission capacity in fiber links, OAM-based fiber imaging, and vortex-mode fiber lasers. In addition, the studies on the all-fiber OAM generator provide alternative choices to commercially available free-space devices. Those all-fiber devices are suitable for applications requiring low insertion loss, good alignment stability, and high integration.

However, the stability of OAM modes in optical fibers needs to be further improved for long-haul, high-capacity data transmission. Although the first-order perturbation theory reveals that increasing the refractive index contrast between core and cladding can split the n_{eff} between the near-degenerate OAM modes, the variation trends of Δn_{eff} among near-degenerate OAM modes with different OAM orders are not consistent. Therefore, there is no mature design scheme for obtaining the optimized combination of guiding mode number and mode stability level. On the other hand, improving the fiber fabrication technique is necessary for improving mode stability. How to keep the hollow structures of AH-RCF and microstructure RCF during fiber fabrication still need to be studied.

For fiber-optic communications, in recent years, a scheme of SDM assisted by low-complexity modular multi-input multi-output (MIMO) equalization was proposed for recovering the signal from the crosstalk among the near-degenerate OAM modes [81]. According to the maximum degeneracy of a fiber mode group, a 4×4 MIMO is sufficient for signal processing. In this scheme, the crosstalk between the near-degenerate OAM modes is no longer the crucial issue; instead, the crosstalk between OAM mode groups becomes the main barrier for increasing the signal quality. Therefore, the fibers suitable for this scheme should achieve low crosstalk between OAM mode groups while having low differential group delay between the near-degenerate OAM modes. The fibers will need no ultra-high refractive index contrast structures such as the air-hole to separate the n_{eff} of near-degenerate OAM modes. Thus, these fibers may have higher fabrication reliability and can be expanded to a multi-core design.

Finally, we should note that active OAM fiber-optic devices such as OAM fiber amplifiers are also important for building long-haul OAM mode fiber links. Although there are many challenges related to mode stability, generation efficiency, multiplexing, demultiplexing, and amplification in the field of OAM, it provides new frontiers for optical communication systems and microscopy techniques.

Author Contributions: Writing–original draft preparation, Z.W.; writing–review and editing, J.T. and C.Y.; investigation and resources, S.G. and Z.L.; supervision, C.L. All authors have read and agreed to the published version of the manuscript.

Funding: This work was supported by National Key R&D Program of China 2019YFA0706300 from Ministry of Science and Technology, China; National Natural Science Foundation of China (NSFC) (Grant No. U1701661 and U2001601); and Guangzhou basic and applied basic research Foundation (Grant No. 202002030327).

Conflicts of Interest: The authors declare no conflict of interest.

References

1. Beth, R.A. Mechanical detection and measurement of the angular momentum of light. *Phys. Rev.* **1936**, *50*, 115–125. [CrossRef]
2. Yao, A.M.; Padgett, M.J. Orbital angular momentum: Origins, behavior and applications. *Adv. Opt. Photonics* **2011**, *3*, 161–204. [CrossRef]
3. Allen, L.; Beijersbergen, M.W.; Spreeuw, R.J.C.; Woerdman, J.P. Orbital angular momentum of light and the transformation of Laguerre-Gaussian laser modes. *Phys. Rev. A* **1992**, *45*, 8185–8189. [CrossRef] [PubMed]
4. Franke-Arnold, S.; Allen, L.; Padgett, M. Advances in optical angular momentum. *Laser Photonics Rev.* **2008**, *2*, 299–313. [CrossRef]
5. Kuga, T.; Torii, Y.; Shiokawa, N.; Hirano, T.; Shimizu, Y.; Sasada, H. Novel optical trap of atoms with a doughnut beam. *Phys. Rev. Lett.* **1997**, *78*, 4713–4716. [CrossRef]
6. Paterson, L.; MacDonald, M.P.; Arlt, J.; Sibbett, W.; Bryant, P.E.; Dholakia, K. Controlled rotation of optically trapped microscopic particles. *Science* **2001**, *292*, 912–914. [CrossRef]
7. Garcés-Chávez, V.; McGloin, D.; Padgett, M.J.; Dultz, W.; Schmitzer, H.; Dholakia, K. Observation of the transfer of the local angular momentum density of a multiringed light beam to an optically trapped particle. *Phys. Rev. Lett.* **2003**, *91*, 093602. [CrossRef]
8. Mair, A.; Vasiri, A.; Weihs, G.; Zeilinger, A. Entanglement of the orbital angular momentum states of photons. *Nature* **2001**, *412*, 313–316. [CrossRef] [PubMed]
9. Mirhosseini, M.; Maga na-Loaiza, O.S.; O'Sullivan, M.N.; Rodenburg, B.; Malik, M.; Lavery, M.P.J.; Padgett, M.J.; Gauthier, D.J.; Boyd, R.W. High-dimensional quantum cryptography with twisted light. *New J. Phys.* **2015**, *17*, 033033. [CrossRef]
10. Jesacher, A.; Fürhapter, S.; Bernet, S.; Ritsch-Marte, M. Shadow effects in spiral phase contrast microscopy. *Phys. Rev. Lett.* **2005**, *94*, 233902. [CrossRef]
11. Fürhapter, S.; Jesacher, A.; Bernet, S.; Ritsch-Marte, M. Spiral interferometry. *Opt. Lett.* **2005**, *30*, 1953–1955. [CrossRef]
12. Gibson, G.; Courtial, M.; Padgett, J.; Vasnetsov, M.; Pas'ko, V.; Barnett, S.M.; Franke-Arnold, S. Free-space information transfer using light beams carrying orbital angular momentum. *Opt. Express* **2004**, *12*, 5448–5456. [CrossRef] [PubMed]
13. Wang, J.; Yang, J.Y.; Fazal, I.M.; Ahmed, N.; Yan, Y.; Huang, H.; Ren, Y.; Yue, Y.; Dolinar, S.; Tur, M.; et al. Terabit free-space data transmission employing orbital angular momentum multiplexing. *Nat. Photonics* **2012**, *6*, 488–496. [CrossRef]
14. Bozinovic, N.; Yue, Y.; Ren, Y.; Tur, M.; Kristensen, P.; Huang, H.; Willner, A.E.; Ramachandran, S. Terabit-scale orbital angular momentum mode division multiplexing in fibers. *Science* **2013**, *340*, 1545–1548. [CrossRef] [PubMed]
15. Willner, A.E.; Huang, H.; Yan, Y.; Ren, Y.; Ahmed, N.; Xie, G.; Bao, C.; Li, L.; Cao, Y.; Wang, J.; et al. Optical communications using orbital angular momentum beams. *Adv. Opt. Photonics* **2015**, *7*, 66–106. [CrossRef]
16. Richardson, D.J.; Fini, J.M.; Nelson, L.E. Space-division multiplexing in optical fibers. *Nat. Photonics* **2013**, *7*, 354–362. [CrossRef]
17. Sueda, K.; Miyaji, G.; Miyanaga, N.; Nakatsuka, M. Laguerre-Gaussian beam generated with a multilevel spiral phase plate for high intensity laser pulses. *Opt. Express* **2004**, *12*, 3548–3553. [CrossRef]
18. Ruffato, G.; Massari, M.; Parisi, G.; Romanato, F. Test of mode-division multiplexing and demultiplexing in free-space with diffractive transformation optics. *Opt. Express* **2017**, *25*, 7859–7868. [CrossRef] [PubMed]
19. Li, W.; Morgan, K.S.; Li, Y.; Miller, J.K.; White, G.; Watkins, R.J.; Johnson, E.G. Rapidly tunable orbital angular momentum (OAM) system for higher order Bessel beams integrated in time (HOBBIT). *Opt. Express* **2019**, *27*, 3920–3934. [CrossRef] [PubMed]
20. Cai, X.; Wang, J.; Strain, M.J.; Johnson-Morris, B.; Zhu, J.; Sorel, M.; O'Brien, J.L.; Thompson, M.G.; Yu, S. Integrated compact optical vortex beam emitters. *Science* **2012**, *338*, 363–365. [CrossRef]
21. Pidishety, S.; Pachava, S.; Gregg, P.; Ramachandran, S.; Brambilla, G.; Srinivasan, B. Orbital angular momentum beam excitation using an all-fiber weakly fused mode selective coupler. *Opt. Lett.* **2017**, *42*, 4347–4350. [CrossRef]
22. Heng, X.; Gan, J.; Zhang, Z.; Li, J.; Li, M.; Zhao, H.; Qian, Q.; Xu, S.; Yang, Z. All-fiber stable orbital angular momentum beam generation and propagation. *Opt. Express* **2018**, *26*, 17429–17436. [CrossRef]
23. Li, S.; Mo, Q.; Hu, X.; Du, C.; Wang, J. Controllable all-fiber orbital angular momentum mode converter. *Opt. Lett.* **2015**, *40*, 4376–4379. [CrossRef] [PubMed]
24. Wu, H.; Gao, S.; Huang, B.; Feng, Y.; Huang, X.; Liu, W.; Li, Z. All-fiber second-order optical vortex generation based on strong modulated long-period grating in a four-mode fiber. *Opt. Lett.* **2017**, *42*, 5210–5213. [CrossRef] [PubMed]
25. He, X.; Tu, J.; Wu, X.; Gao, S.; Shen, L.; Hao, C.; Feng, Y.; Liu, W.; Li, Z. All-fiber third-order orbital angular momentum mode generation employing an asymmetric long-period fiber grating. *Opt. Lett.* **2020**, *45*, 3621–3624. [CrossRef] [PubMed]
26. Zhang, Y.; Bai, Z.; Fu, G.; Liu, S.; Tang, J.; Yu, J.; Liao, C.; Wang, Y.; He, J.; Wang, Y. Polarization-independent orbital angular momentum generator based on a chiral fiber grating. *Opt. Lett.* **2019**, *44*, 61–64. [CrossRef] [PubMed]
27. Detani, T.; Zhao, H.; Wang, P.; Suzuki, T.; Li, H. Simultaneous generation of the second- and third-order OAM modes by using a high-order helical long-period fiber grating. *Opt. Lett.* **2021**, *46*, 949–952. [CrossRef]
28. Zhang, N.; Yuan, C.; Burge, R.E. Extending the detection range of optical vortices by Dammann vortex gratings. *Opt. Lett.* **2010**, *35*, 3495–3497. [CrossRef]
29. Lei, T.; Zhang, M.; Li, Y.; Jia, P.; Liu, G.N.; Xu, X.; Li, Z.; Min, C.; Lin, J.; Yu, C.; et al. Massive individual orbital angular momentum channels for multiplexing enabled by Dammann gratings. *Light Sci. Appl.* **2015**, *4*, e257. [CrossRef]

30. Fontaine, N.K.; Ryf, R.; Chen, H.; Neilson, D.T.; Kim, K.; Carpenter, J. Laguerre-Gaussian mode sorter. *Nat. Commun.* **2019**, *10*, 1865.
31. Lin, Z.; Wen, Y.; Chen, Y.; Yu, S. Transmissive Multi-plane Light Conversion for Demultiplexing Orbital Angular Momentum Modes. In Proceedings of the CLEO, SF1J5, Washington, DC, USA, 10–15 May 2020.
32. Ren, Y.; Huang, H.; Xie, G.; Ahmed, N.; Yan, Y.; Erkmen, B.I.; Chandrasekaran, N.; Lavery, M.P.J.; Steinhoff, N.K.; Tur, M.; et al. Atmospheric turbulence effects on the performance of a free space optical link employing orbital angular momentum multiplexing. *Opt. Lett.* **2013**, *38*, 4062–4065. [CrossRef] [PubMed]
33. Snyder, A.W. Coupled-mode theory for optical fibers. *J. Opt. Soc. Am. A* **1972**, *62*, 1267–1277. [CrossRef]
34. Ramachandran, S.; Kristensen, P.; Yan, M.F. Generation and propagation of radially polarized beams in optical fibers. *Opt. Lett.* **2009**, *34*, 2525–2527. [CrossRef]
35. Gregg, P.; Kristensen, P.; Ramachandran, S. Conservation of orbital angular momentum in air-core optical fibers. *Optica* **2015**, *2*, 267–270. [CrossRef]
36. Gregg, P.; Kristensen, P.; Ramachandran, S. 13.4 km OAM state propagation by recirculation fiber loop. *Opt. Express* **2016**, *24*, 18938–18947. [CrossRef]
37. Zhang, H.; Zhang, W.; Xi, L.; Tang, X.; Zhang, X.; Zhang, X. A new type circular photonic crystal fiber for orbital angular momentum mode transmission. *IEEE Photonics Technol. Lett.* **2016**, *28*, 1426–1429. [CrossRef]
38. Xi, X.M.; Wong, G.K.L.; Frosz, M.H.; Babic, F.; Ahmed, G.; Jiang, X.; Euser, T. G.; Russell, P.S.J. Orbital-angular-momentum-preserving helical Bloch modes in twisted photonic crystal fiber. *Optica* **2014**, *1*, 165–169. [CrossRef]
39. Tu, J.; Liu, Z.; Gao, S.; Wang, Z.; Zhang, J.; Zhang, B.; Li, J.; Liu, W.; Tam, H.; Li, Z.; et al. Ring-core fiber with negative curvature structure supporting orbital angular momentum modes. *Opt. Express* **2019**, *27*, 20358–20372. [CrossRef]
40. Tu, J.; Gao, S.; Wang, Z.; Liu, Z.; Li, W.; Du, C.; Liu, W.; Li, Z.; Yu, C.; Tam, H.; et al. Bend-insensitive grapefruit-type holey ring-core fiber for weakly-coupled OAM mode division multiplexing transmission. *IEEE J. Light. Technol.* **2020**, *38*, 4497–4503. [CrossRef]
41. Marcuse, D. Derivation of coupled power equations. *Bell Syst. Tech. J.* **1972**, *51*, 229–237. [CrossRef]
42. Snyder, A.W.; Love, J.D. *Optical Waveguide Theory*, 1st ed.; Chapman and Hall: New York, NY, USA, 1983; pp. 248–254.
43. Ramachandran, S.; Kristensen, P. Optical vortices in fiber. *Nanophotonics* **2013**, *2*, 455–474. [CrossRef]
44. Zhang, Z.; Gan, J.; Heng, X.; Wu, Y. Li, Q.; Qian, L.; Chen, D.; Yang, Z. Optical fiber design with orbital angular momentum light purity higher than 99.9%. *Opt. Express* **2015**, *23*, 29331–29341. [CrossRef]
45. Chen, S.; Wang, J. Theoretical analyses on orbital angular momentum modes in conventional graded-index multimode fibre. *Sci. Rep.* **2017**, *7*, 3990. [CrossRef] [PubMed]
46. Zhou, W.; Cao, H.; Wang, L.; Wang, J. All-Fiber Orbital Angular Momentum (OAM) Functional Devices for Mode-Division (De)Multiplexing in Conventional Graded-Index Multimode Fiber. In Proceedings of the 2019 Optical Fiber Communications Conference and Exhibition (OFC), Th3D.5, San Diego, CA, USA, 3–7 March 2019.
47. Bozinovic, N.; Golowich, S.; Kristensen, P.; Ramachandran, S. Control of orbital angular momentum of light with optical fibers. *Opt. Lett.* **2012**, *37*, 2451–2453. [CrossRef] [PubMed]
48. Brunet, C.; Ung, B.; Bélanger P.; Messaddeq, Y.; LaRochelle, S.; Rusch, L.A. Vector mode analysis of ring-core fibers: Design tools for spatial division multiplexing. *IEEE J. Light. Technol.* **2014**, *32*, 4046–4057. [CrossRef]
49. Gregg, P.; Kristensen, P.; Golowich, S.E.; Olsen, J.Ø.; Steinvurzel, P.; Ramachandran, S. Stable transmission of 12 OAM states in air-core fiber. In Proceedings of the CLEO, CTu2K.2, San Jose, CA, USA, 9–14 June 2013.
50. Brunet, C.; Vaity, P.; Messaddeq, Y.; LaRochelle, S.; Rusch, L.A. Design, fabrication and validation of an OAM fiber supporting 36 states. *Opt. Express* **2014**, *22*, 26117–26127. [CrossRef]
51. Ramachandran, S.; Gregg, P.; Kristensen, P.; Golowich, S.E. On the scalability of ring fiber designs for OAM multiplexing. *Opt. Express* **2015**, *23*, 3721–3730. [CrossRef]
52. Brunet, C.; Ung, B.; Wang, L.; Messaddeq, Y.; LaRochelle, S.; Rusch, A.L. Design of a family of ring-core fibers for OAM transmission studies. *Opt. Express* **2015**, *23*, 10553–10563. [CrossRef]
53. Kawakami, S.; Ikeda, M. Transmission characteristics of a two-mode optical waveguide. *IEEE J. Quantum Electron.* **1978**, *QE-14*, 608–614. [CrossRef]
54. Ung, B.; Vaity, P.; Wang, L.; Messaddeq, Y.; Rusch, A.L.; LaRochelle, S. Few-mode fiber with inverse-parabolic graded-index profile for transmission of OAM-carrying modes. *Opt. Express* **2014**, *22*, 18044–18055. [CrossRef]
55. Knight, J.C.; Birks, T.A.; Russell, P.S.J.; Atkin, D.M. All-silica single-mode optical fiber with photonic crystal cladding. *Opt. Lett.* **1996**, *21*, 1547–1549. [CrossRef]
56. Russell, P.S.J. Photonic-Crystal fibers. *IEEE J. Light. Technol.* **2006**, *24*, 4729–4749. [CrossRef]
57. Cregan, R.F.; Mangan, B.J.; Knight, J.C.; Birks, T.A.; Russell, P.S.J.; Roberts, P.J.; Allan, D.C. Single-mode photonic band gap guidance of light in air. *Science* **1999**, *285*, 1537–1539. [CrossRef] [PubMed]
58. Couny, F.; Benabid, F.; Light, P.S. Large-pitch kagome-structured hollow-core photonic crystal fiber. *Opt. Lett.* **2006**, *31*, 3574–3576. [CrossRef]
59. Wang, Y.Y.; Wheeler, N.V.; Couny, F.; Roberts, P.J.; Benabid, F. Low loss broadband transmission in hypocycloid-core Kagome hollow-core photonic crystal fiber. *Opt. Lett.* **2011**, *36*, 669–671. [CrossRef] [PubMed]

60. Gao, S.F.; Wang, Y.Y.; Ding, W.; Jiang, D.L.; Gu, S.; Zhang, X.; Wang, P. Hollow-core conjoined-tube negative-curvature fiber with ultralow loss. *Nat. Commun.* **2018**, *9*, 1–6. [CrossRef] [PubMed]
61. Jasion, G.T.; Bradley, T.D.; Harrington, K.; Sakr, H.; Chen, Y.; Fokoua, E.N.; Davidson, I.A.; Taranta, A.; Hayes, J.R.; Richardon, D.J.; et al. Hollow core NANF with 0.28 dB/km attenuation in the C and L bands. In Proceedings of the Optical Fiber Communication Conference, Th4B-4, San Diego, CA, USA, 8–12 March 2020.
62. Wong, G.K.L.; Kang, M.S.; Lee, H.W.; Biancalana, F.; Conti, C.; Weiss, T.; Russell, P.S.J. Excitation of orbital angular momentum resonances in helically twisted photonic crystal fiber. *Science* **2012**, *337*, 446–449. [CrossRef]
63. Chen, C.; Zhou, G.; Zhou, G.; Xu, M.; Hou, Z.; Xia, C.; Yuan, J. A multi-orbital-angular-momentum multi-ring micro-structured fiber with ultra-high-density and low-level crosstalk. *Opt. Commun.* **2016**, *368*, 27–33. [CrossRef]
64. Zhou, G.; Zhou, G.; Chen, C.; Xu, M.; Xia, C.; Hou, Z. Design and analysis of a microstructure ring fiber for orbital angular momentum transmission. *IEEE Photonics J.* **2016**, *8*, 7802512. [CrossRef]
65. Tian, W.; Zhang, H.; Zhang, X.; Xi, L.; Zhang, W.; Tang, X. A circular photonic crystal fiber supporting 26 OAM modes. *Opt. Fiber Technol.* **2016**, *30*, 184–189. [CrossRef]
66. Tandjè, A.; Yammine, J.; Dossou, M.; Bouwmans, G.; Baudelle, K.; Vianou, A.; Andresen, E.R.; Bigot, L. Ring-core photonic crystal fiber for propagation of OAM modes. *Opt. Lett.* **2019**, *44*, 1611–1614. [CrossRef]
67. Pryamikov, A.D.; Biriukov, A.S.; Kosolapov, A.F.; Plotnichenko, V.G.; Semjonov, S.L.; Dianov, E.M. Demonstration of a waveguide regime for a silica hollow-core microstructured optical fiber with a negative curvature of the core boundary in the spectral region >3.5 μm. *Opt. Express* **2011**, *19*, 1441–1448. [CrossRef] [PubMed]
68. Kolyadin, A.N.; Kosolapov, A.F.; Pryamikov, A.D.; Biriukov, A.S.; Plotnichenko, V.G.; Dianov, E.M. Light transmission in negative curvature hollow core fiber in extremely high material loss region. *Opt. Express* **2013**, *21*, 9514–9519. [CrossRef] [PubMed]
69. Poletti, F. Nested antiresonant nodeless hollow core fiber. *Opt. Express* **2014**, *22*, 23807–23828. [CrossRef]
70. Litchinitser, N.M.; Abeeluck, A.K.; Headley, C.; Eggleton, B.J. Antiresonant reflecting photonic crystal optical waveguides. *Opt. Lett.* **2002**, *27*, 1592–1594. [CrossRef] [PubMed]
71. Wang, Z.; Tu, J.; Liu, Z.; Yu, C.; Lu, C. Design of weakly coupled two-mode hollow-core antiresonant fiber with low loss. *IEEE J. Light. Technol.* **2020**, *38*, 864–874. [CrossRef]
72. Koshiba, M.; Saitoh, K.; Kokubun, Y. Heterogeneous multi-core fibers: Proposal and design principle. *IEICE Electron. Express* **2009**, *6*, 98–103. [CrossRef]
73. Takenaga, K.; Arakawa, Y.; Tanigawa, S.; Guan, N.; Matsuo, S.; Saitoh, K.; Koshiba, M. Reduction of crosstalk by trench-assisted multi-core fiber. In Proceedings of the Optical Fiber Communication Conference, OWJ4, Los Angeles, CA, USA, 6–10 March 2011.
74. Tu, J.; Saitoh, K.; Koshiba, M.; Takenaga, K.; Matsuo, S. Design and analysis of large-effective-area heterogeneous trench-assisted multi-core fiber. *Opt. Express* **2012**, *20*, 15157–15170. [CrossRef]
75. Tu, J.; Saitoh, K.; Koshiba, M.; Takenaga, K.; Matsuo, S. Optimized design method for bend-insensitive heterogeneous trench-assisted multi-core fiber with ultra-low crosstalk and high core density. *IEEE J. Light. Technol.* **2013**, *31*, 2590–2598.
76. Tu, J.; Long, K.; Saitoh, K. An efficient core selection method for heterogeneous trench-assisted multi-core fiber. *IEEE Photonics Technol. Lett.* **2016**, *28*, 810–813. [CrossRef]
77. Koshiba, M.; Saitoh, K.; Takenaga, K.; Matsuo, S. Multi-core fiber design and analysis: Coupled-mode theory and coupled-power theory. *Opt. Express* **2011**, *19*, B102–B111. [CrossRef] [PubMed]
78. Koshiba, M.; Saitoh, K.; Takenaga, K.; Matsuo, S. Analytical expression of average power-coupling coefficient for estimating intercore crosstalk in multicore fibers. *IEEE Photonics J.* **2012**, *4*, 1987–1995. [CrossRef]
79. Guerra, G.; Lonardi, M.; Galtarossa, A.; Rusch, L.A.; Bononi, A.; Palmieri, L. Analysis of modal coupling due to birefringence and ellipticity in strongly guiding ring-core OAM fibers. *Opt. Express* **2019**, *27*, 8308–8326. [CrossRef] [PubMed]
80. Zhao, H.; Li, H. Advances on Mode-Coupling Theories, Fabrication Techniques, and Applications of the Helical Long-Period Fiber Gratings: A Review. *Photonics* **2021**, *8*, 8040106. [CrossRef]
81. Liu, J.; Zhu, G.; Zhang, J.; Wen, Y.; Wu, X.; Zhang, Y.; Chen, Y.; Cai, X.; Li, Z.; Hu, Z.; et al. Mode division multiplexing based on ring core optical fibers. *IEEE J. Quantum Electron.* **2018**, *54*, 0700118.

photonics

Article

A Broadband Gold-Coated Photonic Crystal Fiber Polarization Filter with a High Loss Ratio of Both Polarizations at 1550 and 1310 nm

Chao Wang [1,2,3,4,*], Guoxu Zhang [1], Zheng Wu [1], Yajing Zhang [1], Yiyang Zhang [1], Linghong Jiang [1,3,4] and Weihong Bi [2]

[1] Intelligence and Information Engineering College, Tangshan University, Tangshan 063000, China; zhangguoxu@tsc.edu.cn (G.Z.); wuzheng@tsc.edu.cn (Z.W.); zhangyajing@tsc.edu.cn (Y.Z.); zhangyiyang@tsc.edu.cn (Y.Z.); jianglinghong@tsc.edu.cn (L.J.)
[2] School of Information Science and Engineering, Yanshan University, Qinhuangdao 066004, China; bwhong@ysu.edu.cn
[3] Key Lab of Intelligent Data Information Processing and Control of Hebei Province, Tangshan University, Tangshan 063000, China
[4] Key Lab of Intelligent Motion Control System of Tangshan City, Tangshan University, Tangshan 063000, China
* Correspondence: wangchao_198504@sina.com; Tel.: +86-155-1204-8666

Abstract: A new kind of gold-coated hexagonal photonic crystal fiber polarization filter is designed in this paper. The filtering properties can be adjusted through varying the structural parameters. With the 25.60 nm gold film thickness, the losses of the respective modes of Y and X-polarized core mode at 1550 nm are 1024.84 and 0.12 dB/cm with the loss ratio of 8540.33 between two polarizations. However, the losses of Y and X-polarized core mode at 1310 nm are 682.14 and 0.03 dB/cm, and the loss ratio is 22,738 with the gold film thickness of 55.30 nm. That indicates that the proposed filter has a higher loss ratio. Moreover, the crosstalk value with the fiber length of 200 μm at 1550 and 1310 nm are 178.01 and 118.49 dB, respectively. The bandwidths with crosstalk value greater than 20 dB are 640 and 180 nm. The designed polarization filter represents good filtering characteristics and allows great fabrication tolerances. Therefore, the designed hexagonal filter can be well applied in the domain of optical fiber communication.

Keywords: polarization filter; surface plasmon resonance; photonic crystal fiber; gold-coated; crosstalk

Citation: Wang, C.; Zhang, G.; Wu, Z.; Zhang, Y.; Zhang, Y.; Jiang, L.; Bi, W. A Broadband Gold-Coated Photonic Crystal Fiber Polarization Filter with a High Loss Ratio of Both Polarizations at 1550 and 1310 nm. *Photonics* **2021**, *8*, 488. https://doi.org/10.3390/photonics8110488

Received: 21 July 2021
Accepted: 26 October 2021
Published: 1 November 2021

Publisher's Note: MDPI stays neutral with regard to jurisdictional claims in published maps and institutional affiliations.

1. Introduction

Photonic crystal fiber (PCF) has drawn people's attention because of its unique transmission properties, which the conventional optical fibers cannot obtain. The porous structure of PCF cladding makes it possible to fill various functional materials, such as ethanol, graphene, liquid crystal, magnetic fluid, argon, metal, etc. The fusion of functional materials and PCF also greatly broadens the application function of PCF, and provides a new channel for the realization of new PCF devices. In particular, as the air holes of PCF cladding with metal material are served as the defect core, once the phase of the incident light in the PCF core mode (CM) is matched with that of the surface plasmon mode (SPM), the light intensity in the CM would be transferred to the defect core, which is called a surface plasmon resonance (SPR) effect [1]. The confinement loss (CL) of the CM at the phase-matching wavelength, that is, the resonant wavelength, increased rapidly. Based on this, various PCF sensors [2,3], polarizers [4–6] and polarization filters [7–22] on account of SPR have been designed.

Thus far, many researchers have carried out a lot of research on the polarization filters based on PCF combined with metal materials. In 2006, Kuhlmey et al. numerically simulated the conduction characteristics of metal film coated PCF by multipole method [12]. In 2008, Lee et al. realized selective filling of the gold nanowires in the air holes of polarization-maintaining PCF cladding [13]. In 2011, the different gold-filling formats of the PCF are used to realize great polarization filtering transmission by Nagaski et al. [14].

In order to better separate the loss peak of the two orthogonal polarization directions of CM of the polarization filter, the commonly used method is to introduce "asymmetric factor". The asymmetry factor can be introduced by changing the arrangement or structural parameters of pores around PCF core or defective core. That can cause birefringence around the PCF core or defect core, and the mode refractive indices of the two orthogonal polarization directions are no longer equal. Based on this idea, a large number of PCF polarization filters are designed. In 2013, Xue et al. designed a PCF polarization filter with a Y-polarized CM loss of 508 dB/cm at 1310 nm, and the loss ratio of Y-polarized CM to X-polarized CM is 20 [15]. In 2015, Liu et al. proposed a broadband square PCF single polarization filter with an SPR effect. The respective losses of Y-polarized CM are 452.4 and 102 dB/cm at 1310 and 1550 nm, and the relevant losses of X-polarized CM are only 0.9 and 0.8 dB/cm [16]. In 2016, a rhombic plasmonic PCF single polarization filter with the respective losses of 630.20 and 36.90 dB/cm for Y and X-polarized CM around 1550 nm were designed by Dou et al. [17]. In 2017, Wang et al. investigated a gold-filled PCF single polarization single-mode filter. The losses of X-polarized CM can reach to 126.10 and 326.30 dB/cm at 1310 and 1550 nm, while that of X-polarized CM are 0.08 and 1.20 dB/cm, respectively [18]. In 2018, Wang et al. designed a new ultra-wide bandwidth square PCF polarization filter with a Y-polarized CM loss of 718.87 dB/cm at 1550 nm, and the loss ratio of Y-polarized CM to X-polarized CM is 756.71 [19]. In the same year, the respective losses at 1310 and 1550 nm for X-polarized CM based on a gold-filled hexagonal PCF polarization filter are 251.26 and 224.45 dB/cm, while the relevant Y-polarized CM losses of 13.76 and 3.59 dB/cm were obtained by Lu et al. [20]. In 2019, Zhao et al. designed a semi-hourglass plasmonic PCF polarization filter, which reveals the respective losses of 1304.02 and 3.96 dB/cm for Y and X-polarized CM at 1550 nm with 18.70 nm Au film thickness [21]. In 2020, Zhang et al. represented a high intensity polarization filter with a high bandwidth of 990 nm based on a diamond-shaped PCF structure, which can obtain the losses of 563.29 dB/cm and 3.75 dB/cm for the respective Y and X-polarized CM at 1550 nm [22]. According to the current reports, the designed polarization filter using the above method can effectively make the loss peaks of both polarized CM locate at different wavelengths. However, the coupling between the CM and SPM will inevitably occur in both horizontal and vertical polarizations. Therefore, the loss ratio in both polarizations of the reported PCF polarization filter is still not very large, and the filter needs a longer fiber length to achieve a wider bandwidth, which are the main limited factors of the polarization filters widely used in the optical fiber communication system.

In this paper, a new style of gold-coated hexagonal PCF polarization filter is designed. The filtering properties with varying PCF structural values of d_1, d_2, d_3, d_4 and t are numerically simulated by taking advantage of full vector finite element method (FVFEM). The loss of Y-polarized CM around 1550 and 1310 nm are 1024.84 and 682.14 dB/cm, respectively, but that of X-polarized CM is very small with the 25.60 and 55.30 nm gold film thickness. This means that the designed filter has a higher loss ratio, and single polarization filtering at both 1550 and 1310 nm is realized. The Crosstalk (CT) value shows that the presented hexagonal PCF polarization filter has a broadband filtering bandwidth with a shorter PCF length. Moreover, the proposed polarization filter has good fabrication tolerances, because the $\pm 1\%$ deviations of the structural parameters will not affect the properties of the filter.

2. Structure Design

The cladding structure of the presented PCF polarization filter arranged in hexagonal pattern contains four layers air holes as shown in Figure 1. To obtain the much higher birefringence, the asymmetric factor is introduced in the proposed design. It is due to that, that higher birefringence can effectively separate the resonance point of two polarized CM with SPM. A larger loss difference between the Y and X-polarized CM can be achieved for a strong asymmetric PCF structure, which is a benefit for the single polarization filtering. Two large air holes with d_1 are distributed on both sides of the core. The respective structural values of the outer diameter and the thickness of the gold-coated holes are denoted as d_4 and t. The two air holes with d_2 lie between the fiber core and the gold film hole. The two large air holes with d_3 are distributed on both sides of the gold film hole. The spacing between neighboring air holes is Λ = 2 μm. The remaining air holes are identical with d = 1.20 μm.

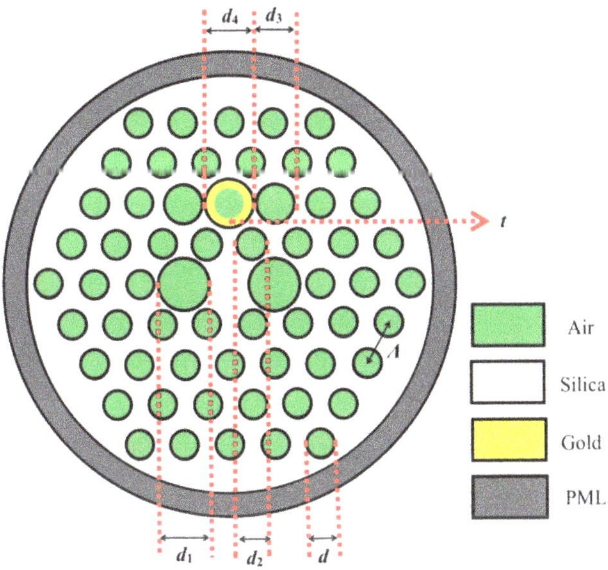

Figure 1. Cross-sectional schematic of the designed PCF polarization filter.

Two large air holes d_1 and d_3 are, respectively, set on both sides of the fiber core and the gold-coated hole region, which can effectively increase the asymmetry of the fiber and improve the resonant coupling between the CM and SPM in a single polarization direction. Two slightly enlarged air holes d_2 of the proposed filter form a bridge between the two kinds of modes. This setting of this structure can form a strong polarization filter transmission.

The silica is chosen as a substrate material of the presented PCF. The refractive index of silica is subjected to its Sellmeier equation, which is represented by [23]:

$$n(\lambda) = \sqrt{1 + \frac{B_1 \cdot \lambda^2}{\lambda^2 - C_1} + \frac{B_2 \cdot \lambda^2}{\lambda^2 - C_2} + \frac{B_3 \cdot \lambda^2}{\lambda^2 - C_3}} \quad (1)$$

here, λ is the light wavelength, and its unit is μm. The Sellmeier coefficients of B_1, B_2, B_3, C_1, C_2 and C_3 are found in Table 1. Using the Sellmerier equation of silica in our simulation analysis, the refractive index of silica varies with wavelength, so the material properties of silica in the designed filter have been considered.

Table 1. The parameters in the Sellmeier equation.

B_1	B_2	B_3	C_1 (μm^2)	C_2 (μm^2)	C_3 (μm^2)
0.6961663	0.4079426	0.8974794	0.00467914826	0.0135120631	97.9340025

Gold is used as filling material, and its relative dielectric constant is expressed by using the Drude–Lorentz model [24]:

$$\varepsilon_{\mathrm{m}} = \varepsilon_{\infty} - \frac{\omega_D^2}{\omega(\omega + \mathrm{j}\gamma_D)} - \frac{\Delta\varepsilon \cdot \Omega_L^2}{(\omega^2 - \Omega_L^2) - \mathrm{j}\Gamma_L\omega} \tag{2}$$

here, ε_{∞} is the high frequency dielectric constant. $\Delta\varepsilon$ is a weighting coefficient. ω is the angular frequency of the light. ω_D and γ_D are the plasma frequency and damping frequency, respectively. Ω_L is the frequency of the Lorentz oscillator. Γ_L is the spectral width of the Lorentz oscillator. The corresponding coefficient parameters of gold are set as shown in Table 2.

Table 2. The parameters of gold.

ε_{∞}	$\omega_D/2\pi$(THz)	$\gamma_D/2\pi$(THz)	$\Omega/2\pi$(THz)	$\Gamma_L/2\pi$(THz)	$\Delta\varepsilon$
5.9673	2113.60	15.92	650.07	104.86	1.09

During the process of numerical simulation, in order to reduce the radiation loss, the perfectly matched layer and scattering boundary conditions are employed. The dispersion distribution between two kinds of modes are analyzed through the FVFEM-based software. The computation area is discretized with triangular subdomains and meshed with 25,360 number of smallest elements. The CL of the fiber mode is calculated by using the imaginary part of effective refractive index (Im(n_{eff})) [25]:

$$\alpha(x, y) = 8.686 \times \frac{2\pi}{\lambda} \times \mathrm{Im}\left(n_{eff}\right) \times 10^4 \tag{3}$$

here, the unit of $\alpha(x, y)$ is dB/cm.

3. Numerical Simulation and Results Analysis

The optimal parameters d_1 = 2.40 μm, d_2 = 1.38 μm, d_3 = 1.80 μm, d_4 = 2.20 μm, t = 25.60 nm and t = 55.30 nm are obtained during a time consuming process. Figure 2 shows the loss and Re(n_{eff}) of CM and SPM with two different t in the left and right arises, respectively. From Figure 2a, when t is 25.60 nm, the Re(n_{eff}) of Y-polarized CM has intersection with that of Y-polarized SPM at 1550 nm, where SPR effect happened due to the phase-matching between them. Furthermore, in the Y-polarized direction, CM and SPM have the same loss, which indicates that complete coupling has occurred [26]. However, the coupling strength of both X-polarized CM and SPM is very low, and the loss spectrum appears flat. The respective losses of both Y and X-polarized CM are 1024.84 and 0.12 dB/cm around 1550 nm with a large loss ratio of 8540.33 in both polarizations. Figure 2b shows that the Re(n_{eff}) of Y-polarized CM has intersection with that of Y-polarized SPM at 1310 nm as t is 55.30 nm. Besides, Y-polarized CM has the resonance peak with maximum loss, when a complete coupling between CM and SPM occurs. Similarly, in the X-polarized direction, the coupling is weak and no loss peak appears. The losses of 682.14 and 0.03 dB/cm for the respective Y and X-polarized CMs are achieved at 1310 nm, and the loss ratio is 22,738.

In order to better illustrate the effectiveness of the exhibited structure in increasing loss ratio of the two orthogonal polarization directions, the loss and distribution of the electric field of CM with different values of d_3 are analyzed. Figures 3 and 4 show the corresponding distribution of the electric field of Y-polarized CM and the losses of two polarized CM with d_3 = 1.40, 1.60, 1.80 μm and d_1 = 2.40 μm, d_2 = 1.38 μm, d_4 = 2.20 μm, t = 25.60 nm. The loss peak wavelength is 1740, 1600 and 1550 nm, respectively. From Figures 3 and 4, as the value of d_3 increases, the distance between the gold film hole and the two air holes with d_3 decreases, the designed structure can effectively prevent the diffusion of the electromagnetic field intensity from the core to the metal surface, and the energy is concentrated in a certain area of the metal surface. Additionally, the mode coupling method has also changed, that is, the CM is no longer coupled with the second-order SPM, but it is only coupled with a part of the second-order SPM. The energy does not spread to other regions. Therefore, the coupling strength in the Y-polarized direction will be stronger. On the other hand, it effectively blocks the coupling channel between the CM and SPM in the X-polarized direction. Therefore, the intensity of the X-polarized loss decreases as d_3 increases. This results in the loss difference and loss ratio of the designed PCF polarization filter being very large.

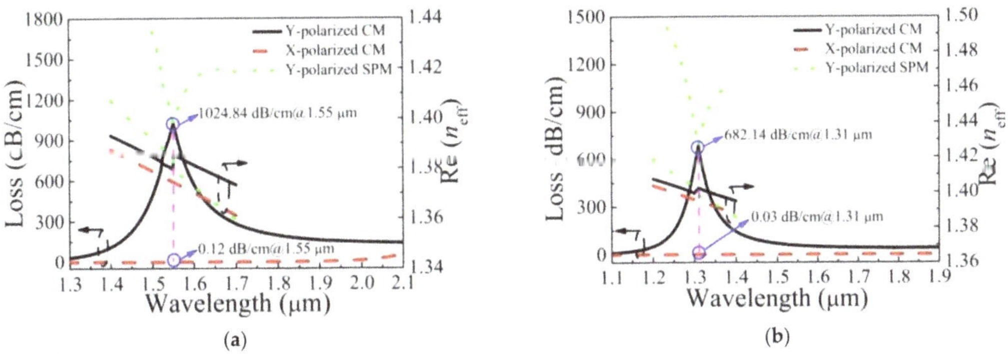

Figure 2. Loss and the real part of effective refractive index (Re(n_{eff})) of CM and SPM as t is (**a**) 25.60 nm and (**b**) 55.30 nm, respectively.

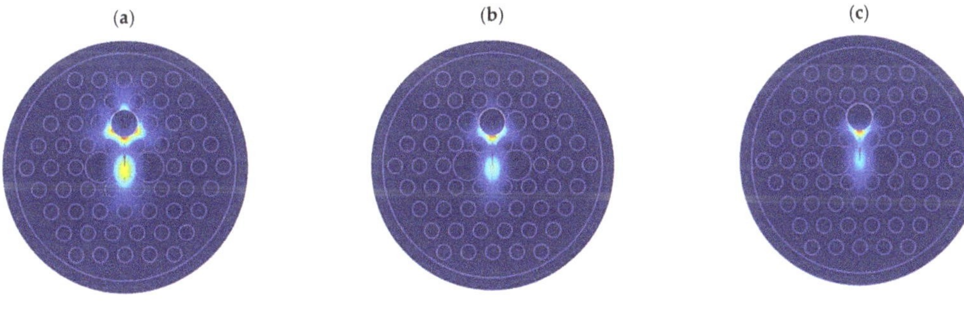

Figure 3. Distribution of the electric field of Y-polarized CM varied with d_3 = (**a**) 1.40, (**b**) 1.60, (**c**) 1.80 μm and d_1 = 2.40 μm, d_2 = 1.38 μm, d_4 = 2.20 μm, t = 25.60 nm.

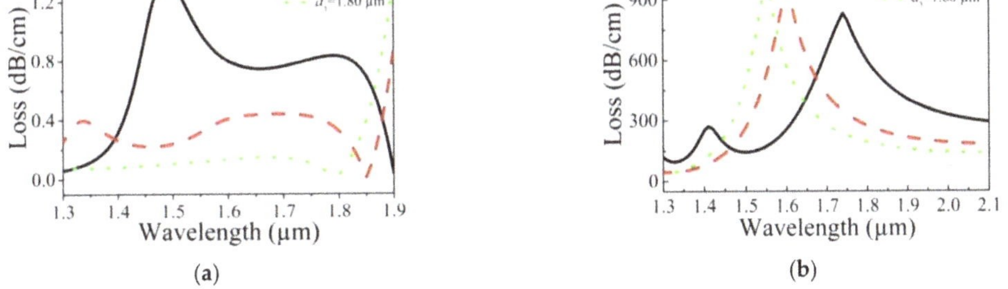

Figure 4. Loss of (**a**) X-polarized and (**b**) Y-polarized CM varied with d_3 = 1.40, 1.60, 1.80 μm and d_1 = 2.40 μm, d_2 = 1.38 μm, d_4 = 2.20 μm, t = 25.60 nm.

Figure 5 shows the distribution of the electric field of X-polarized CM, Y-polarized CM, and Y-polarized SPM at 1550 and 1310 nm with the complete coupling between Y-polarized CM and SPM. From Figure 5, whether at 1310 nm or 1550 nm, the distributions of these three kinds of modes are similar. The electromagnetic field intensity of X-polarized CM is confined to the PCF core, while that of Y-polarized CM is transferred to the SPM. Moreover, the electric field distributions of both Y-polarized CM and SPM are essentially the same. That indicates when the light with the wavelength of 1550 nm or 1310 nm is transmitted for a certain distance, the light intensity distribution in Y-polarized direction is attenuated seriously, but the X-polarized light still retains a high energy. Therefore, the single polarization filter is realized.

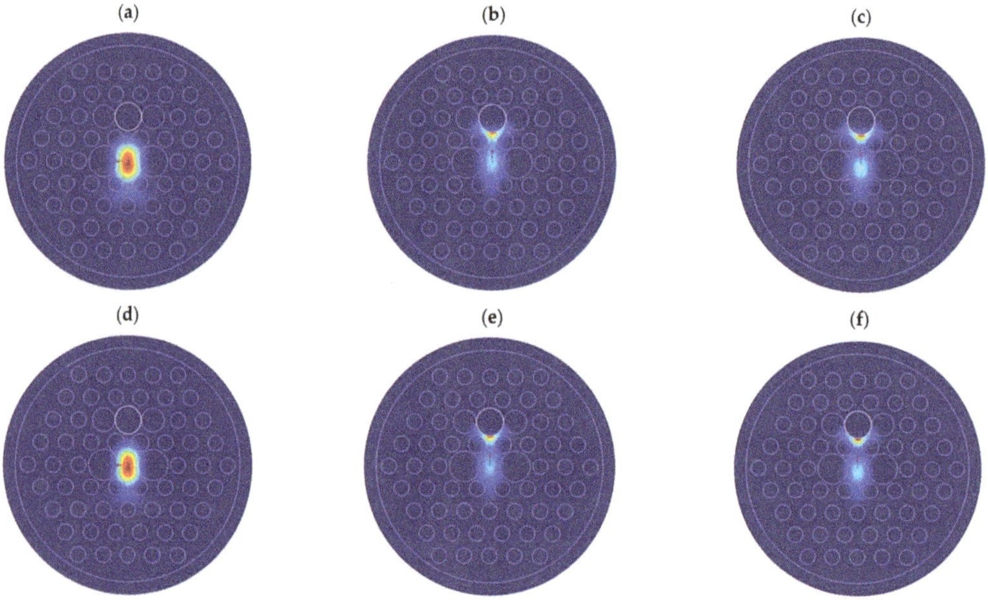

Figure 5. Distribution of the electric field of (**a**,**d**) X-polarized CM, (**b**,**e**) Y-polarized CM, and (**c**,**f**) Y-polarized SPM at 1550 and 1310 nm as the complete coupling occurs.

In the following section, the analysis results in Figure 6 show that the polarization filter characteristics vary with different d_1, d_2, d_3, d_4, and t values. As shown from Figure 6a, the peak wavelength experiences a red shift by increasing d_1 from 2.20, 2.40 to 2.60 μm. It should also be noted that the peak intensity of Y-polarized CM increases firstly and then decreases as d_1 increases. This is because the further increase of d_1 results in the light constrained weakly in the core, which makes the coupling between them weak and the peak intensity reduce. Hence, the peak wavelength and intensity of the loss can be coarsely adjusted by selecting appropriate d_1. Figure 6b shows the loss of Y-polarized CM varied with d_2 = 1.30, 1.35 and 1.40 μm. From Figure 6b, the peak wavelength decreases from 1850 to 1500 nm by increasing d_2 values from 1.30 to 1.40 μm. Furthermore, the peak intensity of Y-polarized CM increases obviously with increasing d_2. Therefore, the peak wavelength and intensity can be coarsely tuned with selecting proper d_2. Figure 6c shows the loss of Y-polarized CM varied with d_3 = 1.60, 1.70 and 1.80 μm. From Figure 6c, the peak wavelength is 1600, 1570 and 1550 nm when d_3 is 1.60, 1.70 and 1.80 μm, respectively. Moreover, the peak intensity of Y-polarized CM increases slightly as d_3 increases. It can be seen from the above analysis that the peak wavelength and intensity of the loss do not obviously change with the increase of d_3. Therefore, the peak wavelength and intensity can be fine-tuned by adjusting d_3 appropriately. Figure 6d shows the loss of Y-polarized CM varied with $d_4 = 2.10$, 2.15 and 2.20 μm. From Figure 6d, the peak wavelength of Y-polarized CM shifts to shorter wavelength, that is, from 1780 to 1550 nm with d_4 increasing from 2.10 to 2.20 μm. Meanwhile, the peak intensity of Y-polarized CM increases from 846.74 to 1024.84 dB/cm with the increase of d_4. Therefore, the peak wavelength and intensity can be coarse adjusted by using the appropriate d_4. Figure 6e shows the loss of Y-polarized CM varied with t = 22, 26 and 30 nm. From Figure 6e, the peak wavelength of CM shifts to shorter wavelength with the increase of t, but the movement speed slows down gradually. This is because the Re(n_{eff}) of SPM decreases with increasing t, but that of CM does not alter distinctly, which generates the blue-shift of the phase-matching point. Furthermore, the increasing t value decreases the peak intensity of Y-polarized CM. That is why the light has difficulty penetrating it as the gold film is too thick, which makes the coupling between CM and SPM weaken. Therefore, the peak wavelength and intensity of the loss can be roughly adjusted by tuning t properly.

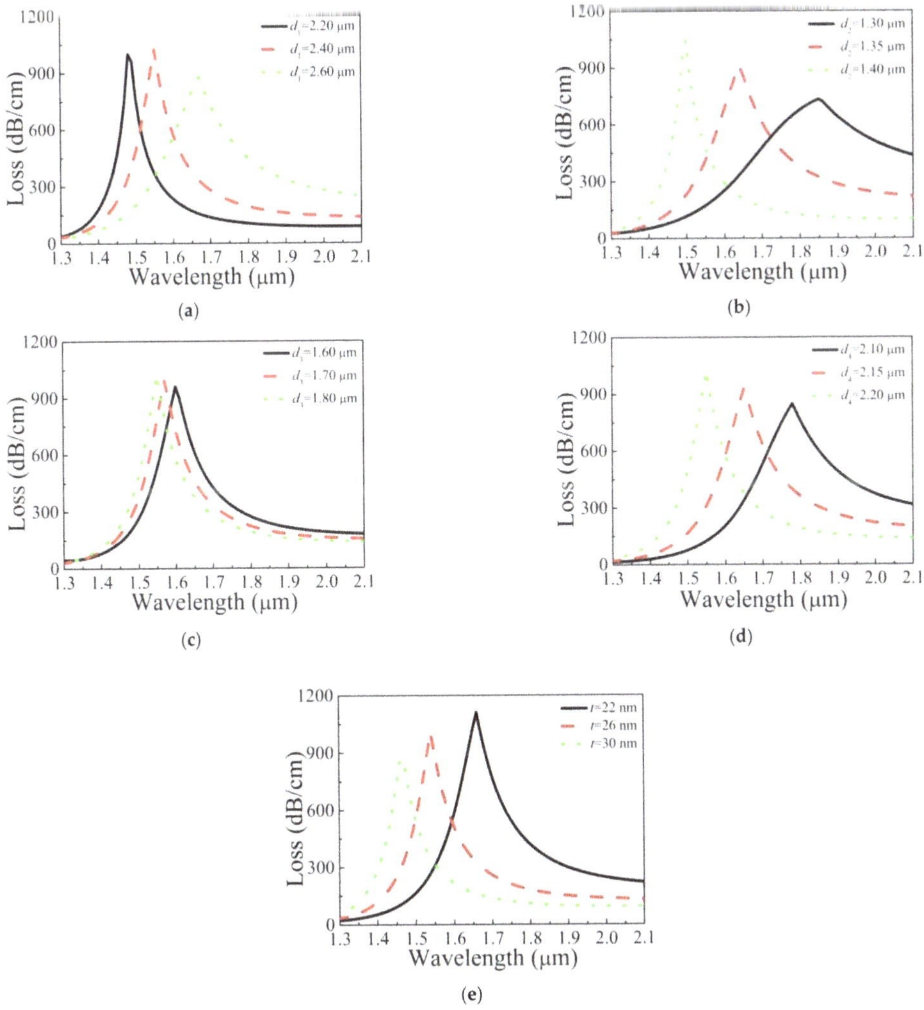

Figure 6. Loss of Y-polarized CM varied with different (**a**) d_1, (**b**) d_2, (**c**) d_3, (**d**) d_4, and (**e**) t, and other parameters kept unchanged.

CT is a key indicator to reflect the property of the filters, which determines the effect on the unnecessary polarization mode. The available bandwidth of the filters is defined as a band with a CT value greater than 20 dB or less than −20 dB. The relationship between the CT value and fiber length can be represented by the following formula [27,28]:

$$\mathrm{CT} = 20\lg\{\exp[(\alpha_y - \alpha_x)L]\} \tag{4}$$

here, L is the PCF length. The unit of the CT and L is dB and μm, respectively.

Figure 7 shows the CT value varied with L = 100, 200, and 300 μm as the PCF is arranged with the optimal structural parameters. From Figure 7, it is clear that the CT value reaches to the peak both at 1550 and 1310 nm. As L is increased from 100 to 300 μm, whether at 1550 nm or 1310 nm, both the peak of the CT value and the bandwidth with a CT value greater than 20 dB are increased. The available bandwidth is 280 nm, and the maximum CT value is 89.01 dB at 1550 nm when L is 100 μm in Figure 7a. The CT value is greater than 20 dB in all bands as L is more than 300 μm. From Figure 7b, as L is 100 μm, the available bandwidth is 90 nm, and the maximum CT value is 59.24 dB around 1310 nm. In contrast to the previously reported filter, the presented filter with shorter lengths reveals higher CT value and wider bandwidth. The results show that the designed polarization filter can achieve a strong filtering performance both at the communication wavelength of 1550 and 1310 nm.

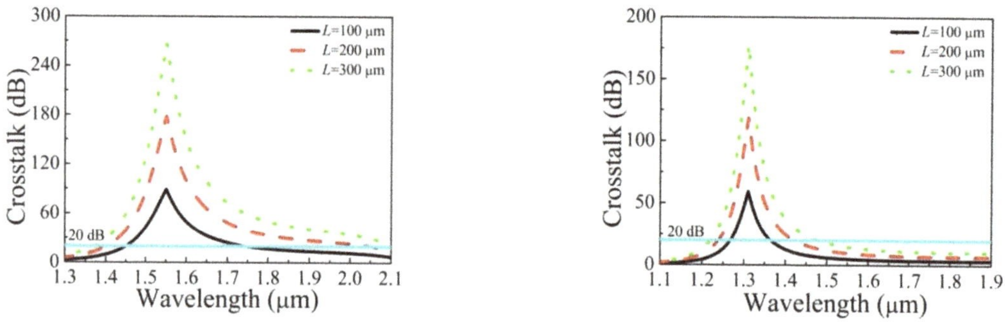

Figure 7. CT value varied with different L when d_1 = 2.40 μm, d_2 = 1.38 μm, d_3 = 1.80 μm, d_4 = 2.20 μm, (**a**) t = 25.60 nm and (**b**) t = 55.30 nm.

Figure 8 shows the CT value varied with different d_1, d_2, d_3, d_4 and t in the case of L = 100 μm. From Figure 8, the structural parameters have the same influence on the CT as loss value. The CT value has the same change rule as the loss by varying d_1. As d_2, d_3 and d_4 increase, the CT value increases. The increase of t causes the CT value to decrease linearly.

To comprehensively demonstrate the performance of the achieved filter, the comparison between the designed filter and the previously reported filters is made. Table 3 shows the results for resonance wavelength, loss, loss ratio, filter length and bandwidth. The simulation results in Table 3 reveal that the realized loss ratio is 8540.33 and 22,738, and the obtained filtering bandwidth at 1550 and 1310 nm is 640 and 180 nm with a fixed length of 200 μm, respectively. The comparison results show that, whether at 1310 or 1550 nm, the designed PCF polarization filter with more miniaturization realizes a higher loss ratio and wider filtering bandwidth. Therefore, the comparison of many parameters shows that our proposed polarization filter has better performance.

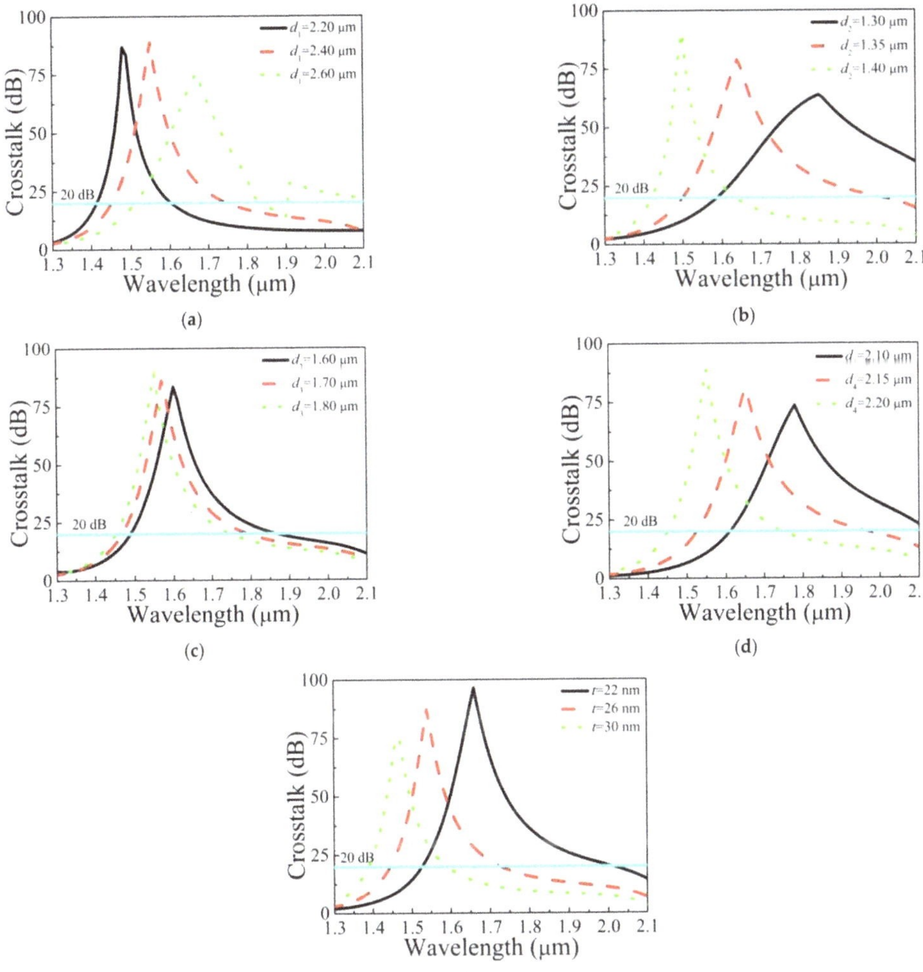

Figure 8. CT value varied with different (**a**) d_1, (**b**) d_2, (**c**) d_3, (**d**) d_4 and (**e**) t.

Table 3. Comparison results between the proposed filter and the previously reported filter.

Ref.	Resonance Wavelength	Y-Polarized Loss	X-Polarized Loss	Loss Ratio	Filter Length	Bandwidth
9	1550 nm	433.65 dB/cm	2.64 dB/cm	164.26	4000 μm	150 nm
10	1420 nm	1.13 dB/cm	692.25 dB/cm	612.61	1000 μm	830 nm
17	1550 nm	630.20 dB/cm	36.90 dB/cm	17.10	2000 μm	220 nm
19	1550 nm	718.87 dB/cm	0.95 dB/cm	756.70	300 μm	360 nm
22	1550 nm	563.29 dB/cm	3.75 dB/cm	150.21	5000 μm	990 nm
our filter	1550 nm	1024.84 dB/cm	0.12 dB/cm	8540.33	200 μm	640 nm
	1310 nm	682.14 dB/cm	0.03 dB/cm	22738	200 μm	180 nm

There are usually two steps to make the proposed hexagonal PCF polarization filter. First of all, the hexagonal PCF is realized based on the improved stack and draw method [29,30]. Secondly, the selective coated gold film in the air hole is carried out by wet chemical deposition, and the gold film thickness can be controlled accurately [4,31]. It is inevitable that the structural parameters change slightly in the process of PCF fabrication. Figure 9 shows the variety of CT value with ±1% tolerances of the main structural parameters. The ±1% deviations of d_2 have a small influence on CT value, and ±1% tolerances of d_1 and t have no influence. Furthermore, the PCF polarization filter still has a high loss ratio, large CT values and a wide filtering bandwidth under ±1% deviations.

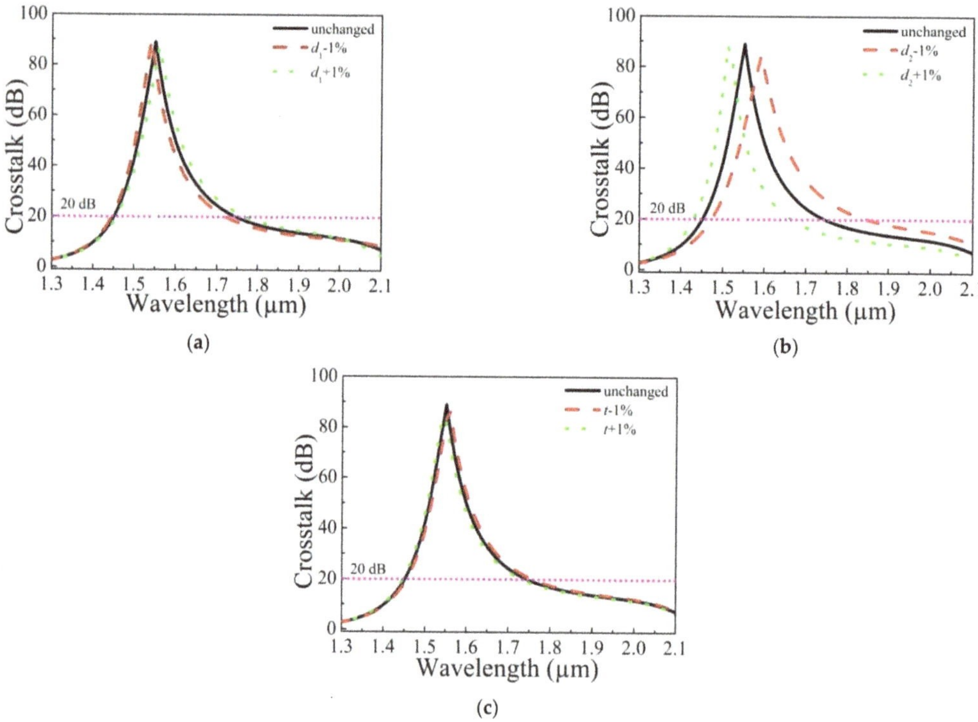

Figure 9. CT value varied with the ±1% tolerances of (**a**) d_1, (**b**) d_2 and (**c**) t.

In view of the application of the proposed PCF single polarization filter. The splice loss between the proposed device and the traditional single-mode fibers (SMFs) is considerable. The splice loss of this designed fiber to a SMF with a mode filed diameter (MFD) of 10 μm is around 7.33 dB at 1.55 μm, which can be further controlled either by the splice-free interconnection techniques [32] or using fusion splicing by repeated arc discharges techniques [33]. According to the reference [34], this splice loss can also be overcome if this PCF structure is coupled with the SMF with the MFD less than 10 μm.

For the spatial optical path coupling in the experiment, a laser beam is emitted from the laser, firstly using one lens to collimate, then using another lens to focus, using a high-precision potential shifting platform and other devices to adjust the position of the proposed PCF polarization filter to improve the coupling efficiency of spatial optical path and achieve the mode field matching.

4. Conclusions

A novel gold-coated hexagonal PCF polarization filter with a high loss ratio is proposed and studied. Numerical simulation results show that d_1, d_2, d_4 and t have greater influence on the peak wavelength and intensity of the loss, but d_3 has smaller impact on that. The CM and SPM can be completely coupled at the wavelength of 1550 nm in

Y-polarized direction, where the respective losses of Y and X-polarized CM are 1024.84 and 0.12 dB/cm with a high loss ratio of 8540.33. Nevertheless, only as t is changed to 55.30 nm, the respective losses of Y and X-polarized CM at 1310 nm are 682.14 and 0.03 dB/cm, and the loss ratio reaches to 22738. The filtering bandwidths with a fixed length of 200 μm at 1550 and 1310 nm are 640 and 180 nm, respectively. The above results also show that the designed PCF polarization filter has larger loss ratio of Y to X-polarized CM, and wider filtering bandwidth with a shorter PCF length. The structure deviation analysis shows that the polarization filter have good fabrication tolerances. Therefore, the designed hexagonal PCF polarization filter will inevitably be extensively put into use in the optical fiber communication system.

Author Contributions: Conceptualization, C.W.; methodology, C.W.; software, C.W.; validation, C.W.; formal analysis, C.W. and G.Z.; investigation, C.W.; resources, C.W.; data curation, C.W; writing—original draft preparation, C.W.; writing—review and editing, Y.Z. (Yajing Zhang) and L.J.; visualization, Z.W. and Y.Z. (Yiyang Zhang); supervision, Z.W., Y.Z. (Yiyang Zhang) and W.B., project administration, C.W. and L.J.; funding acquisition, C.W. All authors have read and agreed to the published version of the manuscript.

Funding: This research is funded by the National Science Foundation of China (Nos. 61405173, 61405172), National Science Foundation of Hebei Province (Nos. F2018105036, F2019105108, F2019203549, 19941708G, 206Z0401G), Science and Technology Project of Hebei Education Department (No. ZD2021409, ZC2021033), PhD Innovation Fund Project of Tangshan University.

Institutional Review Board Statement: Not applicable.

Informed Consent Statement: Not applicable.

Data Availability Statement: Not applicable.

Acknowledgments: We thank Key Lab of Intelligent Data Information Processing and Control of Hebei Province and Key Lab of Intelligent Motion Control System of Tangshan City (Intelligence and Information Engineering College) for the scientific helps and supports throughout this research.

Conflicts of Interest: The authors declare no conflict of interest.

References

1. Schmidt, M.A.; Russell, P.S.J. Long-range spiralling surface plasmon modes on metallic nanowires. *Opt. Express* **2008**, *16*, 13617–13623. [CrossRef]
2. Otupiri, R.; Akowuah, E.K.; Haxha, S.; Ademgil, H.; AbdelMalek, F.; Aggoun, A. A Novel Birefrigent Photonic Crystal Fiber Surface Plasmon Resonance Biosensor. *IEEE Photon. J.* **2014**, *6*, 6801711. [CrossRef]
3. Zhao, Y.; Deng, Z.Q.; Li, J. Photonic crystal fiber based surface plasmon resonance chemical sensors. *Sens. Actuators B Chem.* **2014**, *202*, 557–567. [CrossRef]
4. Zhang, X.; Wang, R.; Cox, F.M.; Kuhlmey, B.T.; Large, M.C.J. Selective coating of holes in microstructured optical fiber and its application to in-fiber absorptive polarizers. *Opt. Express* **2007**, *15*, 16270–16278. [CrossRef]
5. Khaleque, A.; Hattori, H.T. Polarizer based upon a plasmonic resonant thin layer on a squeezed photonic crystal fiber. *Appl. Opt.* **2015**, *54*, 2543–2549. [CrossRef] [PubMed]
6. Zhang, S.Y.; Yu, X.; Zhang, Y.; Shum, P.; Zhang, Y.T.; Xia, L.; Liu, D.M. Theoretical Study of Dual-Core Photonic Crystal Fibers with Metal Wire. *IEEE Photon. J.* **2012**, *4*, 1178–1187. [CrossRef]
7. An, G.W.; Li, S.G.; Zhang, W.; Fan, Z.K.; Bao, Y.J. A polarization filter of gold-filled photonic crystal fiber with regular triangular and rectangular lattices. *Opt. Commun.* **2014**, *331*, 316–319. [CrossRef]
8. Wang, H.Y.; Yan, X.; Li, S.G.; An, G.W.; Zhang, X.N.; Yuan, Z.Y. Design of the polarization filter based on photonic crystal fiber with Au-coated air holes. *Opt. Quant. Electron.* **2016**, *48*, 512. [CrossRef]
9. Shi, M.; Li, S.G.; Chen, H.L.; Wang, G.Y.; Zhao, Y.Y. Surface plasmon resonance effect induced tunable polarization filter based on gold film selectively coated photonic crystal fiber. *Opt. Commun.* **2017**, *396*, 257–260. [CrossRef]
10. Hossen, M.N.; Ferdous, M.; Ahmed, K.; Khalek, M.A.; Chakma, S.J.; Paul, B.K. Single polarization photonic crystal fiber filter based on surface plasmon resonance. *Front. Optoelectron.* **2019**, *12*, 157–164. [CrossRef]
11. Qu, Y.W.; Yuan, J.H.; Zhou, X.; Li, F.; Mei, C.; Yan, B.B.; Wu, Q.; Wang, K.R.; Sang, X.Z.; Long, K.P.; et al. A V-shape photonic crystal fiber polarization filter based on surface plasmon resonance effect. *Opt. Commun.* **2019**, *452*, 1–6. [CrossRef]
12. Kuhlmey, B.T.; Pathmanandavel, K.; McPhedran, R.C. Multipole analysis of photonic crystal fibers with coated inclusions. *Opt. Express* **2006**, *14*, 10851–10864. [CrossRef]

13. Lee., H.W.; Schmidt, M.A.; Tyagi, H.K.; Sempere, L.P.; Russell, P.S.J. Polarization-dependent coupling to plasmon modes on submicron gold wire in photonic crystal fiber. *Appl. Phy. Lett.* **2008**, *93*, 111102. [CrossRef]
14. Nagasaki, A.; Saitoh, K.; Saitoh, M. Polarization characteristics of photonic crystal fibers selectively filled with metal wires into cladding air holes. *Opt. Express* **2011**, *19*, 3799–3808. [CrossRef] [PubMed]
15. Xue, J.R.; Li, S.G.; Xiao, Y.Z.; Qin, W.; Xin, X.J.; Zhu, X.P. Polarization filter characters of the gold-coated and the liquid filled photonic crystal fiber based on surface plasmon resonance. *Opt. Express* **2013**, *21*, 13733–13740. [CrossRef] [PubMed]
16. Liu, Q.; Li, S.G.; Li, H.; Zi, J.C.; Zhang, W.; Fan, Z.K.; An, G.W.; Bao, Y.J. Broadband Single-Polarization Photonic Crystal Fiber Based on Surface Plasmon Resonance for Polarization Filter. *Plasmonics* **2015**, *10*, 931–939. [CrossRef]
17. Dou, C.; Jing, X.L.; Li, S.G.; Liu, Q.; Bian, J. A Photonic Crystal Fiber Polarized Filter at 1.55 μm based on Surface Plasmon Resonance. *Plasmonics* **2016**, *11*, 1163–1168. [CrossRef]
18. Wang, X.Y.; Li, S.G.; Liu, Q.; Wang, G.Y.; Zhao, Y.Y. Design of a Single-Polarization Single-Mode Photonic Crystal Fiber Filter Based on Surface Plasmon Resonance. *Plasmonics* **2017**, *12*, 1325–1330. [CrossRef]
19. Wang, Y.J.; Li, S.G.; Chen, H.L.; Shi, M.; Liu, Y.C. Ultra-wide bandwidth polarization filter based on gold-coated photonic crystal fiber around the wavelength of 1.55 μm. *Opt. Laser Technol.* **2018**, *106*, 22–28. [CrossRef]
20. Lu, X.L.; Chang, M.; Chen, N.; Zhang, X.D.; Zhuang, S.L.; Xu, J. Design of a Metal-Filled Photonic-Crystal Fiber Polarization Filter Based on Surface Plasmon Resonance at 1.31 and 1.55 μm. *IEEE Photon. J.* **2018**, *10*, 7203913. [CrossRef]
21. Zhao, X.T.; Hua, L.; Jiang, G.H.; Cheng, J.R.; Xiong, Q. A Novel Polarization Filter Based on Photonic Crystal Fiber with a Single Au-Coated Air Hole and Semi-Hourglass Structure. *Plasmonics* **2019**, *14*, 1725–1733. [CrossRef]
22. Zhang, Y.X.; Yuan, J.H.; Qu, Y.W.; Zhou, X.; Yan, B.B.; Wu, Q.; Wang, K.R.; Sang, X.Z.; Long, K.P.; Yu, C.X. Design of diamond-shape photonic crystal fiber polarization filter based on surface plasma resonance effect. *Chin. Phys. B* **2020**, *29*, 034208. [CrossRef]
23. Ghosh, G.; Endo, M.; Iwasaki, T. Temperature-Dependent Sellmeier Coefficients and Chromatic Dispersions for Some Optical Fiber Glasses. *J. Lightwave Technol.* **1994**, *12*, 1338–1342. [CrossRef]
24. Islam, M.S.; Sultana, J.; Rifat, A.A.; Ahmed, R.; Dinovitser, A.; Brian, W.H.N.; Heike, E.H.; Abbott, D. Dual-polarized highly sensitive plasmonic sensor in the visible to near-IR spectrum. *Opt. Express* **2018**, *26*, 30347–30361. [CrossRef] [PubMed]
25. Sun, B.; Chen, M.Y.; Zhou, J.; Zhang, Y.K. Surface Plasmon Induced Polarization Splitting Based on Dual-Core Photonic Crystal Fiber with Metal Wire. *Plasmonics* **2013**, *8*, 1253–1258. [CrossRef]
26. Zhang, Z.H.; Shi, Y.F.; Bian, B.M.; Lu, J. Dependence of leaky mode coupling on loss in photonic crystal fiber with hybrid cladding. *Opt. Express* **2008**, *16*, 1915–1922. [CrossRef]
27. Florous, N.; Saitoh, K.; Koshiba, M. A novel approach for designing photonic crystal fiber splitters with polarization-independent propagation characteristics. *Opt. Express* **2005**, *13*, 7365–7373. [CrossRef]
28. Chen, L.; Zhang, W.G.; Zhang, Z.; Liu, Y.J.; Sieg, J.; Zhang, L.Y.; Zhou, Q.; Wang, L.; Wang, L.; Yan, T.Y. Design for a Single-Polarization Photonic Crystal Fiber Wavelength Splitter Based on Hybrid-Surface Plasmon Resonance. *IEEE Photon. J.* **2014**, *6*, 1–9.
29. Mahdiraji, G.A.; Chow, D.M.; Sandoghchi, S.R.; Amirkhan, F.; Dermosesian, E.; Yeo, K.S.; Kakael, Z.; Ghomeishi, M.; Poh, S.Y.; Gang, S.Y.M.; et al. Challenges of solutions in fabrication of silica-based photonic crystal fibers: An experimental study. *Fiber Integr. Opt.* **2014**, *33*, 85–104. [CrossRef]
30. Wang, Y.; Huang, Q.; Zhu, W.J.; Yang, M.H.; Lewis, E. Novel optical fiber SPR temperature sensor based on MMF-PCF-MMF structure and gold-PDMS film. *Opt. Express* **2018**, *26*, 1910–1917. [CrossRef]
31. Liu, B.; Jiang, Y.; Zhu, X.; Tang, X.; Shi, Y. Hollow fiber surface plasmon resonance sensor for the detection of liquid with high refractive index. *Opt. Express* **2013**, *21*, 32349–32357. [CrossRef] [PubMed]
32. Leon-Saval, S.G.; Birks, T.A.; Joly, N.Y.; George, A.K.; Wadsworth, W.J.; Kakarantzas, G. Splice-free interfacing of photonic crystal fibers. *Opt.Lett.* **2005**, *30*, 1629–1631. [CrossRef] [PubMed]
33. Xiao, L.M.; Jin, W.; Demokan, M.S. Fusion splicing small-corephotonic crystal fibers and single-mode fibers by repeated arc discharges. *Opt. Lett.* **2007**, *32*, 115–117. [CrossRef] [PubMed]
34. Asiful Islam, M.; Shah Alam, M. Design Optimization of Equiangular Spiral Photonic Crystal Fiber for Large Negative Flat Dispersion and High Birefringence. *J. Lighwave Technol.* **2012**, *30*, 3545–3551. [CrossRef]

photonics

MDPI

Article

A Novel Liquid Crystal-Filled, Dual-Core Photonic Crystal Fiber Polarization Beam Splitter Covering the E + S + C + L + U Communication Band

Yuwei Qu [1], Ying Han [2,3,*], Jinhui Yuan [1,4], Xian Zhou [4], Binbin Yan [1], Kuiru Wang [1], Xinzhu Sang [1] and Chongxiu Yu [1]

[1] State Key Laboratory of Information Photonics and Optical Communications, Beijing University of Posts and Telecommunications, Beijing 100876, China; quyuwei@bupt.edu.cn (Y.Q.); yuanjinhui81@bupt.edu.cn (J.Y.); yanbinbin@bupt.edu.cn (B.Y.); krwang@bupt.edu.cn (K.W.); xzsang@bupt.edu.cn (X.S.); cxyu@bupt.edu.cn (C.Y.)

[2] School of Information Science and Engineering, Yanshan University, Qinhuangdao 066004, China

[3] Key Laboratory for Special Fiber and Fiber Sensor of Hebei Province, Qinhuangdao 066004, China

[4] Research Center for Convergence Networks and Ubiquitous Services, University of Science & Technology Beijing (USTB), Beijing 100083, China; zhouxian219@ustb.edu.cn

* Correspondence: hanying2020@ysu.edu.cn

Abstract: This paper proposes a novel liquid crystal-filled, dual core photonic crystal fiber polarization beam splitter (LC-DC-PCF PBS) based on the coupled mode theory of DC-PCF. The mode birefringence of odd and even modes, coupling lengths (*CLs*) of the X-polarization (X-pol) and Y-polarization (Y-pol), and the corresponding coupling length ratio (*CLR*) of the proposed LC-DC-PCF PBS filled without LC E7 and with LC E7 are compared. The change rules of the *CLs* of the X-pol and Y-pol and *CLR* of the proposed LC-DC-PCF with wavelengths for different cladding microstructure parameters were investigated. The relationships between the X-pol and Y-pol normalized output powers in core A of the proposed LC-DC-PCF PBS and the propagation length at the wavelength of 1.604 μm are discussed. Finally, by studying the change of extinction ratio (*ER*) with wavelength, the LC-DC-PCF PBS *ER* of 60.3 and 72.2 dB at wavelengths 1.386 and 1.619 μm are achieved, respectively. The final splitting length (L_S) is 94 μm, and the splitting bandwidth is 349 nm (1.352~1.701 μm), covering the whole of the E + S + C + L + U communication bands. The proposed LC-DC-PCF PBS has good beam-splitting performance, such as ultra-short L_S and ultra-wide splitting bandwidth, with potential applications in laser, sensing, and communication systems.

Keywords: liquid crystal; dual-core photonic crystal fiber; polarization beam splitter; extinction ratio

Citation: Qu, Y.; Han, Y.; Yuan, J.; Zhou, X.; Yan, B.; Wang, K.; Sang, X.; Yu, C. A Novel Liquid Crystal-Filled, Dual-Core Photonic Crystal Fiber Polarization Beam Splitter Covering the E + S + C + L + U Communication Band. *Photonics* **2021**, *8*, 461. https://doi.org/10.3390/photonics8110461

Received: 19 August 2021
Accepted: 19 October 2021
Published: 21 October 2021

1. Introduction

In the late 1990s, J.C. Knight et al. proposed and successfully fabricated the first photonic crystal fiber (PCF), whose cladding was composed of micron-sized air holes arranged according to certain rules [1–3]. In the following 20 years, due to the flexible cladding microstructure of PCF and the maturity of fiber post-processing technology, the PCF has experienced incredible development in various fields of optics and photonics [4–10]. Especially after all six optical fiber communication windows O (1260–1360 nm), E (1360–1460 nm), S (1460–1530 nm), C (1530–1565 nm), L (1565–1625 nm), and U (1625–1675 nm) are proposed, PCF has more significant advantages in the field of multi window optical fiber communication.

In recent years, researchers worldwide have conducted a great deal of research on the dual-core photonic crystal fiber (DC-PCF) [11–15]. A large number of studies have shown that a polarization beam splitter (PBS) with excellent performance can be obtained by using the coupled mode theory of DC-PCF and fiber post-processing technology, such as filling with liquid crystal (LC), liquid, or precious metal materials in some fixed air holes of DC-PCF [16–23]. However, the commonly used precious metal filling materials are generally gold or silver materials, which are expensive, and the deposition process is complex. Complex methods such as high-pressure chemical vapor deposition and radio frequency magnetron sputtering are usually needed. Compared with gold and silver materials, LC or liquid filling materials have a lower price and simpler filling process, so many researchers have favored them in recent years. In 2016, Wang et al. proposed a DC-PCF PBS filled with LC E7 in all the air holes of the cladding, where the shortest splitting length (L_S) was 890.5 μm and the widest splitting bandwidth was 150 nm, which almost covers all the S + C + L communication bands [24]. In 2017, Hagras et al. reported an ultra-compact DC-PCF PBS filled with LC E7 in six big air holes of the cladding. The L_S and splitting bandwidth of the DC-PCF PBS were 111.2 μm and 88 nm, respectively [25]. In 2018, Younis et al. designed a DC-PCF PBS whose left core was surrounded by LC E7-filled air holes. At wavelengths 1.3 and 1.55 μm, the splitting bandwidths of the DC-PCF PBS were always 3 nm, and L_S were 5.678 and 7.178 mm, respectively [26]. In 2019, Xu et al. proposed a DC-PCF PBS based on rare metal, Ti, and liquid filling in the cladding air holes. The proposed DC-PCF PBS offered an ultra-short length of 83.9 μm, a high extinction ratio (*ER*) of 44.05 dB, but the splitting bandwidth was only 32.1 nm [27]. However, the above-mentioned DC-PCF PBS, LC, or liquid are always filled in multiple air holes, and some DC-PCF PBS also involves filling multiple materials simultaneously or even filling rare metals, which will lead to increased fabrication difficulty and fabrication cost. In addition, it is difficult for the above-mentioned DC-PCF PBS to obtain a shorter L_S and a wider splitting bandwidth at the same time.

In this paper, we propose a novel LC-filled DC-PCF (LC-DC-PCF) PBS based on the DC-PCF-coupled mode theory. With the full vector-finite element method (FV-FEM), the differences of the effective refractive indices of odd and even modes, mode birefringence of odd and even modes, coupling length (*CL*) of the X-polarization (X-pol) and Y-polarization (Y-pol), and coupling length ratio (*CLR*) in LC-DC-PCF filled without and with LC E7 are simulated and compared. Furthermore, the influences of structural parameters on the *CLs* of the X-pol and Y-pol and *CLR* are analyzed. Finally, a LC-DC-PCF PBS with ultra-short L_S (94 μm) and ultra-wide splitting bandwidth (349 nm) is obtained, covering the whole of the E + S + C + L + U communication bands.

2. Design of the LC-DC-PCF PBS

The cross-sectional structure of the proposed LC-DC-PCF PBS is shown in Figure 1. From Figure 1, the most central air hole with the diameter of d_1 is filled with the LC E7 to obtain high mode birefringence. Near the most central air hole, one air hole is missing in the X-positive direction, and the other is missing in the X-negative direction to form cores A and B, respectively. To obtain higher mode birefringence, there are three different sizes of air holes around cores A and B, with diameters of d_2, d_3, and d_4, respectively. The diameter of the remaining air hole is d_5, and the distance between any two air holes is Λ. The proposed LC-DC-PCF is arranged in a triangular lattice, and the background material is silica. A perfectly matched layer (PML) is added to the outermost layer of the proposed LC-DC-PCF to absorb radiation energy [28]. The refractive indices of the silica material and PML are n_{silica} and n_{silica} + 0.05, respectively. In addition, it also can be seen from Figure 1 that cores A and B of the proposed LC-DC-PCF are two symmetrical cores with the same structure.

The refractive index of the n_{silica} at different wavelengths can be obtained by the Sellmeier equation [29]:

$$n_{\text{silica}}(\lambda) = \sqrt{1 + \frac{A_1\lambda^2}{\lambda^2 - B_1^2} + \frac{A_2\lambda^2}{\lambda^2 - B_2^2} + \frac{A_3\lambda^2}{\lambda^2 - B_3^2}}, \tag{1}$$

where λ is the wavelength of the incident light in the free space. Each coefficient in the Sellmeier equation for n_{silica} can be obtained from Ref. [29].

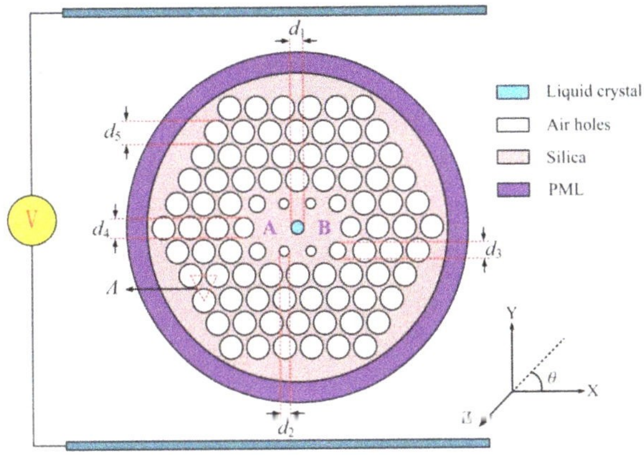

Figure 1. The cross-sectional structure of the proposed LC-DC-PCF PBS.

The ordinary refractive index n_o and the extraordinary refractive index n_e of the LC E7 can be described by the extended Cauchy equation [30]:

$$n_o = D_o + \frac{E_o}{\lambda^2} + \frac{F_o}{\lambda^4}, \tag{2}$$

$$n_e = D_e + \frac{E_e}{\lambda^2} + \frac{F_e}{\lambda^4}, \tag{3}$$

where D_o, E_o, F_o, D_e, E_e, and F_e have some specific functional relationships with temperature, which can be obtained from Ref. [30]. The temperature T is 25 °C D_o = 1.4994, E_o = 0.007 μm^2, F_o = 0.0004 μm^4, D_e = 1.6933, E_e = 0.0078 μm^2, and F_e = 0.0028 μm^4. Here, we set the T to a fixed value of 25 °C.

The relative permittivity of the LC E7 is defined as [31]

$$\varepsilon_r = \begin{pmatrix} n_o^2\sin^2(\theta) + n_e^2\cos^2(\theta) & (n_e^2 - n_o^2)\cos(\theta)\sin(\theta) & 0 \\ (n_e^2 - n_o^2)\cos(\theta)\sin(\theta) & n_o^2\cos^2(\theta) + n_e^2\sin^2(\theta) & 0 \\ 0 & 0 & n_o^2 \end{pmatrix}, \tag{4}$$

where θ is the angle between the direction vector of the LC E7 material and the X-axis. θ can be controlled by an external electric field, such as placing the proposed LC-DC-PCF between two electrodes, as shown in Figure 1. The LC E7 molecules can usually be arranged in three forms by modulating the external electric field: the long axis of the LC E7 molecules is parallel to the X-axis, that is, θ is equal to 0°. θ between the long axis of the LC E7 molecule and the X-axis is 45°. The long axis of the LC E7 molecule is perpendicular to the X-axis; that is, θ is equal to 90°. Here, we set θ to the most commonly used fixed value of 90° [31].

The even mode birefringence (B_e) and odd mode birefringence (B_o) can be defined as [32]

$$B_e = \left| n_{even}^X - n_{even}^Y \right|, \tag{5}$$

$$B_o = \left| n_{odd}^X - n_{odd}^Y \right|, \tag{6}$$

where n_{even}^X, n_{even}^Y, n_{odd}^X, and n_{odd}^Y represent the effective refractive indices of the even and odd modes of the X-pol and Y-pol, respectively.

Because the DC structures of DC-PCF are completely symmetrical and the medium distribution is the same, the dual-core mode coupling equation can be expressed as [33]

$$\frac{da_1(z)}{dz} = i\beta a_1(z) + iKa_2(z), \tag{7}$$

$$\frac{da_2(z)}{dz} = i\beta a_2(z) + iKa_1(z), \tag{8}$$

where β is the propagation constant of the DC, K is the coupling coefficient of the dual core, and z represents the transmission direction and distance.

By setting a series of initial condition parameters, combined with the fact that the DC-PCF has even mode and odd mode in the X-pol and Y-pol, respectively, the mode coupling equation is solved, and the *CLs* of the DC-PCF can be obtained.

The *CLs* of the X-pol and Y-pol of the LC-DC-PCF PBS can be calculated as [34]

$$CL_X = \frac{\lambda}{2|(n^X_{even} - n^X_{odd})|}, \tag{9}$$

$$CL_Y = \frac{\lambda}{2|(n^Y_{even} - n^Y_{odd})|}, \tag{10}$$

where CL_X and CL_Y represent the CL of the X-pol and Y-pol, respectively.

The *CLR* can be calculated by [35]

$$CLR = \frac{CL_Y}{CL_X}, \tag{11}$$

According to the previous studies, when the optimal CLR = 2 or 1/2 and the CL_X and CL_Y are shorter, it is easier for the PBS to have the shortest L_S [36].

Because cores A and B of the proposed LC-DC-PCF have symmetrical and identical structures, only incident light from cores A or B should be considered [37]. Here, we assume that the incident light is incident from core A. The normalized output power (P_{out}) in the X-pol and Y-pol in the core A can be described as [38]

$$P^{X,\,Y}_{out,\,A} = P_{in}\cos^2\left(\frac{\pi}{2}\frac{L_P}{CL_{X,Y}}\right), \tag{12}$$

where P_{in} is the input power of the incident light, and L_P is the propagation length of the LC-DC-PCF PBS. According to Equation (10), L_P will change periodically. According to previous studies, the shortest L_P is also the shortest L_S [39].

The *ER* of the core A can be calculated as [40].

$$ER_A = 10\log_{10}\frac{P^X_{out,\,A}}{P^Y_{out,\,A}}, \tag{13}$$

For the PBS, the *ER* is a significant indicator with which to judge and characterize the performance of the PBS. In practical application, when the *ER* reaches 20 dB, the power of the polarization light is 100 times that of the other, which is enough to separate two orthogonal polarization lights. Generally speaking, the wavelength range with the *ER* greater than 20 dB is the splitting bandwidth of the PBS [41].

3. Simulation Results

The initial structural parameters of the LC-DC-PCF are set as follows: d_1 = 0.85 μm, d_2 = 1.00 μm, d_3 = 1.50 μm, d_4 = 1.60 μm, d_5 = 1.65 μm, and Λ = 2.10 μm. The relationships between the effective refractive indices of the X-pol and Y-pol even and odd modes of the LC-DC-PCF and the wavelength are shown in Figure 2a,b when the LC-DC-PCF is filled without and with LC E7, respectively. From Figure 2a, when the LC-DC-PCF is filled without LC E7, the effective refractive indices of the X-pol and Y-pol even and odd modes decrease with the increase in wavelength. The effective refractive index curves of the X-pol and Y-pol even modes are very close, and those of the X-pol and Y-pol odd modes are virtually overlapped. In other words, the effective refractive index differences between the X-pol and Y-pol even modes or X-pol and Y-pol odd modes are very small. Therefore, the values of the B_e and B_o are also very small. From Figure 2b, when the LC-DC-PCF is filled with LC E7, the effective refractive indices of the X-pol and Y-pol even and odd modes decrease with the increase in wavelength. This change trend is similar to that filled without LC E7. The effective refractive index curves of the X-pol and Y-pol odd modes are also nearly overlapped. However, the effective refractive index values of the X-pol and Y-pol even modes have an obvious change, respectively. In other words, the effective refractive index difference between X-pol and Y-pol odd modes is also very small, but the effective refractive index difference between X-pol and Y-pol even modes is a larger value. Therefore, the B_o will still be small, but the B_e will be large. The B_e and B_o of the LC-DC-PCF are shown in Figure 3 when the LC-DC-PCF is filled without and with LC E7. From Figure 3, the B_e and B_o are kept very small with the wavelength increase when the LC-DC-PCF is filled without LC E7. When the LC-DC-PCF is filled with LC E7, the B_o still keeps a small value, but the B_e increases. B_e varies from 6.7×10^{-3} to 1.2×10^{-2} in the wavelength range of 1.3~1.8 μm, which is a relatively large mode birefringence compared with many previous studies. According to the above analysis, a large B_e can be introduced by filling LC E7 in the most central air hole.

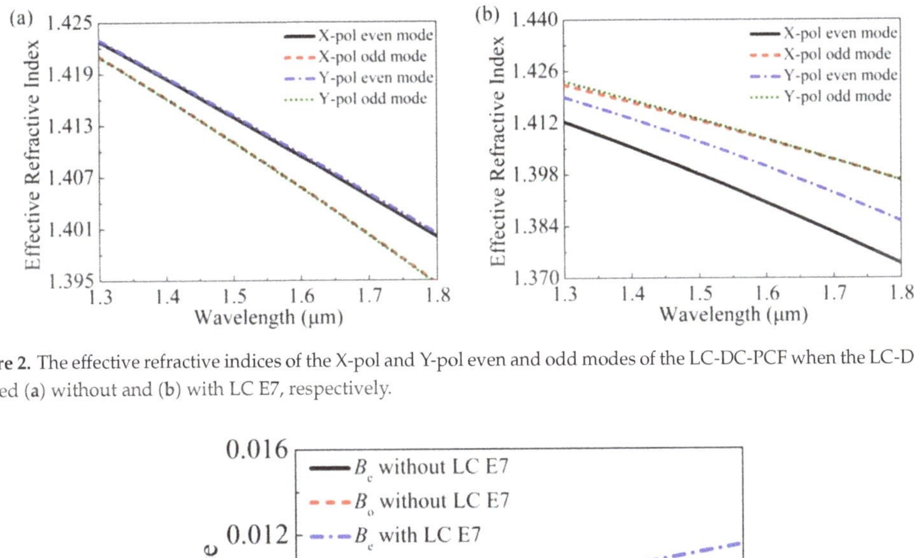

Figure 2. The effective refractive indices of the X-pol and Y-pol even and odd modes of the LC-DC-PCF when the LC-DC-PCF is filled (**a**) without and (**b**) with LC E7, respectively.

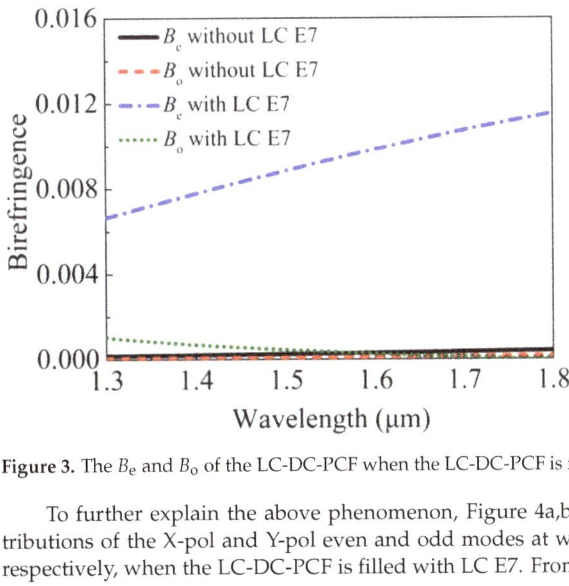

Figure 3. The B_e and B_o of the LC-DC-PCF when the LC-DC-PCF is filled without and with LC E7.

To further explain the above phenomenon, Figure 4a,b shows the mode field distributions of the X-pol and Y-pol even and odd modes at wavelengths 1.3 and 1.8 μm, respectively, when the LC-DC-PCF is filled with LC E7. From Figure 4a,b, when the LC-

DC-PCF is filled with LC E7, the mode field energies of the X-pol and Y-pol odd modes are mainly distributed in cores A and B at wavelengths 1.3 and 1.8 μm, respectively. According to previous work, this is the same as that of conventional DC-PCF. However, no matter at 1.3 or 1.8 μm, most of the mode field energy of the X-pol and Y-pol even modes is still distributed in cores A and B, but a small part is distributed in the most central air hole filled with LC E7. This is also why B_e is larger when the LC E7 is filled into the most central air hole. Therefore, the most central air hole filled with LC E7 can be regarded as a modulation core. At wavelengths 1.3 and 1.8 μm, the energy percentages of mode field energy of the X-pol and Y-pol even modes transmitted in the modulation core are 6.79%, 1.23%, and 6.96%, 3.46%, respectively. It can be seen from Figure 4a,b that the energy transmitted in cores A and B is much greater than that in modulation core. Therefore, the CL_X, CL_Y, and CLR can be analyzed using the DC-PCF coupling mode theory [42] and Equations (9)–(11).

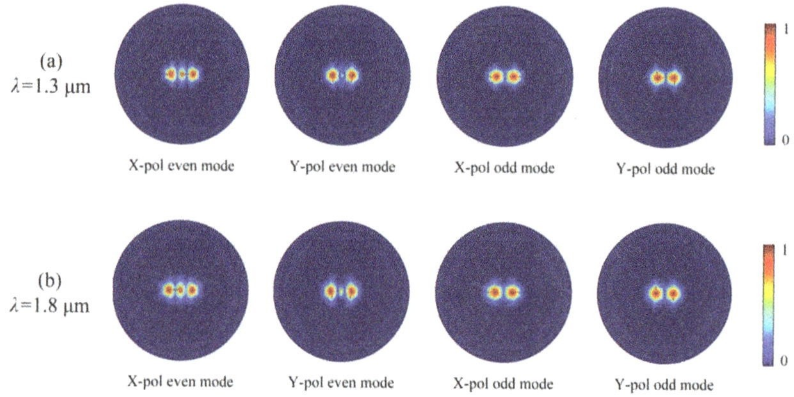

Figure 4. The mode field distribution of the X-pol and Y-pol even and odd modes at (**a**) 1.3 μm and (**b**) 1.8 μm when the LC-DC-PCF is filled with LC E7.

Figure 5a,b shows the CL_X, CL_Y, and CLR of the LC-DC-PCF when the LC-DC-PCF is filled without and with LC E7, respectively. From Figure 5a, when the LC-DC-PCF is filled without LC E7, the CL_X, CL_Y, and CLR decrease approximately linearly with the wavelength increase. However, the difference between CL_X and CL_Y is small, which leads to smaller values and smaller changes of CLR. This phenomenon can be deduced from the results shown in Figure 2a and Equations (9)–(11). In addition, at a wavelength of 1.8 μm, the minimum values of the CL_X and CL_Y are 151 and 167 μm, respectively. At wavelengths of 1.3 and 1.8 μm, the largest and smallest CLR are 1.102 and 1.125, respectively. From Figure 5b, when the LC-DC-PCF is filled with LC E7, although the CL_X and CL_Y decrease approximate linearly with the increase in wavelength, the difference between the CL_X and CL_Y decreases gradually. However, the minimum difference is also much larger than when the LC-DC-PCF is filled without LC E7. With the increase in wavelength, CLR has a trend of increasing first and then decreasing. This is because the introduction of the modulation core produces a large B_e, leading to a large change in the effective refractive index of the X-pol and Y-pol even modes, which leads to the corresponding change of the CL_X, CL_Y, and CLR. At 1.3 μm, the maximum values of the CL_X and CL_Y are 65 and 148 μm, respectively. Additionally, in the wavelength range of 1.3~1.8 μm, the minimum and maximum values of CLR are 2.057 and 2.395, respectively. Therefore, when the LC-DC-PCF is filled with LC E7, the maximum CL_X and CL_Y are smaller than the minimum CL_X and CL_Y when the LC-DC-PCF is filled without LC E7, and the CLR value in the range of 1.3~1.8 μm is closer to the optimal $CLR = 2$. In other words, the LC-DC-PCF is filled with LC E7; if $CLR = 2$ is obtained at a certain wavelength in the communication band and the CL_X and CL_Y have a small value, the proposed LC-DC-PCF PBS may obtain better beam splitting characteristics.

Therefore, in the following, we will discuss the change rule of the CL_X, CL_Y, and *CLR* of the proposed LC-DC-PCF with the change of each structural parameter.

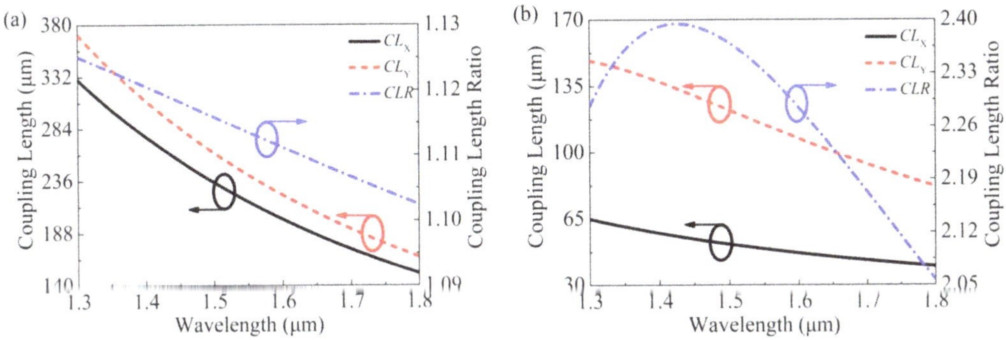

Figure 5. The CL_X, CL_Y, and *CLR* of the LC-DC-PCF when the LC-DC-PCF is filled (**a**) without and (**b**) with LC E7, respectively.

The change rules of the CL_X, CL_Y, and *CLR* of the proposed LC-DC-PCF with wavelengths for different d_1 are shown in Figure 6a–c. From Figure 6a, when d_1 increases from 0.65 to 1.05 μm, the CL_X increases gradually in the whole wavelength range of 1.3~1.8 μm. However, the increased extent of the CL_X at a short wavelength is larger than that at a long wavelength. From Figure 6b, when d_1 increases from 0.65 to 0.75 μm, the CL_Y also increases gradually in the whole wavelength range of 1.3~1.8 μm. However, when d_1 increases from 0.75~0.95 μm, the CL_Y decreases gradually at the short wavelength, and the decreasing range is larger, while the CL_Y increases gradually at the long wavelength and the increasing range is smaller. When d_1 increases from 0.95 to 1.05 μm, the CL_Y decreases gradually in the wavelength range of 1.3~1.8 μm, while the CL_Y has a larger decrease at a short wavelength and a smaller decrease at a long wavelength. On the whole, the change range of the CL_Y in the short wavelength is larger than that in the long wavelength when d_1 increases from 0.75 to 1.05 μm. From Figure 6c, the *CLR* also increases in the whole wavelength range of 1.3~1.8 μm with the increase in d_1 from 0.65 to 0.75 μm, then decreases in the short wavelength and increases in the long wavelength with the increase in d_1 from 0.75 to 0.95 μm, and finally decreases in the whole wavelength range of 1.3~1.8 μm with the increase in d_1 from 0.95 to 1.05 μm. On the whole, the *CLR* occurs to fluctuate in the range from 0.84 to 2.74. When d_1 is 0.85 μm, the minimum and maximum of the *CLR* are 2.3948 and 2.0568, respectively, and the difference between them is 0.338.

The change rules of the CL_X, CL_Y, and *CLR* of the proposed LC-DC-PCF with wavelengths for different d_2 are shown in Figure 7a–c. From Figure 7a, when d_2 increases from 0.80 to 1.20 μm, the CL_X maintains a relatively stable increase in the whole wavelength range of 1.3~1.8 μm. However, overall, the changing amplitude of the CL_X is not very large. From Figure 7b, when d_2 increases from 0.80 to 1.00 μm, the CL_Y changes little at the short wavelength but increases steadily at the long wavelength. The CL_Y also maintains a relatively stable increase in the whole wavelength range of 1.3~1.8 μm when d_2 increases from 1.00 to 1.20 μm. However, on the whole, the changing amplitude of the CL_Y is slightly larger than that of the CL_X. According to the change rule of the CL_X and the CL_Y with d_2, the *CLR* decreases gradually at the short wavelength and increases gradually at the long wavelength when d_2 increases from 1.00 to 1.20 μm. In addition, it is found that the larger the d_2, the smaller the difference between the minimum and maximum *CLR*. When d_2 is 1.20 μm, the minimum and maximum of the *CLR* are 2.3994 and 2.1534, respectively, and the difference between them is only 0.246.

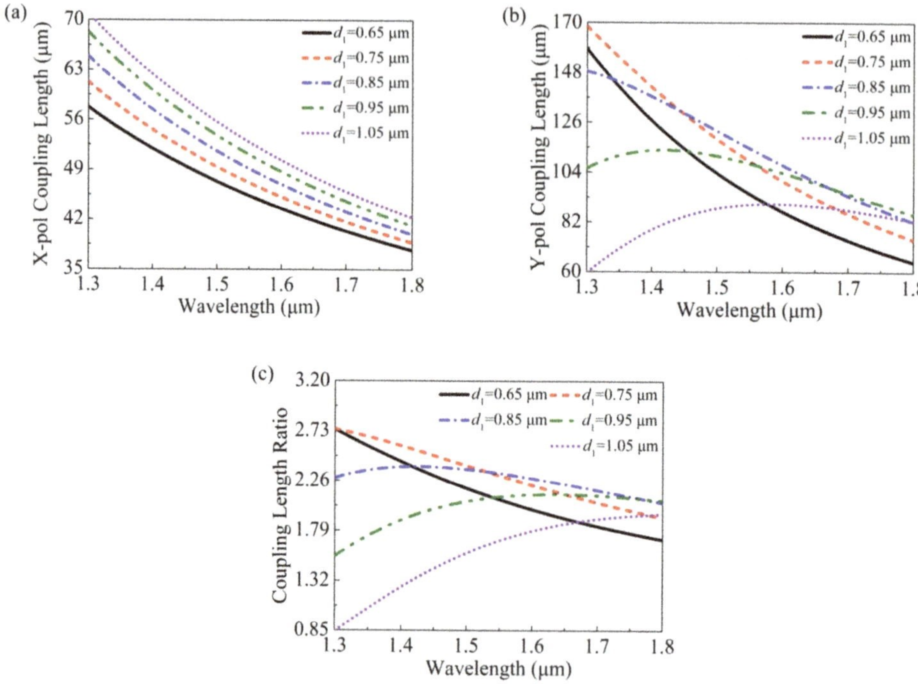

Figure 6. The change rules of the (**a**) CL_X, (**b**) CL_Y, and (**c**) CLR of the LC-DC-PCF with wavelengths for different d_1.

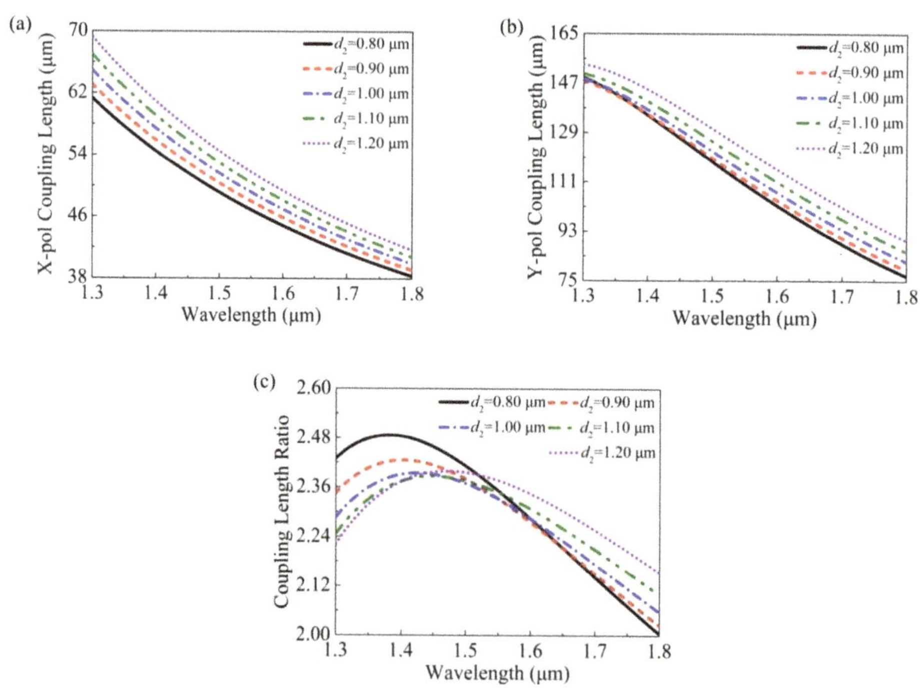

Figure 7. The change rules of the (**a**) CL_X, (**b**) CL_Y, and (**c**) CLR of the LC-DC-PCF with wavelengths Figure 2.

The change rules of the CL_X, CL_Y, and CLR of the proposed LC-DC-PCF with wavelengths for different d_3 are shown in Figure 8a–c. From Figure 8a,b, the CL_X and CL_Y decrease gradually when d_3 increases from 1.30 to 1.70 μm in the whole wavelength range of 1.3~1.8 μm. The decreasing amplitude of the CL_Y is slightly larger than that of the CL_X. Thus, it can be seen from Figure 8c that the CLR maintains a relatively stable decrease when d_3 increases from 1.30 to 1.70 μm in the whole wavelength range of 1.3~1.8 μm. In addition, we also find that the larger the d_3, the smaller the difference between the minimum and maximum CLR. When d_3 is 1.70 μm, the minimum and maximum of the CLR are 2.2541 and 1.9409, respectively, and the difference between them is only 0.3132.

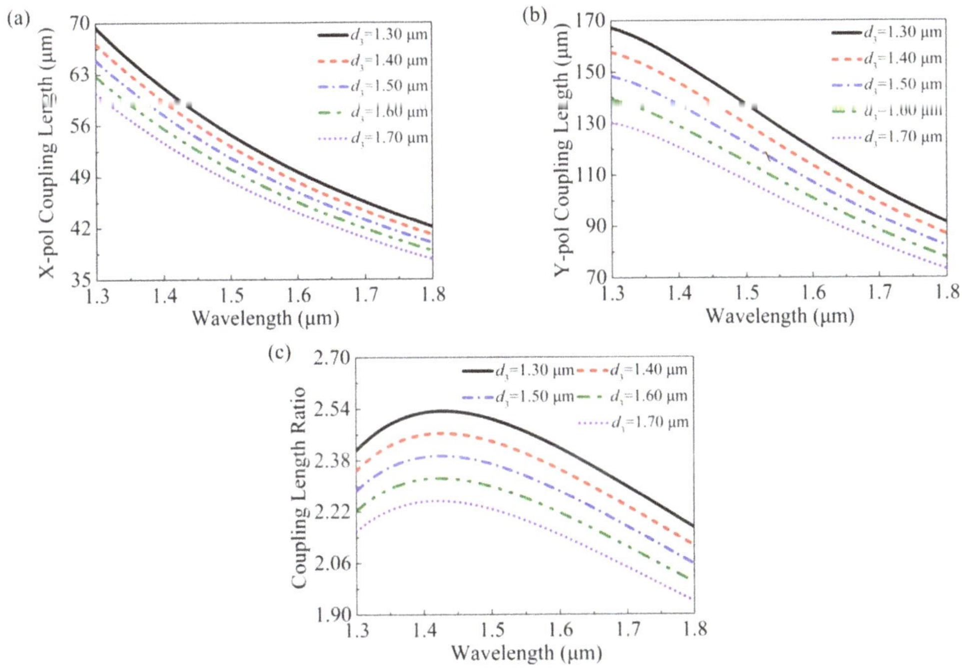

Figure 8. The change rules of the (**a**) CL_X, (**b**) CL_Y, and (**c**) CLR of the LC-DC-PCF with wavelengths for different d_3.

The change rules of the CL_X, CL_Y, and CLR of the proposed LC-DC-PCF with wavelengths for different d_4 are shown in Figure 9a–c. From Figure 9a,b, the CL_X and CL_Y gradually decrease when d_4 increases from 1.40 to 1.80 μm in the whole wavelength range of 1.3~1.8 μm. The decreasing amplitude of the CL_X and CL_Y is nearly the same. Therefore, as shown in Figure 9c, the CLR has a small change when d_4 increases from 1.40 to 1.80 μm in the whole wavelength range of 1.3~1.8 μm. In addition, when d_4 is 1.40 μm and 1.8 μm, the minimum and maximum of the CLR are 2.0436, 2.3835, and 2.0653, 2.3996, respectively. The maximum difference of the CLR is only 0.0217.

The change rules of the CL_X, CL_Y, and CLR of the proposed LC-DC-PCF with wavelengths for different d_5 are shown in Figure 10a–c. From Figure 10a,b, the curves under different d_5 are virtually coincident, that is to say, the CL_X and CL_Y do not change when d_5 increases from 1.40 to 1.80 μm in the whole wavelength range of 1.3~1.8 μm. Certainly, the CLR will not change with the change of d_5 in the whole wavelength range of 1.3~1.8 μm.

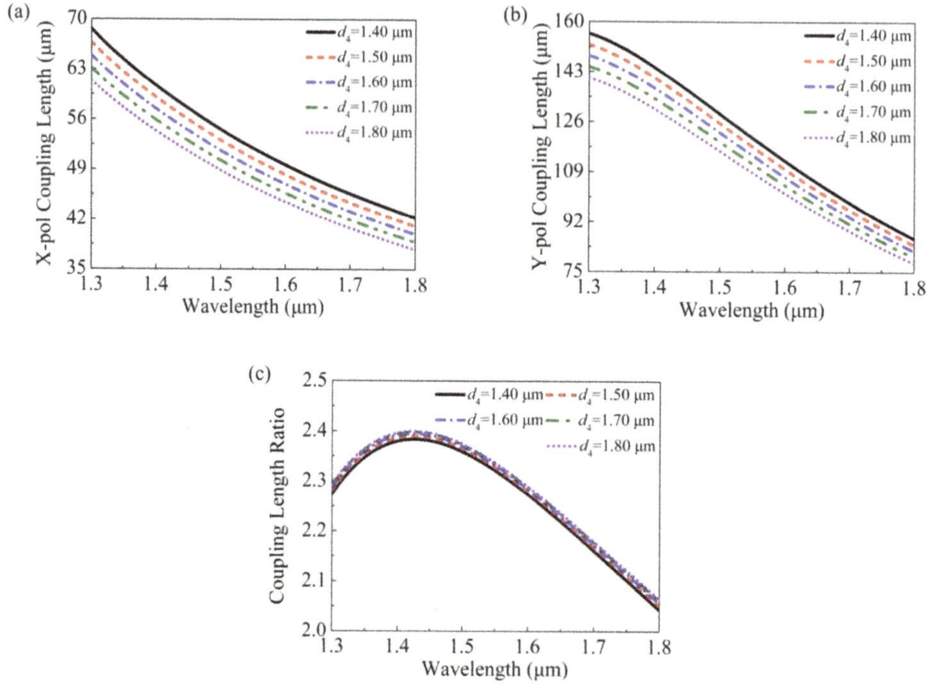

Figure 9. The change rules of the (**a**) CL_X, (**b**) CL_Y, and (**c**) CLR of the LC-DC-PCF with wavelengths for different d_4.

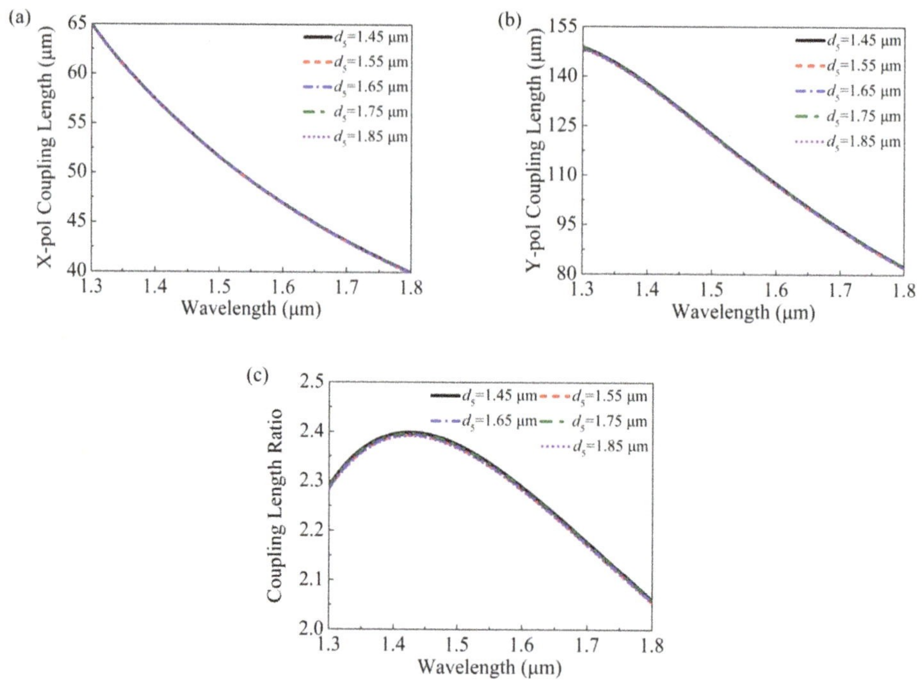

Figure 10. The change rules of the (**a**) CL_X, (**b**) CL_Y, and (**c**) CLR of the LC-DC-PCF with wavelengths for different d_5.

The change rules of the CL_X, CL_Y, and CLR of the proposed LC-DC-PCF with wavelengths for different Λ are shown in Figure 11a–c. From Figure 11a it can be seen that the CL_X increases gradually in the whole wavelength range of 1.3~1.8 μm when Λ increases from 1.90 to 2.30 μm. From Figure 11b, although the CL_Y also increases gradually in the whole wavelength range of 1.3~1.8 μm when Λ increases from 1.90 to 2.30 μm, the increasing amplitude of the CL_Y is larger at the shorter wavelength and smaller at the longer wavelength. Hence, when Λ increases from 1.90 to 2.30 μm, the CLR increases gradually, but the increased amplitude of the CLR is larger at the shorter wavelength and smaller at the longer wavelength. When Λ is 1.90 μm, the maximum and minimum of the CLR are 2.0778 and 1.9035, respectively, and the difference between them is only 0.1743. When Λ is 2.30 μm, the maximum and minimum of the CLR are 2.2349 and 2.8137, respectively, and the difference between them can reach 0.5877. That is to say, the difference between the maximum and minimum of the CLR becomes larger when Λ increases from 1.90 to 2.30 μm.

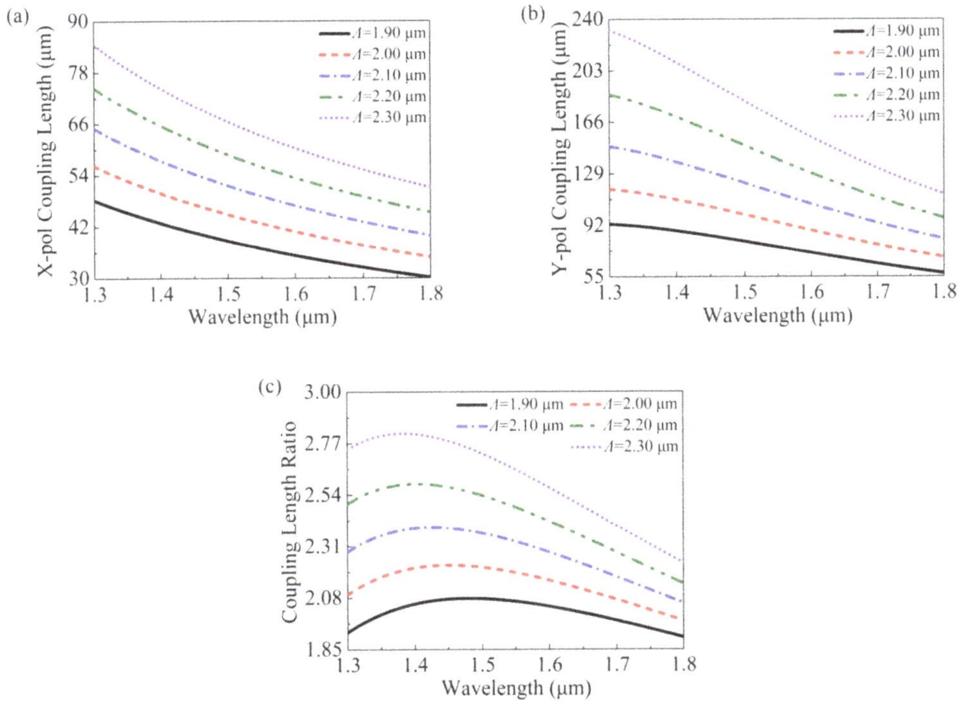

Figure 11. The change rules of the (**a**) CL_X, (**b**) CL_Y, and (**c**) CLR of the LC-DC-PCF with wavelengths for different Λ.

4. Discussion

According to the above analysis, considering the influences of structural parameters on the CL_X, CL_Y, and CLR of the proposed LC-DC-PCF, a set of final parameters is set as d_1 = 1.00 μm, d_2 = 0.90 μm, d_3 = 1.50 μm, d_4 = 1.65 μm, d_5 = 1.70 μm, and Λ = 2.10 μm. The relationships of the CL_X, CL_Y, and CLR of the proposed LC-DC-PCF with wavelengths obtained under the final parameters are shown in Figure 12. From Figure 12, the CL_X decreases with the increase in wavelength, while the CL_Y first increases and then decreases with the increase in wavelength. At the same time, we notice that the CL_X and CL_Y have a relatively small length, and the maximum length is only 98.4 μm. With the increase in wavelength, the CLR first presents a gradual upward trend, but the upward trend gradually decreases and finally has a nearly flat change. When the wavelength is equal to 1.604 μm, the CLR is equal to 2. Before the wavelength of 1.604 μm, the range of the CLR change is 1.22~2.0. After the wavelength of 1.604 μm, the range of the CLR change is only 2.0~2.036.

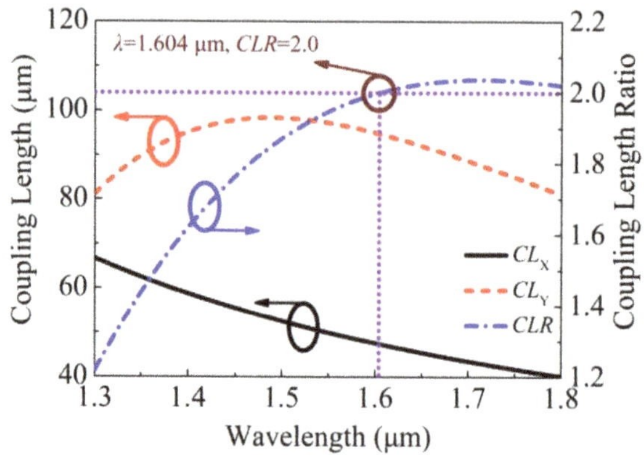

Figure 12. The relationships of the CL_X, CL_Y, and CLR of the proposed LC-DC-PCF with wavelengths.

We mentioned earlier that, when the CLR = 2 and the CL_X and CL_Y are shorter, the proposed PBS is easier to obtain the shortest L_S. Therefore, the relationships between the X-pol and Y-pol P_{out} in the core A of the proposed LC-DC-PCF PBS and L_P at the wavelength of 1.604 μm are shown in Figure 13. From Figure 13, when the L_P is 0 μm, P_{out} of the X-pol and Y-pol is the maximum because the initial incident light enters the core A. When L_P is 94 μm, P_{out} of the X-pol reaches the maximum again, while P_{out} of the Y-pol reaches 0 for the first time. This shows that only X-pol light exists in core A, and Y-pol light only exists in core B. When L_P is 188 μm, P_{out} of the X-pol and Y-pol is once again maximum. It indicates that both the X-pol and Y-pol light completely exist in core A again, just as in the initial incident. That is enough to show that P_{out} of the X-pol and Y-pol changes periodically with the increase in L_P. That is to say, the X-pol and Y-pol light propagate periodically in cores A and B. When the X-pol and Y-pol light are separated to cores A and B for the first time, respectively, the corresponding L_P = 94 μm is the shortest L_S.

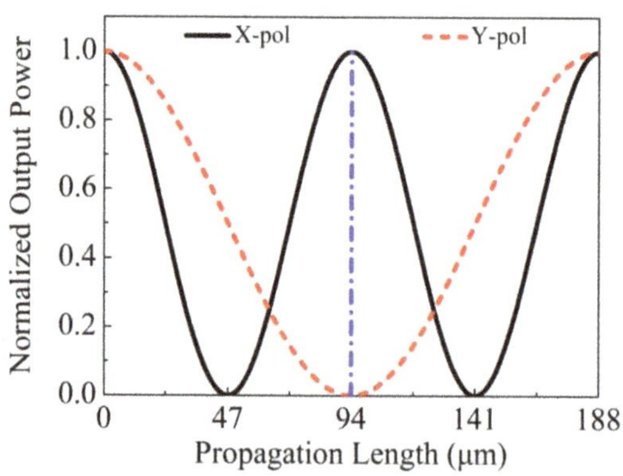

Figure 13. The relationships between the X-pol and Y-pol P_{out} in the core A of the proposed LC-DC-PCF PBS and the L_P at 1.604 μm.

When L_P is 94 μm, the relationship of the ER in the core A of the proposed LC-DC-PCF with wavelengths is shown in Figure 14. It can be seen from Figure 14 that with the increase in wavelength, the ER first increases and then reaches the first maximum peak value of 60.3 dB at the wavelength of 1.386 μm. Then, as the wavelength continues to increase, the *ER* first decreases and then increases; a relatively gentle *ER* valley is formed between 1.386 and 1.619 μm, and then reaches the second maximum peak value of 72.2 dB at the wavelength of 1.619 μm. After 1.619 μm, the *ER* decreases gradually with the increase in wavelength. Between 1.352 and 1.701 μm, the *ER* is greater than 20 dB. In other words, the splitting bandwidth of the proposed LC-DC-PCF PBS can reach 349 nm (1.352~1.701 μm), which can cover the whole of the E + S + C + L + U communication bands.

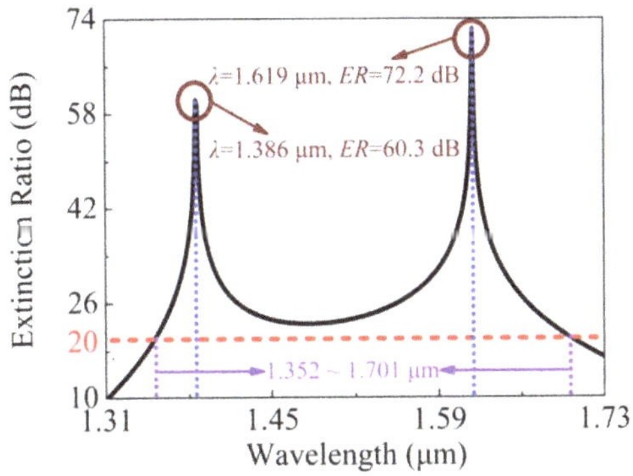

Figure 14. The relationship of the *ER* in the core A of the proposed LC-DC-PCF with wavelengths.

The comparisons between the proposed LC-DC-PCF PBS and reported DC-PCF PBS are shown in Table 1. From Table 1, since the cost of liquid crystal E7 is significantly lower than that of gold, from the perspective of structural fabrication and cost reduction, the proposed LC-DC-PCF PBS in this work is only filling liquid crystal E7 in one air hole, does not fill other materials, and does not contain elliptical air holes, which not only reduces the fabrication difficulty but also reduces the fabrication cost. In addition, only L_S of the DC-PCF PBS reported in Ref. [27] is slightly shorter than that of the proposed LC-DC-PCF PBS, but the splitting bandwidth of the proposed LC-DC-PCF PBS is more than ten times that of the DC-PCF PBS reported in Ref. [27]. Moreover, the maximum *ER* of the DC-PCF PBS reported in Refs. [16,38,39] is slightly higher than that of the proposed LC-DC-PCF PBS, but the proposed LC-DC-PCF PBS has the widest splitting bandwidth and the shortest L_S compared with other results reported in Refs. [16,38,39]. Because the splitting bandwidth of the DC-PCF PBS is the wavelength range corresponding to the *ER* greater than 20 dB, even if the maximum ER of a certain wavelength is large, the final splitting bandwidth is not necessarily the widest, so it is only necessary to compare the L_S and splitting bandwidth. To summarize, it can be seen that the proposed LC-DC-PCF PBS in this work can obtain a shorter L_S and wider splitting bandwidth at the same time and has a lower cost and simpler fabrication process.

Table 1. Comparisons between the proposed LC-DC-PCF PBS and reported DC-PCF PBS.

Ref.	Structural Characteristics	Splitting Bandwidth	L_S	Max *ER*
[16]	Filling liquid crystal E7 in one air hole	250 nm	175 μm	80.7 dB
[24]	Filling liquid crystal E7 in all air holes	150 nm	890.5 μm	45 dB
[25]	Filling liquid crystal E7 in six air holes and filling As_2S_3 in one air hole	88 nm	111.2 μm	55 dB
[26]	Filling liquid crystal E7 in six air holes	3 nm	5678 μm	<30 dB
[27]	Filling liquid (ethanol) in six air holes and filling Ti in two air holes	32.1 nm	83.9 μm	44.05 dB
[38]	Filling elliptical gold wire in one air hole	70 nm	1079 μm	174.92 dB
[39]	Coating gold film in one air hole	318 nm	188 μm	<82dB
[40]	Filling silver wire in one air hole	250 nm	577.5 μm	42 dB
This work	Filling liquid crystal E7 in one air hole	349 nm	94 μm	72.2 dB

The fabrication methods of the PCF mainly include the stack-and-draw method, femtosecond laser drilling method, 3D-printing method, etc., [3,43–45]. In addition, the technology of selectively filling LC in some or all of the air holes has also been very mature; in the past three years, there have been many reports on the practical fabrication of PCFs selectively filled with LC material [46–50]. In this work, the fabrication steps of the proposed LC-DC-PCF are as follows: first, the proposed LC-DC-PCF is fabricated by the stack-and-draw method. With this method, the high-purity silica glass tubes of different diameters and wall thicknesses are arranged according to the designed structure. Two high-purity silica solid rods with the same diameter are used to replace the high-purity silica glass tubes on the left and right sides of the most central high-purity silica glass tubes, respectively, to form the LC-DC-PCF preform, and then the LC-DC-PCF preform is drawn into the designed LC-DC-PCF. In the fabrication process, parameters such as temperature, air pressure, and traction speed must be controlled simultaneously to obtain the LC-DC-PCF with an excellent final structure. Finally, the LC E7 is selectively filled in the central air hole of the proposed LC-DC-PCF by femtosecond laser-assisted selective infiltration technology. Among them, the technology mainly includes several steps, such as UV curable, normal splicing, laser cutting, capillary infiltration, etc.

In the actual fabrication process for the proposed LC-DC-PCF, some unnecessary tolerances will inevitably occur in the diameter of the air hole, which will lead to a ±1% change in the final L_S. When the final L_S changes ±1%, the relationships of the *ER* in the core A of the proposed LC-DC-PCF with wavelength are shown in Figure 15. From Figure 15, when the final L_S changes -1% and +1%, the wavelength ranges with *ER* greater than 20 dB are 1.346~1.713 μm and 1.357~1.689 μm, respectively, which can cover the whole of the E + S + C + L + U communication band. This indicates that the proposed LC-DC-PCF has good fabrication fault tolerance.

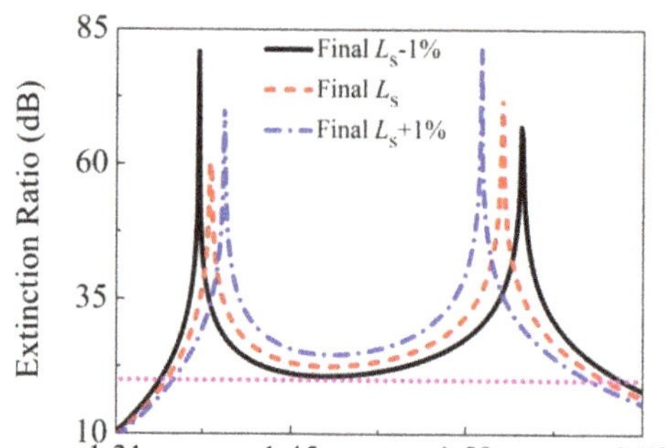

Figure 15. The relationship of the *ER* in the core A of the proposed LC-DC-PCF with wavelengths when the final L_S changes ±1%.

5. Conclusions

In summary, a novel LC-DC-PCF PBS based on the DC-PCF-coupled mode theory is proposed. Using the FV-FEM, the effective refractive indices of the X-pol and Y-pol even and odd modes, the B_e and B_o, the CL_X, CL_Y, and *CLR* of the LC-DC-PCF filled without and with LC E7 are analyzed and compared. By optimizing the influences of cladding microstructure parameters on the CL_X, CL_Y, and *CLR* of the LC-DC-PCF, the *CLR* is exactly equal to 2 at a wavelength of 1.604 μm. The relationships between the X-pol and Y-pol P_{out} in core A of the proposed LC-DC-PCF PBS and L_P at the wavelength of 1.604 μm are investigated. The *ER*s in core A achieve 60.3 and 72.2 dB at wavelengths 1.386 and 1.619 μm, respectively. The final L_S is 94 μm, and the splitting bandwidth can reach 349 nm (1.352~1.701 μm), covering the whole of the E + S + C + L + U communication bands. The proposed LC-DC-PCF PBS has an ultra-short L_S and ultra-wide splitting bandwidth, so it can be applied in laser, sensing, and communication systems.

Author Contributions: Conceptualization, Y.Q.; methodology, Y.Q. and Y.H.; software, Y.Q.; validation, J.Y.; formal analysis, Y.Q., Y.H. and J.Y.; investigation, Y.Q.; resources, Y.H. and J.Y.; data curation, Y.Q. and Y.H.; writing—original draft preparation, Y.Q.; writing—review and editing, Y.H. and J.Y.; visualization, X.Z. and B.Y.; supervision, K.W., X.S., and C.Y.; project administration, X.Z. and B.Y.; funding acquisition, J.Y. All authors have read and agreed to the published version of the manuscript.

Funding: This research was funded by the National Key Research and Development Project of China (2019YFB2204001), Program of the Natural Science Foundation of Hebei Province (F2021203002), and the Important Research Project of Hebei Province Application Foundation Plan (17961701D).

Institutional Review Board Statement: Not applicable.

Informed Consent Statement: Not applicable.

Data Availability Statement: Not applicable.

Acknowledgments: We thank the State Key Laboratory of Information Photonics and Optical Communications (Beijing University of Posts and Telecommunications of China) for the scientific helps and supports throughout this research.

Conflicts of Interest: The authors declare no conflict of interest.

References

1. Knight, J.C.; Birks, T.A.; Russell, P.S.J.; Atkin, D.M. All-silica single-mode optical fiber with photonic crystal cladding. *Opt. Lett.* **1996**, *21*, 1547–1549. [CrossRef]
2. Birks, T.A.; Knight, J.C.; Russell, P.S.J. Endlessly single-mode photonic crystal fiber. *Opt. Lett.* **1997**, *22*, 961–963. [CrossRef]
3. Knight, J.C.; Broeng, J.; Birks, T.A.; Russell, P.S.J. Photonic band gap guidance in optical fibers. *Science* **1998**, *282*, 1476–1478. [CrossRef]
4. Islam, R.; Habib, M.S.; Hasanuzzaman, G.K.M.; Ahmad, R.; Rana, S.; Kaijage, S.F. Extremely high-birefringent asymmetric slotted-core photonic crystal fiber in THz regime. *IEEE Photon. Technol. Lett.* **2015**, *27*, 2222–2225. [CrossRef]
5. Xu, Z.L.; Lim, J.L.; Hu, D.J.J.; Sun, Q.Z.; Wong, R.Y.N.; Li, K.; Jiang, M.; Shum, P.P. Investigation of temperature sensing characteristics in selectively infiltrated photonic crystal fiber. *Opt. Express* **2016**, *24*, 1699–1707. [CrossRef]
6. Hu, D.J.J.; Ho, H.P. Recent advances in plasmonic photonic crystal fibers: Design, fabrication and applications. *Adv. Opt. Photonics* **2017**, *9*, 257–314. [CrossRef]
7. Mollaha, M.S.; Yousufalia, M.; Rifat Bin Asif Faysala, M.; Rabiul Hasanb, M.; Biplob Hossainc, M.; Amiri, I.S. Highly sensitive photonic crystal fiber salinity sensor based on Sagnac interferometer. *Results Phys.* **2020**, *16*, 103022. [CrossRef]
8. Maidi, A.M.; Abas, P.E.; Petra, P.I.; Kaijage, S.; Zou, N.Y.; Begum, F. Theoretical considerations of photonic crystal fiber with all uniform-sized air holes for liquid sensing. *Photonics* **2021**, *8*, 249. [CrossRef]
9. Paul, A.K.; Mollah, M.A.; Hassan, M.Z.; Gomez-Cardona, N.; Reyes-Vera, E. Graphene-coated highly sensitive photonic crystal fiber surface plasmon resonance sensor for aqueous solution: Design and numerical analysis. *Photonics* **2021**, *8*, 155. [CrossRef]
10. Haider, F.; Aoni, R.A.; Ahmed, R.; Chew, W.J.; Mahdiraji, G.A. Plasmonic micro-channel based highly sensitive biosensor in visible to mid-IR. *Opt. Laser Technol.* **2021**, *140*, 107020. [CrossRef]
11. Wang, K.; Qu, Y.W.; Yuan, J.H.; Qiu, S.; Zhou, X.; Yan, B.B.; Wu, Q.; Liu, B.; Wang, K.R.; Sang, X.Z.; et al. Ultra-short polarization beam splitter based on dual-core photonic crystal fiber with surface plasmon resonance effect. *Opt. Eng.* **2021**, *60*, 076104. [CrossRef]
12. Wang, C.L.; Shum, P.P.; Hu, D.J.J.; Chen, Y.C.; Xu, Z.L.; Liu, S.H.; Zhang, Y.N.; Zhu, Y.W.; Zheng, Y.; Li, B.C.; et al. Two-core photonic crystal fiber with selective liquid infiltration in the central air hole for temperature sensing. *OSA Contin.* **2020**, *3*, 2264–2276. [CrossRef]
13. Zhang, Y.X.; Qu, Y.W.; Yuan, J.H.; Wang, H.Y.; Zhou, X.; Huo, J.H.; Yan, B.B.; Wu, Q.; Wang, K.R.; Sang, X.Z.; et al. Polarization beam splitter based on the gold wire-filled dual-core photonic crystal fiber at the communication wavelengths. *Fiber Integr. Opt.* **2021**, *40*, 1–14. [CrossRef]
14. Qiu, S.; Yuan, J.H.; Zhou, X.; Qu, Y.W.; Yan, B.B.; Wu, Q.; Wang, K.R.; Sang, X.Z.; Long, K.P.; Yu, C.X. Highly sensitive temperature sensing based on all-solid cladding dual-core photonic crystal fiber filled with the toluene and ethanol. *Opt. Commun.* **2020**, *477*, 126357. [CrossRef]
15. Qu, Y.W.; Yuan, J.H.; Zhou, X.; Feng, L.; Yan, B.B.; Wu, Q.; Wang, K.R.; Sang, X.Z.; Long, K.P.; Yu, C.X. Surface plasmon resonance-based silicon dual-core photonic crystal fiber polarization beam splitter at the mid-infrared spectral region. *J. Opt. Soc. Am. B* **2020**, *37*, 2221–2230. [CrossRef]
16. Chen, H.L.; Li, S.G.; Fan, Z.K.; An, G.W.; Li, J.S.; Han, Y. A novel polarization splitter based on dual-core photonic crystal fiber with a liquid crystal modulation core. *IEEE Photo. J.* **2014**, *6*, 2201109. [CrossRef]

17. Hameed, M.F.O.; Balat, R.T.; Heikal, A.M.; Abo-Elkhier, M.M.; Abo el Maaty, M.I.; Obayya, S.S.A. Polarization-independent surface plasmon liquid crystal photonic crystal multiplexer-demultiplexer. *IEEE Photo. J.* **2015**, *7*, 4801110. [CrossRef]
18. Zi, J.C.; Li, S.G.; Wang, G.Y.; An, G.W.; Fan, Z.K. Design of ultra-short polarization beam splitter based on liquid-filled photonic crystal fiber. *Opt. Quant. Electron.* **2016**, *48*, 233. [CrossRef]
19. Wang, J.S.; Pei, L.; Weng, S.J.; Wu, L.Y.; Ning, T.G.; Li, J. Ultrashort polarization beam splitter based on liquid-filled dual-core photonic crystal fiber. *Appl. Opt.* **2018**, *57*, 3847–3852. [CrossRef]
20. Wang, H.Y.; Yan, X.; Li, S.G.; Zhang, X.N. Tunable surface plasmon resonance polarization beam splitter based on dual-core photonic crystal fiber with magnetic fluid. *Opt. Quant. Electron.* **2017**, *49*, 368. [CrossRef]
21. Wang, J.S.; Pei, L.; Weng, S.J.; Wu, L.Y.; Huang, L.; Ning, T.G.; Li, J. A tunable polarization beam splitter based on magnetic fluids-filled dual-core photonic crystal fiber. *IEEE Photo. J.* **2017**, *9*, 2200410. [CrossRef]
22. Jiang, L.H.; Zheng, Y.; Hou, L.T.; Zheng, K.; Peng, J.Y.; Zhao, X.T. An ultrabraoadband polarization splitter based on square-lattice dualcore photonic crystal fiber with a gold wire. *Opt. Commun.* **2015**, *351*, 50–56. [CrossRef]
23. Li, P.; Zhao, J.L. Polarization-dependent coupling in gold-filled dual-core photonic crystal fibers. *Opt. Express* **2013**, *21*, 5232–5238. [CrossRef] [PubMed]
24. Wang, E.L.; Jiang, H.M.; Xie, K.; Chen, C.; Hu, Z.J. Polarization splitter based on dual core liquid crystal-filled holey fiber. *J. Appl. Phys.* **2016**, *120*, 114501. [CrossRef]
25. Hagras, E.A.A.; Heikal, A.M.; Hamed, M.F.O.; El-Azab, J.M.; El-Nozahi, A.M.; Obayya, S.S.A. Ultra compact soft glass liquid photonic crystal polarization splitter with As_2S_3 core. *Opt. Quant. Electron.* **2017**, *49*, 55. [CrossRef]
26. Younis, B.M.; Heikal, A.M.; Hameed, M.F.O.; Obayya, S.S.A. Highly wavelength-selective asymmetric dual-core liquid photonic crystal fiber polarization splitter. *J. Opt. Soc. Am. B* **2018**, *35*, 1020–1028. [CrossRef]
27. Xu, Q.; Luo, W.L.; Li, K.; Copner, N.; Lin, S. Design of polarization splitter via liquid and Ti infiltrated photonic crystal fiber. *Crystals* **2019**, *9*, 103. [CrossRef]
28. Rahman, M.T.; Datto, S.; Sakib, M.N. Highly sensitive circular slotted gold-coatedmicro channel photonic crystal fiber basedplasmonic biosensor. *OSA Contin.* **2021**, *4*, 1808–1826. [CrossRef]
29. Qiu, S.; Yuan, J.H.; Zhou, X.; Feng, L.; Wang, Q.W.; Qu, Y.W.; Yan, B.B.; Wu, Q.; Wang, K.R.; Sang, X.Z.; et al. Hollow-core negative curvature fiber with high birefringence for low refractive index sensing based on surface plasmon resonance effect. *Sensors* **2020**, *20*, 6539. [CrossRef]
30. Hameed, M.F.O.; Heikal, A.M.; Younis, B.M.; Abdelrazzak, M.; Obayya, S.S.A. Ultra-high tunable liquid crystal-plasmonic photonic crystal fiber polarization filter. *Opt. Express* **2015**, *23*, 7007–7020. [CrossRef]
31. Jiang, L.H.; Zheng, Y.; Yang, J.J.; Hou, L.T.; Li, Z.H.; Zhao, X.T. An Ultra-broadband single polarization filter based on plasmonic photonic crystal fiber with a liquid crystal core. *Plasmonics* **2017**, *12*, 411–417. [CrossRef]
32. Bao, Y.J.; Li, S.G.; Zhang, W.; An, G.W.; Fan, Z.K. Designing of a polarization beam splitter for the wavelength of 1310 nm on dual-core photonic crystal fiber with high birefringence and double-zero dispersion. *Chin. Phys. B* **2014**, *23*, 104218. [CrossRef]
33. Snyder, A.W.; Love, J.D. *Optical Waveguide Theory*; Chapman&Hall: London, UK, 1983; pp. 542–552.
34. Dou, C.; Jing, X.L.; Li, S.G.; Wu, J.J.; Wang, Q.B. A compact and low-loss polarization splitter based on dual-core photonic crystal fber. *Opt. Quant. Electron.* **2018**, *50*, 255. [CrossRef]
35. Jiang, H.M.; Wang, E.L.; Zhang, J.; Hu, L.; Mao, Q.P.; Li, Q.; Xie, K. Polarization splitter based on dual-core photonic crystal fiber. *Opt. Express* **2014**, *22*, 30461–30466. [CrossRef]
36. Zhang, Y.Z.; Liu, H.; Chen, C.; Bai, B.B.; Tang, S.F. Temperature-controlled and multi-functional splitter based on dual-core photonic crystal fiber. *Results Phys.* **2020**, *19*, 103578. [CrossRef]
37. Chu, L.H.; Liu, M.; Shum, P.; Fu, Y.B. Simultaneous achievement of an ultrashort length and a high extinction ratio polarization splitter based on the dual-core photonic crystal fiber with $Ge_{20}Sb_{15}Se_{65}$ glass. *Appl. Opt.* **2019**, *58*, 7892–7896. [CrossRef]
38. Wang, X.Y.; Li, S.G.; Liu, Q.; Fan, Z.K.; Wang, G.Y.; Zhao, Y.Y. High-extinction ratio and short-length polarization splitter based on microstructured optical fiber with tellurite glass. *Opt. Mater.* **2017**, *66*, 542–546. [CrossRef]
39. Qu, Y.W.; Yuan, J.H.; Qiu, S.; Zhou, X.; Feng, L.; Yan, B.B.; Wu, Q.; Wang, K.R.; Sang, X.Z.; Long, K.P.; et al. A novel gold film-coated V-shape dual-core photonic crystal fiber polarization beam splitter covering the E + S + C + L + U band. *Sensors* **2021**, *21*, 496. [CrossRef]
40. Sun, B.; Chen, M.Y.; Zhang, Y.K.; Zhou, J. Polarization-dependent coupling characteristics of metal-wire filled dual-core photonic crystal fiber. *Opt. Quant. Electron.* **2015**, *47*, 441–451. [CrossRef]
41. Zou, H.; Xiong, H.; Zhang, Y.S.; Ma, Y.; Zheng, J.J. Ultra-broadband polarization splitter based on graphene layer-filled dual-core photonic crystal fiber. *Chin. Phys. B* **2017**, *26*, 124216. [CrossRef]
42. Zhao, Y.Y.; Li, S.G.; Wang, X.Y.; Wang, G.Y.; Shi, M.; Wu, J.J. Design of a novel multi channel photonic crystal fiber polarization beam splitter. *Opt. Commun.* **2017**, *400*, 79–83. [CrossRef]
43. Li, Y.; Itoh, K.; Watanabe, W.; Yamada, K.; Kuroda, D.; Nishii, J.; Jiang, Y.Y. Three-dimensional hole drilling of silica glass from the rear surface with femtosecond laser pulses. *Opt. Lett.* **2001**, *26*, 1912–1914. [CrossRef] [PubMed]
44. Bertoncini, A.; Liberale, C. 3D printed waveguides based on photonic crystal fiber designs for complex fiber-end photonic devices. *Optica* **2020**, *7*, 1487–1494. [CrossRef]
45. Feng, X.; Mairaj, A.K.; Hewak, D.W.; Monro, T.M. Nonsilica glasses for holey fibers. *J. Lightw. Technol.* **2005**, *23*, 2046–2054. [CrossRef]

46. Huang, Y.J.; Wang, Y.; Zhang, L.F.; Shao, Y.; Zhang, F.; Liao, C.R.; Wang, Y.P. Tunable electro-optical modulator based on a photonic crystal fiber selectively filled with liquid crystal. *J. Lightw. Technol.* **2019**, *37*, 1903–1908. [CrossRef]
47. Lin, J.D.; Chiu, C.Y.; Mo, T.S.; Lee, C.R. All-optical directional control of emission in a photonic liquid crystal fiber laser. *J. Lightw. Technol.* **2020**, *38*, 5149–5156. [CrossRef]
48. Liu, Q.; Xue, P.S.; Wu, Q.; Zhao, C.Y.; Ng, W.P.; Fu, Y.Q.; Binns, R. Electrically sensing characteristics of the sagnac interferometer embedded with a liquid crystal-infiltrated photonic crystal fiber. *IEEE Trans. Instrum. Meas.* **2021**, *70*, 9511209.
49. Tian, S.; Yang, T.Y.; Zhang, J.X.; Xie, K.; Ma, J.J.; Hong, L.; Luo, Y.H.; Hu, Z.J. Multi-band thermal optical switch based on nematic liquid crystal filled photonic crystal fiber. *J. Lightw. Technol.* **2021**, *39*, 3297–3302. [CrossRef]
50. Li, Y.; Wang, L.N.; Chen, Y.Z.; Yi, D.; Teng, F.; Hong, X.M.; Li, X.J.; Geng, Y.F.; Shi, Y.; Luo, D. High-performance fiber sensor via Mach-Zehnder interferometer based on immersing exposed-core microstructure fiber in oriented liquid crystals. *Opt. Express* **2020**, *28*, 3576–3586. [CrossRef]

MDPI
St. Alban-Anlage 66
4052 Basel
Switzerland
Tel. +41 61 683 77 34
Fax +41 61 302 89 18
www.mdpi.com

Photonics Editorial Office
E-mail: photonics@mdpi.com
www.mdpi.com/journal/photonics

www.ingramcontent.com/pod-product-compliance
Lightning Source LLC
Chambersburg PA
CBHW040801250926
51850CB00007BA/1048
* 9 7 8 3 0 3 6 5 5 1 2 6 5 *